Value-Free Science?

Robert N. Proctor

Value-Free Science?
Purity and Power in Modern Knowledge

Harvard University Press
Cambridge, Massachusetts, and London, England   1991

Copyright © 1991 by the President and Fellows
  of Harvard College
All rights reserved
Printed in the United States of America
10 9 8 7 6 5 4 3 2 1

This book is printed on acid-free paper, and its binding
materials have been chosen for strength and durability.

Library of Congress Cataloging in Publication Data

Proctor, Robert N., 1954–
  Value-free science? : purity and power in modern knowledge /
Robert N. Proctor.
    p.   cm.
  Includes bibliographical references and index.
  ISBN 0-674-93170-X (acid-free paper)
  1. Science—Social aspects.   2. Science—Philosophy.
3. Science—History.   I. Title.
Q175.5.P785   1991
303.48′3—dc20   90-28895
                                                         CIP

For Londa

Contents

Preface ix

Introduction. The Dilemma of Science Policy 1

Part One. "Pure Science" and the Baconian Critique

1 The Cosmos as Construct 17
2 Baconian Caveats, Royalist Compromise 25
3 The Devalorization of Being 39
4 Secondary Qualities and Subjective Value 53

Part Two. The Politics of Neutrality in German Social Theory

5 The German University and the Research Ideal 65
6 Empirical Science and Specialized Expertise 75
7 The *Werturteilsstreit*, or Controversy over Values 85
8 The Social Context of German Social Science 99
9 Neutral Marxism 121
10 Max Weber and *Wertfreie Wissenschaft* 134

Part Three. The Legacy of Neutrality: Positivism and Its Critics

11	Catholicism without Christianity	157
12	Logical Positivism	163
13	Positive Economics	182
14	Emotivist Ethics	201
15	Social Theory of Science	209
16	Realism versus Moralism	224
17	Critiques of Science	232

Conclusion. Neutrality as Myth, Mask, Shield, and Sword — 262

Notes — 273
Bibliography — 311
Index — 321

Preface

The question of the place of values in science has vexed philosophers and social theorists for more than a century. Several different issues are at stake. Foremost among these is the question of the moral responsibilities of scientists—but personal morality is not the only issue involved. There are also larger, political, questions. How should a democratic society establish research priorities? Can social goals—such as environmental responsibility or cultural diversity—be incorporated into science without compromising intellectual freedom? What does scientific freedom mean in an era of Big Science—when research and development may involve military clearance, corporate patents, and billions of dollars in taxpayer support?

There was a time when philosophers debated the relations of science and ethics in terms of the "fact-value problem." Since the 1960s, however, the ground has shifted, partly because it was realized that "facts" suffer from many of the foibles once attributed only to "values." The shift is, to some degree, a consequence of the changed political climate since the 1960s; but much also has to do with the fact that science itself has changed. In an era of multibillion-dollar research projects it is hard to deny that science has both value implications and value origins. Science is as important to economic and military power as to soul-searching questions of cosmology or cosmogony. Science lies close to the roots of many forms of power: power to create or destroy, to heal or to harm, to feed or let hunger, to enlighten or obscure.

My first book, *Racial Hygiene: Medicine under the Nazis,* was originally framed as an epilogue to the present work; my goal there was to understand how scientists came to support a political order unprecedented in its cruelty. Here I return to the larger, historical, question: Why did scientists articulate the ideal of value-neutrality in the first place? Why have scientists feared (and often continue to fear) the intrusion of values into their work?

These are questions that require philosophical and political analysis. The fact-value problem, central to positivist philosophy only a couple of decades ago, foundered for having been treated as a purely semantic problem—a problem of the meanings of words. The relation of science and values is certainly more than this, especially if we admit that the boundary between science and other parts of society is a shifting and permeable one. This is one of the assumptions of the present book: that the larger conditions of life shape the kind and quality of science people produce. Whether a nation is at war or at peace, in triumph or defeat, boom or bust—these things shape the kinds of questions scientists ask, or at least the kinds of questions they are likely to be able to answer. So can the politics of the party in power, or the religion, gender, ethnicity, or ideology of researchers or those who fund research. Politics, in other words, is more than government. The politics of science has to do with how government or industry encourages or discourages research, but also with how power relations influence what kind of science gets done and what kind doesn't get done. There is politics in the power of professions, in the style of scientific rhetoric, in the structure of scientific priorities. There is a micropolitics of science, a political and sexual economy of science. There is also a politics of nature.

The thesis of this book is that value-neutrality, far from being a timeless or self-evident principle, has a distinctive geography: "value-freedom" has meant different things to different people at different times. Slogans like "science must be value-free" or "all knowledge is political" must be understood in light of specific fears and goals that change over time. Value-neutrality may be a response to state or religious suppression of scientific ideas; value-neutrality may be a way to guard against personal interests obstructing scientific progress. Value-neutrality may reflect the desires of scholars to professionalize or to secularize; value-neutrality may conceal the fact that science has social origins and social consequences. Neutrality may provide a path along which one retreats or a platform from which one launches an offensive.

It has become fashionable to claim that "all science is social," and there are many good reasons for this fashion. Still, we should not forget that there are some good reasons that scholars have tried (and continue to try) to distance their work from questions of values or politics. The idea that science is and should remain value-free has complex historical roots. This complexity is obscured if we simply declare that all facts are theory-laden or that knowledge is a social product. We need to know much more about how social goals can and should shape scientific practice. Who is served by science and who is ignored? Who benefits and who suffers? How might science look different were its methods or personnel or priorities to change? What can be done? What should be done?

These are ultimately moral questions—questions that have to be asked, and not simply assumed to have been answered. My hope is that the problem of whether science is or ought to be value-free will remain an open question, a question of how properly to politicize or depoliticize science, to enliven science education, to bring science more closely and humanely to bear on matters of vital human interest.

I would like to thank the Fulbright Commission for a fellowship to the Free University in West Berlin (1980–81), where I did much of my research; I would also like to thank the Charlotte Newcombe Foundation and the Howard Foundation for helping support my early research. My earliest interest in the politics of knowledge was sparked by my father, Neel Proctor; I must also give credit to Jim Adler for providing me with invaluable advice and counsel at nearly every stage of this project. Friends and colleagues who read and commented on the manuscript in one form or another include: I. Bernard Cohen, Robert S. Cohen, James Farr, Steve Fuller, Peter Galison, Stephen Jay Gould, Loren Graham, Gillian Hadfield, David Hollinger, Evelyn Fox Keller, Richard Levins, Everett Mendelsohn, Robert Merton, John Proctor, Margaret Schabas, Alan Sica, and Peter Taylor. In matters such as these it is often as difficult to separate friends and colleagues as to separate values from facts; let me simply note that to each of these I owe a debt that is both personal and profound. I would also like to acknowledge my debts to Londa Schiebinger, my companion in all things, to whom I dedicate this book.

Value-Free Science?

Introduction

The Dilemma of Science Policy

Vain is the word of the philosopher which heals no suffering of man. —EPICURUS

The inhumanity humans show toward one another is probably no greater now than in other times in history; the crimes of the Nazis against the Jews, or the Turks against the Armenians, or Americans against the Vietnamese are not the first such crimes in history. What is new in our times is that we have greater means to effect such crimes—and these we owe largely to the growth of science and technology. Today's powers are, of course, the product of centuries of technological progress. Francis Bacon celebrated the compass, gunpowder, and printing for having revolutionized navigation, warfare, and access to knowledge. But consider also that, in the century prior to Bacon, the compass allowed the Spanish to send their armies into the Americas; gunpowder allowed them to subdue the natives; and printing allowed records of these acts to survive (while at the same time Spanish priests burned the Mayan libraries).

In recent years, more sophisticated, modern technologies promise to raise the stakes of destruction. The first hydrogen bomb, exploded on Bikini atoll in 1952, released more destructive power than had been loosed in all previous wars in history. And by the 1980s, the time required to complete a "nuclear exchange" between New York and Moscow had dwindled to something on the order of half an hour, less if one considers nuclear submarines lying outside national waters. Fears are translated into a barbaric will to survive: as "survival groups" take target practice to endure the holocaust, doctors explore the horrors of "catastrophe medicine" and governments stockpile opium to soothe the last minutes of the mutilated.[1] "Preparedness" has taken on an almost comic air, as bureaucrats ponder the price of post-holocaust postage and morticians practice the art of embalming radioactive bodies. Most terrifying of all, perhaps, is that there are those who have lost their fear. It is not so long since Colin Gray and Keith Payne assured us (in harmony with former vice-president George Bush) that "victory is possible" in a war in which 20 million Americans are killed.[2]

Language has been developed to soften the brutality. Departments of war have become departments of defense, recalling Orwell's forecast of "Ministries of Peace." War has been given the language of peace, and even art. We speak of "theater weapons," "arenas of defense," and "windows of vulnerability." In classic doublespeak, the U.S. Department of Defense described its mission in Vietnam as waging "peacefare" as well as "warfare"; American army officials described their program of defoliation and deployment of anti-personnel weapons as "pacification" and as "bringing peace to the countryside." The horrors of war have been disguised with the sanitary terms of medicine, business, or even sex and birth—there is talk of "surgical strikes" and "penetration aids," of nuclear bombs as "assets" or "babies" or keepers of the peace.[3]

Language, of course, is only part of the problem. Since the 1970s we have seen the rapid militarization of science in the United States, as the proportion of government funds for military research and development rose from 50 percent under President Carter to more than 70 percent under Reagan and Bush. There remains today a large and shadowy secret scientific community, with secret scientific journals and secret scientific conferences. Those who have studied this phenomenon estimate that as many as a quarter or a third of all American scientists and engineers have some form of military security clearance. This figure itself, of course, is not open to public scrutiny.

Still we often hear that however foul its application, science itself is pure. Science may be political in its application, but not in its origin or structure. And certainly it is true that science and technology alone are hardly a threat to world peace. Politics and moralities stand behind our sciences and give them life; science can be used for good or for evil. This is one sense of the "neutrality" of science—that science (or technology) "in itself" is neither good nor bad; that science may be *used,* or it may be *abused.* This is hardly a new idea. Plato goes to some lengths to show that those most capable of healing are also those most capable of harming, that those most competent to tell the truth are those most able to tell a lie.

Yet this supposed neutrality describes only the simplest technologies, the most abstract principles. The seven simple machines, perhaps, or the rules of arithmetic, may be neutral in this sense. But an abstract truth often conceals a concrete lie. "Guns don't kill people, people kill people." Yet is it surprising that a society that surrounds itself with guns will use them? "A newly sharpened sword," reads the African proverb, "marches by itself to the next village." Tools, we realize, have alternative uses; the knife bought for cooking might be used for killing. Yet knives or levers are not what modern science-based technology is all about. A nuclear power plant, cruise missile, or linear accelerator can hardly be used for ends other than those for which they are designed. Science-based technologies are increasingly *end-specific:* the means constrain the ends; it is no longer so easy to separate the origins of a tool from its intended use. What does it mean to "abuse" a cruise missile or a neutron bomb?

These abstract problems have historical origins. The nineteenth-century distinction between science "pure" and "applied" was designed to recognize that whereas technical invention or innovation is almost always goal-oriented, there is also pure science—science pursued "for its own sake," not just for its applications. Especially in recent years, however, the boundaries separating science and its application have become blurred. Governments have recognized the "usefulness of useless research." The consequence has been not just a shortened time lag between discovery and application, but entirely new relations of science and industry. In the modern industrial laboratory discoveries are sought in areas of anticipated applicability; the establishment of biotechnology firms to explore recombinant DNA techniques is a recent example of industries sponsoring "basic research" with the expectation of applying the knowledge gained to medicine or agriculture.

For these and other reasons, it has become increasingly difficult to separate pure and applied science along traditional lines. It is no longer so easy to separate a passive, disinterested "science" from a transformative, engaged "technology." Applied science works increasingly with principles; pure science is increasingly dependent upon large-scale instrumentation. As most practicing scientists are aware, much of science consists of pure technology. The traditional distinction between science pure and applied has been modified from two directions: in the degree that applied science works with principles, but also in the degree that pure science has social origins and consequences. Appreciation of these origins and consequences, especially in recent years, has made science a vital aspect of both government and industry. Today, governments or industries that ignore R and D are doomed to languish in the backwaters of economic competitiveness.

According to one common sense of science, however, politics or values are only to be found external to the practice of science—in its uses but not its origins, in its failures but not its triumphs, in the exceptional or the peripheral but not in the everyday and fundamental. This is the ideology of *pure* or *value-free* science, the belief that science "in itself" is pure, and that values or politics enter only as contamination.

The image of pure science in the modern West is associated with a hierarchy from pure to applied, from theory to practice. And yet the myth confronts a reality: science throughout the modern world is recognized as a vital part of industrial and military production. Science is planned, and increasingly so. Priorities are established and goals are set. When monies are given, there are expectations of returns—perhaps not soon, and certainly not for every dollar spent, but still there are expectations. History has shown that pure science can yield substantial practical rewards. In World War I (the "chemists' war"), chemists proved that battles could be won by transforming chemical knowledge into poison gas or powerful munitions. In World War II (the "physicists' war"), the proximity fuse, radar, and the atom bomb proved the value of esoteric physics for the waging of war. Science, so it seemed, was good for war; war, some even claimed, was good for science.

And so, it appeared, was industry. In the spring of 1964, physicists Arno A. Penzias and Robert W. Wilson working at Bell Telephone Laboratories at Crawford Hill, New Jersey, discovered a curious background microwave radiation that seemed to have the character of electronic noise coming with equal intensity from all directions of the sky, independent of time of day or season of the year. In the course of conversation with a colleague at the Massachusetts Institute of Technology, Penzias learned that cosmologists had predicted background radiation of this sort from the "big bang" model of the origin of the universe. The contribution of these industrial scientists to the relatively pure field of cosmology provided further proof that industry, alongside government and the academy, could serve to advance basic science.

Industrial support for science is not in itself a new phenomenon. What *is* new is its magnitude and its penetration into university environments. In the early 1990s many industrial firms have research budgets in excess of $500 million; IBM, the largest in this sphere, funds R and D to the tune of nearly $7 billion per year and has won to date a total of five Nobel Prizes. In the universities, as federal funds for basic research no longer grow as rapidly as they once did, scientists have turned to industrial laboratories for support. Industry in turn has looked to universities for a place to invest. In the early 1980s, the Whitehead Foundation of Revlon promised $127 million for the establishment of a microbiology research facility at MIT; Hoechst Chemicals offered comparable funds ($70 million) to the Massachusetts General Hospital. Efforts to link industry and the academy have entered a new and more intensive phase; the claim can now be made (though not without strong criticism) that the interests of industry and of government are those of science itself.

Also new is the fact that research efforts can absorb hundreds of millions or even billions of dollars. The effort to map the human genome will cost on the order of $3 billion, and the Hubble Space Telescope about $2 billion. The Superconducting Supercollider is projected to cost as much as $8 billion—making it (according to Harvard physicist Sheldon Glashow) "one of the most ambitious projects ever conceived by our species."[4] Even this is dwarfed, though, by the behemoth of Star Wars, the engineering nightmare that staggered the imagination when figures as high as $1 trillion were bandied about for total costs through deployment. (Star Wars funds have been trimmed, although funding is still going ahead for the Pentagon's supersecret MILSTAR, the $20 billion satellite system designed to choreograph World Wars III and IV.) The plant and equipment involved in such projects are a far cry from the "tools" or "arts" of the classical world. They are institutions, with highly trained personnel, specialized transport and maintenance facilities, proving grounds, computing centers, and so forth. They are systems, communities, or societies, which simply would not exist without massive funding from government or corporate treasuries.[5] Again, for projects of such magnitudes, there are expectations of returns.

And yet science in the West has for much of its history demanded a certain distance, or detachment, from society.[6] Socrates in *The Statesman* praised the science of number for its purity, unalloyed with the affairs of practical life. Even Francis Bacon, herald of the utilitarian ideal of knowledge, warned against the disregard of "science for its own sake" (see Chapter 2). Deep within the Western philosophical tradition is an appreciation of the free and unhampered pursuit of knowledge, a freedom based on a distinction between the ideal of *theory*, on the one hand, and the pursuit of personal gain or social need, on the other. Any attempt to turn science toward the "social good" or any other social or political goal invites the charge of illegitimately "politicizing science"; it is often held that if science has anything but itself as a guide, its progress will be impeded. Any attempt to "plan" science (in this view) can only result in its destruction.[7]

Science is forced on the horns of a dilemma. On the one hand, there are calls for scientific *freedom,* born of fears for the survival of science in the face of political tyranny. On the other hand, there are calls for *accountability* in science, for sciences more in tune with practical human needs and desires. There is the sense that science is objective, that objectivity is an essential quality of science; yet there is also the growing sense that science serves specific social interests. There is a perception that "wishing does not make it so," that the earth will circle the sun regardless of whether we live under socialism, communism, or fascism; yet there are also calls for "alternatives" in the sciences, for science more in tune with human needs, for broader participation in and "social responsibility" of the sciences.

It is the purpose of this study to explore the origins of one key aspect of this dilemma—the ideal of neutral or value-free science—as expressed across a number of different historical, political, and philosophical contexts. Much of our attention will be focused on debates surrounding the nature of *social science*. The *Werturteilsstreit*, the "controversy over values" that raged in German academic circles in the early 1900s, was a battle fought over the nature of social theory; most subsequent critics of the ideal of value-neutrality have been social theorists. This is partly because, for many years, it was widely assumed that value-neutrality was the *sine qua non* of scientific objectivity. Physics, according to this view, was neutral; could the social sciences achieve the same rigorous detachment? The "value-problem" became a problem partly because many assumed that, unless social science could be value-free (as physics supposedly was), it could not be made truly scientific.

But is neutrality the hallmark of objectivity? Does advocacy preclude objectivity? Why have scholars defended their work as value-free? How has neutrality come to dominate the world of science—and how does one distinguish boast from reality? As we shall see, the ideal of value-neutrality is not a single notion but rather a collection of loosely associated ideals that emerged at different times to serve different social functions. Neutrality has meant very different things to different people; these changing meanings must be understood, if one is to understand why people continue to believe that science is and/or ought to be value-free.

My argument may be outlined as follows. Part I traces three basic ways the neutrality of science has been defended prior to the nineteenth century. In the ancient world, Plato and his followers saw contemplative thought as superior to practical action. The unexamined life, Socrates proclaimed, is not worth living; the life of principle should be prized over the unreflective life of manual labor and toil. Science in the Platonic vision is the product of reflective leisure; science is the privilege of a society that has satisfied its physical needs and is free to entertain the luxury of contemplative inquiry.

The Platonic separation of theory and practice has continued since ancient times to serve as a central element in the defense of what we today consider "neutral science." Yet the contemplative ideal is only part of what we mean by neutrality; and in fact, it is possible to argue that the contemplative ideal has long been supplanted by very different concerns. The rise of modern science in the seventeenth century is associated with a new vision of the importance of uniting *theoria* with *praxis,* a vision that grants a new prominence to human agency and labor. The practical impulse of Baconian science posted a challenge to the notion of science as something isolated from practical concerns. At the same time, however, early modern philosophers argued that ethical concerns should be excluded from natural philosophy because they bias us in our search for knowledge, much as the "idols" named by Bacon distort the true nature of things. This is our second sense of neutrality—that moral knowledge acts to spoil or "taint" the knowledge of nature.

There is a third sense of neutrality in early modern science. In the mathematico-mechanical conception of the universe, the world of spirit is radically divorced from the world of matter. As a result, the rise of modern science is associated with a new ethical ontology, where the ancient conception of a world suffused with spirit and purpose is replaced by a "devalorized" conception of the cosmos as everywhere the same and devoid of purpose. Value for the moderns does not inhere in the earth or its creatures but rather is rooted in human needs and desires (Hobbes, Petty). The source of value is not God or nature but the industry and utility of men—value in use, value in exchange. The virtuous or beneficial attributes of a thing are secondary, subjective qualities, as opposed to the hard, primary (and objective) qualities of matter in motion (Galileo, Descartes, Locke, Hume). Science in this view is neutral because nature itself has been *devalorized* (Koyré)—secularized and *disenchanted* (Weber)—stripped of its qualities, of teleology, of magical resonances and meanings. It was this that Thomas Huxley had in mind when he declared, in his Romanes lecture of 1891, that "the Cosmic process has no sort of relation to moral ends."

There is also an institutional aspect to the early exclusion of values from science. Recent scholarship suggests that the rise of apolitical science in seventeenth-century Europe must be understood in terms of a young and fragile science seeking autonomy with respect to church and state. In the "royalist compromise" of seventeenth-century English science, natural philosophers of the Royal Society of London promised not to meddle in matters of "Divinity, Metaphysics, Moralls [or] Politicks," in exchange for rights to publish without censorship, to correspond freely with other members, to pursue science with the blessing and support of the state. Neutrality was part of the bargain struck, the price science had to pay, for social legitimation in the eyes of church and state.

Each of these early conceptions of the place of science in society—the primacy of theory over practice, the devalorization of being, the notion of morality as a contamination of objective scholarship, and the royalist compromise—reemerge in nineteenth- and twentieth-century justifications of science as value-free. Beginning especially in the nineteenth century, however, a series of further defenses emerge. Part II explores the exclusion of politics and morals from science with the rise of academic specialization, industrial research, and (especially) the new science of sociology. At one level, value-neutrality is defended as a consequence of an ontological dualism between the true and the good. In the twentieth-century formulation, propositions about "what ought to be" can never be derived from propositions about "what is"; facts cannot be derived from values. Hume is said to have discovered this principle; it is one that appeals to social philosophers as diverse as Whitehead, Poincaré, Weber, Sombart, the positivists of the Vienna School, Oxford economists in the tradition of Robbins, emotivists in the tradition of Moore and Ayer, and Chicago economists in the tradition of Frank Knight and Milton Friedman.

Value-neutrality is also used, however, to deny that the real is necessarily rational or good. Weber and Poincaré argue that if the world of ethical ideals transcends what is empirically true, then empirical science cannot provide the grounds for moral claims. Neutrality represents a critique of the Panglossian optimism that ours is "the best of all possible worlds," also an argument against those who seek to preserve some *status* or *fluxus quo,* whether that be based on the will of God, the structure of our genes, or invariant laws of history. Neutrality represents a critique of those who try to promote certain values in the guise of science, those who proclaim a certain social or moral order to be the natural one, that which is most "fit" or "functional" or otherwise ideal or predetermined. Science must be neutral, in this view, because ours is not in fact the best of all possible worlds; scholars and nonscholars alike should be wary of efforts to confuse the empirically real and the ontologically possible. Science must be *value-free,* in order to guarantee that values remain *science-free.*

Again, I want to stress that the political significance of value-neutrality—in its various forms—has changed according to use and circumstance. There are times when critics have pointed to the neutrality of science (or technology) to show that technology may be wrongly applied. The very same point has also been used to argue that science (or even technology) should not be subject to moral or political criticism. Today one commonly hears that criticism of science must be restricted to criticism of its application, its "use" or "abuse."

Beginning in the nineteenth century, it becomes popular to suggest that the scientific attitude is exceptionally fitted to the resolution of social conflict. Science in this view is a great and neutral arbiter, an impartial judge to whom social problems may be posed and from whom "balanced" answers will be forthcoming. Science provides a neutral ground upon which people of all creeds and colors might unite, on which all political contradictions might be overcome. Science is to provide a balance between opposing interests, a source of unity amidst diversity, order amidst chaos.

Finally, there is the so-called subjectivist defense. Liberal economists and social theorists from Walras, Jevons, and Menger to Weber, Friedman, and Robbins call for the exclusion of value judgments from science on the grounds that values are subjective and that science therefore can make no claims concerning their validity. Neoclassical economists argue that alternative distributions of property should not enter into scientific discussions of economics, on the grounds that doing so confuses science and politics. Marginalists thus advocate the disinterested study of market behavior without questioning the origins of or alternatives to this form of behavior. Neutrality here is part of the intellectual apparatus of liberal capitalism, part of a world view designed to ensure that questions concerning the origins, nature, and distribution of property are not raised in scholarly discourse.

In Part III we move away from the problem of the origins of neutrality to consider four twentieth-century ways of thinking about science: theories of the logic of science, of scientific change, of the social origins of science, and, finally, the critique of science. My purpose here is to explore "the legacy of neutrality," especially in its positivist embodiment, to see how recent social theorists and social critics have understood the politics of knowledge.

How are we to understand the term *science*? The danger of reification (making something more real than it really is) is a pervasive one in both historical and philosophical writing. When we speak of science among the Greeks *(theoria, episteme, philosophia, peri physeos historia)*, we mean something very different from science as it is practiced today; elucidating some of this variety is one task of this study. What is true for science in general, however, is equally true for the abstract ideals of neutrality or purity. The concepts and movements we shall explore are not the product of any single, well-defined historical tradition. The point in classifying these traditions is not to discover some transhistorical principle of neutrality, but rather to explore the several paths by which we have come to where we are today.

In view of the diverse origins of the ideal of value-free science, it is important that the ideal be understood not simply as an abstract filiation of ideas but also as a response to certain broader changes in the social and economic context of science: changes in its institutional and professional locus; changes associated with the rise of industrial, military, and state-supported science; changes associated with the rise of political movements (such as feminism or socialism) challenging both the autonomy of science and the power of those who hold its purse strings.

The problem of value judgments in the sciences has been the object of much attention, both among philosophers and philosophically minded scientists. At the level of advertisement, the tide has turned in favor of the debunkers. Especially in recent years, we hear a lot about "the myth of neutrality"; value-free research is described as "malignant nonsense," "logically incoherent," a mask for "liberal ideology."[8] Abraham Edel in 1988 called the idea of value-neutrality "bizarre"; Sandra Harding in 1989 called it "a delusion."[9] Yet relatively little attention has been given to the origins of neutrality and the variety of its forms, to neutrality as a phenomenon that has emerged in response to particular historical circumstances and whose political significance has changed according to historical time and place. Among philosophers in the Anglo-American analytic tradition, the problem for many years was cast as one of determining the precise linguistic relations of "ought" and "is," "fact" and "value," independent of changing historical relations of science and society, independent of the variety of ways that science has been defended or attacked.

The problem of neutrality is a philosophical problem, but it cannot be approached in the abstract, in terms of the meanings of words; instead, it must be understood in terms of the concrete uses of ideas in particular historical settings. The approach must be both historical and comparative. It must be historical, for the meaning of words and ideas changes with time. (As we shall see, the ideal of value-free science means something very different for German sociologists in 1911, Jewish philosophers in 1937, and American sociobiologists in 1975.) It must be comparative, because one wants to know what is the same or different about ideals of science in Plato's Greece, Bacon's England, and Weber's Germany.

I should also mention a philosophical problem and a historiographic problem. Philosophers in the analytic tradition may object to any attempt to contextualize the problem of neutrality on the grounds that doing so jeopardizes the notion of scientific objectivity. Two responses are in order. First, one must point out that neutrality and objectivity are not the same thing. Neutrality refers to whether a science takes a stand; objectivity, to whether a science merits certain claims to reliability. The two need not have anything to do with each other. Certain sciences may be completely "objective"—that is, valid—and yet designed to serve certain political interests. Geologists know more about oil-bearing shales than about many other rocks, but the knowledge is thereby no less reliable. Counterinsurgency theorists know how to manipulate populations in revolt, but the fact that their knowledge is goal-oriented does not mean it doesn't work.

The appropriate critique of these sciences is not that they are not "objective" but that they are partial, or narrow, or directed toward ends which one opposes. In general, knowledge is no less objective (that is, true or reliable) for being in the service of interests. This has been a stumbling block for much social theory of science. Abstractly, the assumption that bias compromises objectivity, or that "what you see depends upon where you look," grasps only part of the problem. The more interesting and fruitful question is: How is science involved in patterns of dominance and/or exclusion? The purely epistemological question—"How do we know?"—often diverts attention from more fundamental (and ultimately political) questions such as: Why do we know this and not that? Why are our interests here and not there? Who gains from knowledge of this and not that? What is to be done—or undone?

The recognition especially in recent years of the contingency of scientific knowledge (that knowledge may change, that facts are theory-laden, that knowledge is contextual) and the fact that science may serve certain interests (the "politics" of knowledge) has generated a relativistic critique of science which denies the possibility of scientific objectivity and/or truth independent of a knowing subject. We shall examine this further in Chapter 16; here let me simply suggest that this epistemological relativism (like older forms of subjectivism or skepticism) stands traditional epistemology on its head without challenging the problem-field that generated interest in epistemology in the first place. The abstract relativist, like the positivist, assumes that the question of whether science serves interests is equivalent to the question of whether science is objective. Such a view confuses not just present-day science with all possible science, but more importantly science and nature. Both positivists and abstract relativists conflate the contextuality of knowledge with the possibility of its objectivity.

An alternative approach is to argue that the objectivity of science actually depends upon certain aspects of its contextuality—this is part of what Lenin meant when he claimed that absolute truth is the sum of all possible (subjective) truths; it is also what Popper and Polanyi meant when they held that the objectivity of science derives from the social nature of its verification. Sandra Harding argues that objectivity may even *require* an atmosphere of politicized criticism.[10]

Finally, there is the question of policy and political action. Both positivists and relativists confuse the problem of objectivity and the question of whether science might be practiced in ways that are different from established canons of method or focus. The failure to distinguish these two senses of contextuality—that knowledge may serve political interests and that knowledge is necessarily contingent—has hampered investigations into the political philosophy of science. Both positivists and their relativist critics ironically share this confusion—both see the subordination of (objective) science to (political or moral) interests as a compromise of either scientific objectivity or social responsibility; both assume that the objectivity of science stands or falls with its neutrality.

The point, then, of understanding the origins of the ideal of value-free science is not to sabotage the mission of science in its search for truth but to bring to light certain bright and dark spots in that search. (This is why I emphasize the origins and effects of neutrality and not simply its abstract coherence or validity). The fact that meanings are situated does not mean that they are arbitrary, nor that they are meaningless outside local contexts. It does mean, though, that the fullness of that meaning is better grasped when viewed both with telescope and microscope.

The analyst of the ideal of neutrality is confronted with a curious paradox, given that it is difficult to avoid taking sides: to remain neutral assumes one side; not remaining neutral implies the other. If, however, it is wrong to believe that advocacy necessarily compromises objectivity, then this paradox may be resolved by cultivating the dual methodological principles of *sympathy* and *critique*. This brings us to the question of historiography.

One of the traditional goals of historical understanding, as many philosophers have observed, is a kind of sympathy or identification with the past. This was Vico's conception of the historical sciences, and it is a conception that has remained central to the philosophy of history ever since. Historical understanding implies a certain closeness to, as well as distance from and respect for, the objects of one's attention. It implies a certain suspension of one's own judgment in order to appreciate the judgments of others. The ideal of historical understanding is intended to guard against the danger that we interpret the past through our own categories (what George Stocking called *presentism*), or that we look to history merely to find the forerunners of present-day thought, ranking these as heroes or villains according to how they stand up against modern conceptions of what is good or true (what Herbert Butterfield called *whiggism*).[11] History should not be seen as a series of victories or blunders along the high road to the glorious present.

Understanding is not the only goal of history, however. The purpose of writing philosophical history is not just to reproduce the past but to add something to it. There is thus a second obstacle in the path of critical historical understanding, one that, especially among historians, is as great a danger as presentism. This is the danger of *antiquarianism*, where history speaks to the past but not to the present, where historical study is pursued for "its own sake" with little or no concern to relate the problems under study to contemporary problems or interests.

Sympathy, in other words, is not enough. To know something is to understand its contingency as well as its necessity, to know the forms of its opposites as well as the variety of its appearance. To know a thing is to know what it excludes or precludes as well as what it encompasses. It is, after all, usually in reaction to certain ideas that others gain favor, and we do not comprehend the one without understanding the other. Understanding a doctrine or movement requires a sense not just of its meaning for its advocates but also of its effects—its abuse as well as its use, its legacy and not just its origins. In historical writing, then, there are two kinds of justice—justice to the past and justice to the present. Fairness to the one implies a certain sympathy with the past; fairness to the other requires a willingness to engage the past for the sake of the present.

A prudent combination of sympathy and critique has been made difficult through the separation, in the present century, of history and philosophy. Philosophy (at least in its mainstream forms) explores the timelessly true, history, the eternally changing. The result has been a substantial degree of trivialization in both disciplines—in history, through a collapse into antiquarianism; and in philosophy, through a tedious obsession with logical form and analytic precision.

What we need is a political philosophy of science, a philosophy that focuses on the forms of power in and around the sciences. The question for the political philosopher of science is not "how do we know?"—as it is traditionally formulated in the merry-go-round between realism and relativism, internalism and externalism, skepticism and productivism; it is, rather, Why do we know what we know—and why don't we know what we don't know? Who benefits from knowledge (or ignorance!) of a particular sort, and who suffers? How might the practice of science be different? How should the practice of science be different?

These are ultimately political and ethical questions, questions for the activist-moralist scholar. Such questions take us beyond both positivist and social constructivist science theory, beyond also the abstract ontogeny of science proposed by Thomas Kuhn and others.[12] The interesting question for the political philosopher of science is not the abstract relation of theory and experiment or the morphology of scientific change, but rather how and where the boundaries between knowledge and ignorance are established, how power is wielded in and around the sciences.

A political philosophy of science requires a broader conception of science than is commonly understood. Science is, after all, many different things. Science is a body of knowledge and instrumental traditions; there are scientific styles of speech and writing. Science is also a set of institutions, with implicit or explicit codes of conduct and rules of membership. Science can be an instrument of public relations; there is a political and a sexual economy of science. Science is (as Latour paraphrases Clausewitz) "politics by other means"; knowledge is (as Lyotard has put it) "a question of government."[13] Science can be a key to human wonder or an assistant in public terror.

I would therefore suggest that it is not enough to supplement the *epistemological* question (How do we know?) only by the *social contextual* question (What are the origins of knowledge?). We must also ask the *political, ethical, and activist* questions: Why do we know what we know and why don't we know what we don't know? What *should* we know and what shouldn't we know? How might we know differently?

Part One

"Pure Science" and the Baconian Critique

We have as yet no natural philosophy that is pure.
—FRANCIS BACON

What good is a newborn baby?
—BENJAMIN FRANKLIN, ASKED ABOUT THE PRACTICAL VALUE OF A BALLOON

1

The Cosmos as Construct

In his discussion with Protarchus in Plato's *Philebus,* Socrates proposes to divide knowledge into that which is technical, concerned with the arts and crafts, on the one hand and that which is concerned with education and culture on the other. It is this latter, Socrates tells us, that is of a "purer" sort than the former. The two are not unlike two species of metals, one of which is pure, the other alloyed with base and foreign elements. And within the trades themselves, it is the mathematical elements we esteem to be the highest, for outside of numbering, weighing, and measuring in an art there is nothing left but guesswork and rule of thumb.[1] Those trades which make use of numbers—such as the building of ships or houses—are to be ranked higher than those without numbers, such as music. And within the class of those who speak of numbers, we can distinguish two kinds of persons: those who speak of two cows or three armies or some other particular thing large or small, and philosophers, who speak of "number" alone in its pure sense, apart from any objects to which it might refer. But what is it for the Greek philosopher about number that raises an art or science above the "merely practical"?

In Plato's *Statesman,* the Stranger explains to the Young Socrates that mathematics is to be sharply distinguished from the more practical arts. The science of number, in that it has nothing to do with practical activity, is worthy of the name "pure." But the products of the arts *(techne)* are of a different sort; these do not exist but for the plans of man and from the operation of his tools. In the arts, the objects desired are those which can be put to use, objects which by their very nature are constantly becoming and passing away. The objects of the true sciences, in contrast, are eternal: witness Socrates' declaration to Protarchus, in the *Philebus,* that "we find fixity, purity, truth, and what we have called perfect clarity, either in those things that are always, unchanged, unaltered, and free of all admixture, or in what is most akin to them; everything else must be called inferior and of secondary importance."[2] It was in this sense that the Stranger could argue that geometry "is no art, for it concerns itself not with the opinions and desires of men, nor with the generation and service of that which grows and can be put together, but with that which is true by the ideal and perfect forms of things."

Knowledge for the Greek philosopher was "pure" in the degree to which it makes no appeal to practical arts. Thus Plato criticized Eudoxus, Archytas, and Menaechmus for trying to solve the classical problems of geometry—the squaring of the circle, the trisection of the arc, and the doubling of the cube—using the mechanical instruments and aids available to any Greek artisan. Aristotle was more the empiric, but no less a purist. Natural knowledge, he writes in his *Nicomachean Ethics,* is by definition that science whose object is eternal, unchanging truth. For that which changes cannot be known with certainty, and certainty is the object of scientific knowledge *(episteme)*. Wisdom is the knowledge of certain principles and causes: we admire the man of wisdom more than the man of experience, for the man of wisdom will know the *why,* yet the man of experience will merely know the *what.* The master-worker is wiser than the mechanic, and the man of theoretical knowledge is wiser than the man of productive knowledge. In each case, while the former makes use of principle, the latter merely seeks utility. Knowledge pursued for its own sake is superior to that pursued for its results; indeed, it was Aristotle's view that genuine knowledge arose only after men's practical needs had been satisfied: the mathematical arts first arose in Egypt, for it was there that the priestly class was first allowed to live in leisure.[3]

Pure knowledge, for Plato, was not possible "in the company of the body." Pure knowledge "consists in separating the soul as much as possible from the body," that it might dwell alone in the realm of ideas. True philosophers thus "abstain from all bodily desires and withstand them and do not yield to them." Wisdom itself is a kind of purification men achieve by casting off lowly emotions of the body.[4] Men first philosophized, says Plato, not for any utilitarian end, or to seek pleasures in this world, but to escape from ignorance, to revel in knowledge for its own sake. Just as he is free who exists for his own, rather than any other man's, sake, so we pursue science freely, for its own sake alone. Knowledge is a luxury, but it is the most necessary of luxuries: it is what distinguishes man from animals, man from woman, mind from body.

Some historians have argued that the philosopher's contempt for manual labor derived in part from the gulf that separated the producing and the consuming classes in ancient Greek society. Athens was a slave-owning republic—a republic of some 30,000 (adult, male) citizens and perhaps three or four times this many slaves, most of whom were captured in battle. It was slaves who manned the army and worked the crafts; small wonder that the Greek philosopher could cherish the idea of a bow while disdaining the art required to make one.[5]

Others have pointed out that the Platonic revulsion for the body was the product of an intellectual community that excluded and depreciated women. The Greeks of Periclean Athens rejected the fertility cults of previous centuries, replacing these with the rational gods and goddesses of the Athenaeum—a process dramatized in Euripides' *Iphigeneia in Tauris*.[6] Even for the earlier Pythagoreans, however, women and the feminine were linked with a host of qualities inferior to men and the masculine. Women were the realm of passive matter, men the realm of active spirit. Women were cold and dark, men hot and light. The feminine was vague, unbounded, and indeterminate; the male clear, bounded, and determinate. Plato subsequently incorporated these distinctions into his mind-matter dualism. In the *Timaeus*, Plato derives the souls of women from the fallen souls of men lacking in reason. For both Plato and Aristotle, males play the dominant role in human or animal reproduction: males provide the form, females the matter; generation begins with male sperm (semen) imposing form on female matter (menses). The Platonic ideal of knowledge (mind) as escape from the body (matter) thus resonated with gender divisions that would persist into modern times.

The "Greek" vision is not, of course, a uniform one. On the question of labor, dissenting views are easy to find. The early Greek writers—the philosophers of the Ionian isles, centered around Miletus—extolled the virtues of the practical arts. Herodotus praised Anacharsis of Scythia for his invention of the anchor, the bellows, and the potter's wheel; he further extolled Theodorus of Samos for his invention of the level, the square, and the rule, along with new techniques for polishing stones, draining swamps, and casting bronze. Aeschylus praised Prometheus in similar terms for having brought fire to man from the Gods. Even Socrates is accused, in several dialogues, of drawing too many of his arguments from the practical crafts. Alcibiades in Plato's *Symposium* describes Socrates as ridiculous, talking always as he does about "brass-founders, leather-cutters, and skin-dressers"; Callicles in the *Gorgias* complains that Socrates is always talking of "cobblers and pedlars and cooks and doctors, as if this had to do with our argument." Xenophon, too, has Critias ask Socrates to refrain from speaking of "those shoemakers and smiths." Socrates was a man of unbounded interest in all things human; his point was to contrast the expert knowledge of the trades with our ignorance in matters of virtue.[7]

Philosophical contempt for manual labor seems to have emerged sometime in the fifth century B.C., when Greek society had become sufficiently stratified that protection of the unique status of the citizen (nonslave resident males) became an important task for intellectuals.[8] The issue was a legislative as well as a philosophical one: in Sparta, it became illegal for citizens to practice the lower arts. The stigma attached to the practical arts continued into the Christian era, despite the substantial advance of those arts. If we can believe Plutarch's *Life of Marcellus,* Archimedes expressed contempt even for his own practical inventions. And the Roman philosopher Seneca, in his *Epistulae Morales,* chastised Posidonius for daring to include the discovery of the arch and the invention of metals among the contributions of philosophy. The true philosopher, Seneca argued, teaches neither the building of arches nor the use of metals but rather the formation of the soul: "In my own time, there have been inventions of this sort, transparent windows, tubes for diffusing warmth equally through all parts of a building, shorthand, which has been carried to such perfection that a writer can keep pace with the most rapid speaker. But the invention of such things is a drudgery for the lowest slaves; philosophy lies deeper. It is not her office to teach men how to use their hands. The object of her lessons is to form the soul." And as if this were not enough, writes Seneca, "we shall next be told that the first shoemaker was a philosopher."[9]

A number of ancient writers believed that the stigma attached to labor derived from occupational hazards faced by those practicing the arts. An Egyptian papyrus of 1100 B.C. tells of a father advising his son to learn to write in order to release himself from the ravages of physical labor, such as that suffered by the "metalworker at the mouth of his furnace, with fingers like a crocodile," stinking worse than fish spawn. Xenophon in his *Oeconomicus* has Socrates declare that the mechanical arts "carry a social stigma and are rightly dishonored in our cities. For these arts damage the bodies of those who work at them or who have charge of them, by compelling the workers to a sedentary and indoor life, by compelling them, indeed, in some cases to spend the whole day by the fire. This physical degeneration results also in the deterioration of the soul." Still others recognized that sciences could be used to reinforce the social order. Plutarch linked the practice of geometry with a political interpretation of Plato's expression that "God ever geometrizes":

> Lycurgus is said to have banished the study of arithmetic from Sparta, as being democratic and popular in its effect, and to have introduced geometry, as being better suited to a sober oligarchy and constitutional monarchy. For arithmetic, by its employment of number, distributes things equally; geometry, by the employment of proportion, distributes things according to merit. Geometry is therefore not a source of confusion in the State, but has in it a notable distinction between good men and bad, who

are awarded their portions not by weight or lot, but by the difference between vice and virtue. This, the geometrical, is the system of proportion which God applies to affairs . . . he protects and maintains the distribution of things according to merit, determining it geometrically, that is in accordance with proportion and law.[10]

Purity was thus a political concept. Pure knowledge—knowledge of the fixed and final forms of things—separated one from the filth of the furnaces, from the practical arts and trades. Pure knowledge transcended the body, the feminine, the world of the slaves, the world of production and reproduction. This was the world that the moderns of sixteenth- and seventeenth-century Europe would overthrow—as both sterile and corrupt.

The rise of modern science in the sixteenth and seventeenth centuries was part of a more fundamental shift in the relation of nature and art, *theoria* and *praxis,* the mathematical sciences and the manual crafts. Nature was for Aristotle that which creates itself—that which is essential, as opposed to that which is incidental or the product of human artifice. With the rise of seventeenth-century science—and earlier, in the works of German and Italian artist-engineers—this essentialist, respectful view of nature begins to decline. Nature is increasingly seen as raw material, as stuff waiting to be transformed by the artifice of man. The calling of humankind is now a productive one—to carry forth God's creation in the form of works and deeds; to bring to perfection the work that God has left undone. The Book of Daniel provides the watchwords of the age, that "Many shall go to and fro, and knowledge shall be increased"; humans are called upon to bring forth inventions of all kinds, to seek out evidence of Grace through works. A new vision of knowledge emerges, supplanting the ancient call to transcend or *escape* from nature by the modern call to *dominate* it.

With the moderns, we also find a new conception of the nature and source of value—what is good in a thing. The goodness of God is copied in the fruits of men's labors. Value in nature or in the eyes of God becomes value in the eyes of man: value in use, value in exchange. It is through the labors of men that wool becomes cloth or sod becomes a house. William Petty demonstrates the glory and power of England in the anatomy of her economy: there is the value of nature "raw," in the fruits we pick and the fish we catch; beyond this, however, there is the value of nature "multiplied and transformed" by human labors—ores forged into cannon, land tilled to yield its crops.

The place of humans in such a world is changed. It is commonly argued that one of the signal elements in the transition to modern science was the replacement of the earth- and human-centered world of antiquity by first the heliocentric and ultimately the centerless universe of modern cosmology. The moral usually drawn from this is that nothing is "special" about humans and their place in the cosmos. The earth is a rather ordinary planet circling a rather ordinary star, somewhere (or rather nowhere in particular) in the endless expanse of space. The Copernican Revolution is, in this view, a humbling revolution, the first of several showing the contingency and fragility of human existence.[11]

Yet the opposite can as easily be argued. The "special place" of humans lost in the Copernican world is compensated for by a new confidence, confidence that the future of the world is for humans to create. God makes laws, but humans can discover and master them. Laws found by human sense and reason rule the world. Furthermore, the laws of modern science are universal: laws valid here and now will be true everywhere and for all time to come. Scientists make claims of great sweep and extent. The powers given to those who, through science, become "masters and possessors of nature" are vast and unprecedented.

In the science of the moderns, there arises a curious reversal of the order of art and nature. Art becomes the standard against which nature is judged. Francis Bacon's "nature in distress"—nature distraught by experiment—is as genuine as nature left alone. "The artificial," writes Bacon, "does not differ from the natural either in form or in essence, but only in the efficient." Art *improves* on nature: art is "nature with man to help." The new philosophy celebrates the Renaissance artisan's vision that glory is to be found not in nature but in works; this was what Leon Battista Alberti had meant when, speaking of Florentine art, he wrote that "the power to achieve widespread glory in art or science lies more in the pains we take and the precision we achieve than in the gifts of nature or of time."[12] New meaning is given to Marcus Terentius Varro's ancient claim that "God made the country, but man's skills made the towns."

The new appreciation of the value of artifice is reflected in the rise of mechanical models of the earth, the cosmos, and living beings. Copernicus in 1543 spoke of the "machine of the world founded by the best and most regular artificer"; William Harvey modeled the circulation of the blood on the mechanical pump. Descartes localized the seat of the soul in the pineal gland and explained the production of animal heat through mechanical friction (the heart, for him, was a "furnace"). Hobbes, in his *Leviathan*, asked "what is the *heart,* but a *spring;* and the nerves, but so many strings; and the joints, but so many *wheels,* giving motion to the whole body, such as was intended by the artificer?" For others man was clockwork, or bridgework, or some variety of automaton or *bête machine;* an Italian manuscript of the late sixteenth century depicts man as a distillation apparatus. French physician-philosopher Julien de La Mettrie would later carry this tradition to an extreme in speaking of *l'homme machine* (man a machine), and *l'homme plant* (man a plant).[13]

The conception of the cosmos as artifice (and God as the "great watchmaker") emerges with the rise of the crafts to a respected position in society. According to Edgar Zilsel, science in the modern sense emerges in the sixteenth century from the integration of three social groups: the literary humanists, who provided the tradition of letters; the Aristotelian scholastics, who provided the tradition of logic and mathematics; and the empirical artist-engineers, who provided the tradition of practical experiment and application.[14] The first artist-engineer workshops grew out of the specialized craft shops created to build Gothic cathedrals. Unlike in the guilds, where innovation had been frowned upon, in the workshops of the artist-engineers innovation was encouraged as independent shops competed to produce the more beautiful and complex cathedrals.

The turning point for Zilsel in the origins of modern science was the joining of the skills of the practical crafts and trades with the abstract principles of philosophy and mathematics and the literary skills of the Renaissance humanists. The craftsmen of the medieval guilds were almost entirely illiterate. Skills were passed on from master to apprentice, by word of mouth or example. In the fourteenth century, as technological innovation became increasingly a part of a craftsman's reputation, architects and painters began to write down their methods for distribution amongst their colleagues. The origins of these writings are not entirely clear—unlike humanist or scholarly writings, most were never intended to be preserved.[15]

The first known writing of an artist-engineer is the sketchbook of the French architect Villard de Honnecourt, from about 1235. Written in the form of an encyclopedia of the skills and technology of the time, the work was intended to instruct colleagues on the construction of siege machines, waterlocks, and automata. Villard's and similar writings were generally in the vernacular rather than in Latin; they are more like blueprints or building charts than systematic treatises. According to Zilsel, it was first among these literate craftsmen that we find the ideal of directional perfectibility so central in the modern ideal of scientific progress. The idea of continuous and indefinite perfectibility (of knowledge or skill) was generalized from the craft ideal of progressive technical improvement, and develops first in the works of military engineers and architects, painters such as Albrecht Dürer, and practical mathematicians such as Niccolò Tartaglia and Simon Stevin. The ideal of progress emerges from the marriage of two previously distinct traditions—the humanist ideal of perfection and the technical ideal of stepwise improvement by trial and error. The cyclical, nondirectional ideal of the humanists is rendered directional and practical through combination with the methods and ideals of the artisans; the artisans' conception of stepwise improvement is made readable and systematic through combination with the humanist literary tradition.[16]

The incorporation of craft skills into the armory of science and learning represented more than the creation of a new profession: it also represented a shift in the ontology of the moderns. This was what Otto Bauer meant when he claimed that the origins of materialism lay not in metaphysics but in the material demands of industrial civilization.[17] The central question asked in industrial life is: How can something be made? How can it be taken apart, that it might be put back together again? What are its elements, and the rules by which these can be combined? What are the forces that bind these elements, and what are the means by which these forces might be dissolved? If for the ancients the central problem of philosophy was *being*, the central problem for the moderns was that of *making*. *Homo faber* would analyze and move the earth, in ways others had scarcely imagined, by virtue of the newfound tools and perspectives. It was in this spirit that George Santayana argued that Filippo Brunelleschi's design and construction of the dome of the cathedral in Florence marked the beginning of modern science for it was then, in the 1480s, that humans first demonstrated that large blocks of stone would behave exactly as mathematicians said they would.

The conception of the cosmos as construct and craft, as "heavenly clockwork," emerges with a dramatic shift in the relation of humans to their natural environment. In the modern age, the nature to which humans must adapt is increasingly that of their own design. In 1400 cities still stood as islands amidst the wilderness. By the eighteenth century, when even the landscape has been engineered, trees and forests exist only by the grace of man. It is in this context that labor and the machine emerge in cosmological models.

The rise of the mechanical conception of nature, the conception of nature as the product of artifice, challenged the Greek ideal of theory. *Theoria* in Greek philosophy originally implied a contemplative spectator who maintained a certain distance from his or her object. The *theoros* was the representative sent by Greek city-states to public celebrations. Through *theoria*—that is, through looking on—he abandoned himself to the sacred events. The new ideals that emerged from the amalgamation of the learning of artist-engineers, literary humanists, and academic scholars in fifteenth- and sixteenth-century Europe repudiated this contemplative ideal, incorporating a practical dimension into the pursuit of theoretical knowledge. Science in the mechanical view implied that humans could control the world around them; knowledge becomes inextricably intertwined with the realization of practical human ends. This practical vision of knowledge is expressed most fully in the philosophy of Francis Bacon.

2

Baconian Caveats, Royalist Compromise

The rise of modern science represents a celebration of the gifts and agency of humans. And no single man can be considered a greater architect of that vision than Francis Bacon (1561–1626). "It seems to me," he begins his *Great Instauration*, "that men do not rightly understand either their store or their strength, but overrate the one and underrate the other." From an extravagant estimate of the arts they possess they seek no further; from a failure to imagine their actual powers they waste their time on trivial pursuits.[1] Bacon's self-appointed task was to create such methods as would allow the generation of great sciences, sciences that would be as useful as they were instructive.

Bacon's task was not a modest one: "I have taken all knowledge to be my province." For the glory of God and the relief of man's estate, Bacon proposed a vision of human understanding "entirely different from any hitherto known . . . in order that the mind may exercise over the nature of things the authority which properly belongs to it." The pursuit of knowledge was a divine task: Daniel's prophecy foretold that "knowledge shall be increased." For Bacon, it was "the glory of God to conceal a thing, but it is the glory of a King to find a thing out."[2]

Bacon was also of the view, however, that progress in the realm of knowledge had been hindered by the distance scholars had kept from the works of ordinary men. Bacon is an empiricist, promising to admit nothing to knowledge "but on the faith of eyes." Knowledge is to be gained not by sight alone but by muscle: scientific work is to be "like mines, where the noise of new works and further advances is heard on every side."[3] Science is to take instructions from spinning and weaving, from the crafts and trades, the mines and mills. Practical knowledge often precedes knowledge of principles; this was the message of Celsus, Bacon reports, that "medicines and cures were first found out, and then after the reasons and causes were discoursed; and not the causes first found out, and by light from them the medicines and cures discovered."[4]

Thomas Sprat in his 1667 *History of the Royal Society of London* celebrated this Baconian vision of the academy as partner with the world of the shops. Fellows were to learn not just from philosophers, but from "the Shops of *Mechanics;* from the Voyages of *Merchants;* from the Ploughs of *Husbandmen;* from the Sports, the Fishponds, the Parks, the Gardens of *Gentlemen.*" Royal Society fellows were to help construct models for "Houses, Roofs, Chimnies, Conduits, Wharfs, and Streets," helping also to perfect new techniques of "Graving, Timing, Statuary, Coining, and all the works of Smiths, in Iron, or Steel, or Silver." Sprat quotes Dr. Jonathan Goddard, reporting for the Royal Society on the possibility of making wine from sugar cane, that however unlikely such a project might seem "this is one of the greatest *powers* of the true, and unwearied *Experimenter,* that he often rescues things, from the jaws of those dreadful Monsters, *Improbability,* and *Impossibility.*"[5]

Method was to be the key to the New Learning. Genuine philosophy for Bacon had two parts—the rational and the moral—corresponding to the two faculties of the human mind: the reason and the will. Genuine science combined these faculties, satisfying itself neither with the passive light of reason nor the dull act of experience alone. For Sprat, too, utilization of proper methods would show that the secrets of nature need not be secrets at all. Sprat asserted that even if "all the Authors or Possessors of extraordinary inventions should conjure to conceal all that was in their power from them, yet the *Method* which they take will quickly make apparent abundant reparation for that defect. If they cannot come at Nature in its particular *Streams,* they will have it in the *Fountain.*" This fountain was the fountain of experience, tempered with rational and empirical methods. Knowledge was also to be augmented by the cooperative nature of the new society: knowledge was to be derived from the industry, use, and observation of men "united in Assemblies, where it is true for science, as for the passions, that the wits of men are sharper, their apprehensions lessened."[6]

This dependence on methods that anyone may use inspired a new politics of science. Baconian science was supposed to be "popular" science. For both Bacon and Descartes, the methods of natural philosophy were to work "as if by machinery," helping thereby to "level men's wits" and put men on an equal footing. Science was the fruit of good method, not of good men; the methods of science were to be like *machines* that any man might own and use. The consequences Bacon drew from this were radical: "however various are the forms of civil polities, there is but one form of polity in the sciences, and that always has been and always will be popular."[7] For Sprat, too, science was to be open to all, regardless of creed. Royal Society members were to be "freely admitted men of different Religions, Countries, and Professions of Life." The new philosophy was to be neither English nor Scotch, popish nor Protestant, but would combine the skills of each. The ideal philosopher would have the industry and inquisitive humor of the Dutch, French, and English, yet also the cold and wary disposition of the Italians and Spaniards. Science would provide a common language, uniting men above their petty differences.[8]

It is important, of course, to distinguish rhetoric from reality. Bacon may have intended his method to be popular, but this was by no means the result. Nicholas Hans has noted that among 680 scientists listed in the British *Dictionary of National Biography* born before 1665, 52 percent were from the aristocracy, a class that comprised less than 1 percent of the population. According to the charter of the Royal Society, anyone with the title of baron or above had the right to attend meetings and could be nominated for membership in the society. Furthermore, of the 200 members of the Royal Society in 1677, all were men. Democratic principles notwithstanding, the Royal Society did not have a single woman member until 1945, when two women (Kathleen Lonsdale and Marjory Stephenson) were elected to the society.[9]

Bacon's own views were far from democratic on certain issues. Scientists in the *New Atlantis* reserved the right to decide which discoveries would be revealed to the state and which would remain secret. Neither Francis Bacon nor William Petty were terribly democratic when it came to overseas affairs. Both spoke of the need to suppress the rebellious Irish; Bacon, for example, advocated the destruction of their "barbarous laws, customs, their brehon laws, habits of apparel, their poets or heralds that enchant them in savage manners, and sundry other dregs of barbarism and rebellion." William Petty suggested that "the Irish might in rigor deserve to be extirpated" for their risings.[10]

In the Baconian vision, science was to celebrate the glory of God, but it would also augment the comforts of man. Others echoed this aspect of Bacon's philosophy. In part six of his *Discourse on Method*, René Descartes described how his science was "to promote as far as possible the general good of mankind." Opposing the speculative philosophy taught in the schools, Descartes champions what he called a practical philosophy—one through which we may come to understand the nature and behavior of fire, water, air, and the stars "as well as we now understand the different skills of our workers." This would not only be desirable in bringing about "the invention of an infinity of devices to enable us to enjoy the fruits of agriculture and all the wealth of the earth without labor, but even more so in conserving health, the principle good and the basis of all other goods in this life."[11] Science, Descartes assures us, will make us "masters and possessors of nature."

The Baconian vision of the importance of the practical arts for science was a central aspect of the stance taken by the moderns in their quarrel with the ancients over the question of *progress*. In the late sixteenth and early seventeenth century, defenders of the ancients bemoaned the demise of classical arts and civilization; some even warned that trees and men had become smaller than in earlier times. Morality, wits, and even nature itself were said to have declined since the glorious days of Greece and Rome: who could claim that the works of Shakespeare surpassed those of Homer? How could one say that the thoughts of modern philosophers equalled those of Plato, Socrates, or Aristotle?

For the moderns, however, the proof of progress lay not in nature or even genius but in art, not in wits but in works. The signs of progress were most clearly evident in science and invention. The poetry and theater of the ancients may have been superior, but who among the ancients had ever traveled to India or to China? Ptolemy, writes Richard Eden in 1574, though an excellent man, had certainly never heard of America, nor of gunpowder, nor printing. "Our Age may seem not only to contend with the Ancients, but also in many inventions of art and wit, far to exceed them." George Hakewill, in his *Apologie*, agreed with Eden. The inventions of the ancients were good, but only "toyes and trifles" compared with "those most useful inventions, which these *latter ages* challenge as due and proper to themselves, *Printing, Gunnes,* and the *Mariner's Compasse* . . . All antiquity can boast of nothing equall to these three." These were new things, the kind of things that humanists of the Renaissance recognized as beyond the power of the ancients even to imagine. What, asks Lorenzo Valla, is the Latin for *compass,* or *cannon,* or *quadrant* or *windmill?* What do we call in Latin the air-filled ball, the tallow candle, or the chime clock? The moderns celebrate such things as proofs of progress, progress that depends not on men's character but on their method. Individual men of the past may well have been greater, but we moderns, by the cumulative effects of our works, are like "dwarfs standing on the shoulders of giants," so that even if small of stature, we still may see farther than those who have gone before.[12]

Interestingly, few early champions of modern science considered the mission of science as something that might continue indefinitely. Both Bacon and Descartes, for example, forecast the completion of their project (that is, the discovery of all the principles of natural philosophy!) if not in their own lifetimes then shortly thereafter. George Hakewill, first to launch the battle cry of the moderns, assumed his own age to be the last great age of the world, the mission of man on earth accomplished but for "the subversion of Rome and the conversion of the Jews." Charles Perrault, best known today as an author of children's fairy tales, wrote in a similar vein in 1688–89 that, on the basis of the arts and sciences, "our age has, in some sort, arrived at the summit of perfection." Perrault believed that the rate of progress had already begun to slow, and that it was therefore "pleasant to think that probably there are not many things for which we need envy future generations."[13]

For growing numbers of others, however, it was not out of the question that progress might continue forever. Bernard de Fontenelle, for example, in his 1688 *Digression on the Ancients and Moderns* distinguished science from poetry on the grounds that, whereas it was hardly possible to speak of progress in literature or the arts, progress in the sciences was both possible and without end. The Abbé de Saint-Pierre in his *Observations on the Continuous Progress of Universal Reason* (1737) forecast the improvement of human reason and wisdom for a hundred thousand years, if not blocked by war, superstition, and the jealousy of rulers.[14]

Associated with the spirit of progress is a new respect for that which is *new* in the world. Thus Tartaglia in 1537 writes a book with the title *Nova scientia;* Galileo three quarters of a century later proudly writes of *Two New Sciences,* not long after Johannes Kepler describes his *Astronomia Nova.* Giambattista Vico continues this tradition for the historical sciences with (in the eighteenth century) his *Nova Scienza.* Science, together with the voyages of discovery and the progress of inventions, was widely recognized as a leading beacon in creating the new. When Ben Jonson wrote about "News from the New World," he spoke not of America but of the "New World" of the heavens, discovered by the instruments of science.[15] Thomas Sprat in his apology for the Royal Society reasoned that it was "not an offence to foster the introduction of *New Things,* unless that which is introduced prove pernicious in itself." Nor should we be frightened of a thing simply because it is new: "If to be the *Author* of *new things,* be a crime; how will the first Civilizers of *Men,* and makers of *Laws,* and Founders of *Governments* escape? Whatever now delights us in the Works of *Nature,* that excells the rudeness of the first Creation, is *New.*"[16]

Science in the Baconian vision was to combine both practical and theoretical knowledge, challenging the Greek conception of the "purity" of theoretical knowledge unalloyed with practical concerns. Samuel Johnson, in good Baconian fashion, compared the "speculist" to one who, secure on solid ground, lectures on navigation as if the sea were always smooth and the winds favorable.[17] Natural philosophers were supposed to get their hands dirty. Typical of Bacon himself (or at least the Baconian legend) is that he died from a cold caught while trying to stuff a chicken with snow, apparently part of an experiment on the preservation of meat.[18] The values of the new science—novelty, democracy, utility, clarity, modesty, confidence in the possibility of progress, trust in "the faith of eyes"—were values foreign to classical ideals of scholarship.

Bacon did not, however, abandon entirely the ideal of theory separated from practice. As we shall see, he warned against two dangers: the danger of "harvesting the fruits of science before they are ripe" and the danger of the "moist heat of the passions" infecting the "dry light of reason." Both of these dangers—of knowledge directed too forcefully towards utility and of moral knowledge "tainting" the knowledge of nature—were to become important elements in subsequent justifications of the neutrality of science.

No revolution is a perfect one; continuities bind the past to the present. In its break with the scholastic heritage, Baconian science preserved an interest in truth "for its own sake"; the genius of Bacon's synthesis lay in his recognition of what we today might call the "usefulness of useless research." Bacon's ethic was a utilitarian one: knowledge is power, and the measure of our knowledge of nature is our ability to command her. Yet nature has limits, and laws, and will bend only so far. Nature, per Bacon's dictum, "to be commanded must be obeyed."[19]

Nature in Bacon's view must be imitated, but not without care, for there are two sorts of potential error. Nature imitated too closely reveals only the bare facts of observation; hence the follies of the "empyrick," who gathers facts without a guiding design or principle. And nature ignored as a result of the dogmas of the "reasoners" also yields no fruit: "The men of experiment are like the ant, they only collect and use; the reasoners resemble spiders, who make cobwebs out of their own substance. But the bee takes a middle course: it gathers its material from the flowers of the garden and of the field, but transforms and digests it by a power of its own." The prudent middle course is to combine the virtues of both the "empiric" and the "reasoner." Bacon supposes himself to have established "a true and lawful marriage between the empirical and the rational faculty, the unkind and ill-starred divorce and separation of which has thrown into confusion all the affairs of the human family."[20]

Nature for Bacon is of three kinds: *free,* as in the movement of the heavens; *in error,* as in the production of monsters; and *in bondage,* as in nature wrought by men into experiments. Nature reveals her secrets only in the third of these forms—not in freedom or in error, but only "under constraint and vexed"—that is, squeezed by human arts into shapes more convenient for study, according to the stringent demands of experiments.[21] Experimental science forces nature into new and unfamiliar forms. In this sense, modern science originates as much with the abandonment of common sense as with its cultivation.

What concerns us here, however, are the qualifications Bacon raises against his general view that knowledge is power and dependent on practice. In several of his works, Bacon cautions against the disregard for "knowledge for its own sake." In the parable of Atalanta, repeated three times in the *New Organon* and explicated at length in his book of fables, Atalanta stops to pick up the apples cast in her path and thereby loses the race. The moral of the story is that it is dangerous to seek the quick and easy profit, to harvest the fruits of science before they are ripe. It is not that the practical arts are ignoble pursuits. The sun "enters the sewer no less than the palace, yet takes no pollution" therefrom. Indeed, "the roads to human power and to human knowledge lie close together, and are very near the same." Yet we must beware as well, for there are those who will too hastily try to bring their researches to useful ends, forgetting that the Lord in his method did material work only after a day wherein He created only light. However eager we may be for the fruits of science, let us not ruin the project but rather "wait for harvest-time" and not attempt to "mow the moss" or "reap the green corn."[22]

Sprat, too, warned against the corruption of learning produced when men stoop "to consult profit too soon." It can only weaken the sciences when "not the best, but the most gainful of them flourish." The seeking of early profit busies a man about possessing "some pretty prize, while Nature itself, with all its mighty Treasures, slips from them," like foolish guards who, while picking up coins that a prisoner has dropped "let the prisoner himself escape, from whom they might have got a great ransom."[23]

Science is to be of use, but not planned exclusively for that use. In the *New Organon,* Bacon distinguishes between experiments of fruit *(experimenta fructifera)* and experiments of light *(experimenta lucifera):* experiments which are of immediate use, and experiments which are not. Science should be pursued "for its own sake" and not for whatever short-run utilitarian ends may result. He laments the fact that natural philosophy "has scarcely ever possessed, especially in these later times, a disengaged and whole man (unless it were some monk studying in his cell, or some gentleman in his country house)." Science has been like a servant, attendant upon medicine or mathematics but never courted "with honest affection and for her own sake." Like happiness, utility is something found only when not sought. Science is thus like fortune, having "somewhat of the nature of a woman, that if she be too much wooed she is the farther off."[24]

This, then, is the first Baconian caveat: that though knowledge is power, and should be based on and contribute to practical knowledge, the power of science is not to be gained directly, or immediately, but only through the patient accumulation of knowledge through experiments. There is also a further caveat, however, one that becomes central to ideologies of subsequent centuries. This is the idea that progress in the sciences has been hampered by the taint of moral knowledge.

In his *New Organon*, Bacon proposes that progress in the knowledge of nature has been hindered by various "idols" that distort our knowledge, mingling into the true nature of things the colorations of our own human qualities. Idols of the *tribe* are those frailties of human nature which warp our vision of things, as a false mirror bends the true light of nature. Idols of the *cave* are those particular foibles of the individual which further refract the light of nature, according to one's own particular education and experience. Idols of the *marketplace* are those frailties derived from human association, as when the use of ill and unfit words forces and overrules the understanding. And finally, idols of the *theater* are those philosophical dogmas preached in the learned schools that prejudice one's thoughts and opinions.[25]

Distortions can also arise, however, from inquiries into the nature of morals. Bacon separates moral and natural knowledge as belonging to separate spheres. The "proud knowledge of good and evil" is the product of faith and communion with God, whereas the "pure knowledge of nature and universality" is the product of observation and communion with nature. It is important, he says, not to mix these two kinds of knowledge. Bacon warns that moral concerns can easily distort the perceptions of the mind—for human understanding "is no dry light," but receives a harmful infusion from the will and affections. For what a man had rather were true, Bacon writes, he more readily believes. A man thus rejects difficult things "from impatience of research," sober things "because they narrow hope," and the deeper things of nature "from superstition." The affections color and infect one's understanding in innumerable ways. Concern for moral knowledge, based on these affections, Bacon considers to have been a hindrance to the progress of science since ancient times. The seven wise men of ancient Greece (all but Thales) applied themselves not to natural philosophy but to morals and politics. As a consequence, even after Socrates had drawn down philosophy from heaven to earth, "moral philosophy became more fashionable than ever, and diverted the minds of men from the philosophy of nature." It is directly after this passage that we hear Bacon's complaint that philosophy has never possessed "a disengaged and whole man," practicing science "for its own sake."[26]

Both of these caveats—the danger of knowledge applied too soon and the taint of moral knowledge—are important in subsequent conceptions of the neutrality of science (see Part II). An element of fear was also involved. In his *Great Instauration*, Bacon writes that while knowledge of nature is safe, knowledge of morals invites danger. It was not "pure and uncorrupted natural knowledge" that gave occasion to the fall of man, it was "the ambitious and proud desire of moral knowledge to judge of good and evil to the end that man may revolt from God and give laws to himself."[27] Moral knowledge is thus dangerous knowledge. It was recognition of this danger that resulted in the rejection of morals as the proper object of study for the later Baconians, in circumstances surrounding the "royalist compromise" of early English science.

Baconian science in the years of the Interregnum (1640–1660) under Cromwell was an integral part of larger struggles for political, religious, and educational reform. Science was to serve the glory of God and the good of man; early natural philosophers were optimistic for the prospects of their new endeavors. As Thomas Macaulay tells it, philosophers were enchanted as "dreams of perfect forms of government made way for dreams of wings with which men were to fly from the Tower to the Abbey, and of double-keeled ships which were never to founder in the fiercest storm."

But the scope of science is often constrained by the social forms into which it is organized. With the restoration of Charles II to the throne in 1660, science was restricted in the ground that it might cover. The king's decree granting fellows of the Royal Society of London rights to assemble, to correspond with foreign members, to publish free from censorship, to dissect bodies of executed criminals, and so forth, was granted with the understanding that the "Business and Design" of the society, in the words of Robert Hooke, was "To improve the knowledge of natural things, and all useful Arts, Manufactures, Mechanics, Practices, Engynes and Inventions by Experiments (not meddling with Divinity, Metaphysics, Moralls, Politicks, Grammar, Rhetoric, or Logick)." Sir Robert Moray, the "soul" of the early Royal Society by one account, in a letter of 1665 to Christiaan Huygens pointed out that the *Philosophical Transactions* of the society would be "much more philosophical" than its French counterpart, the *Journal des Savants,* but still that it would not "interfere with legal or theological matters." Thomas Sprat, too, proclaimed that the subjects investigated by the Royal Society could include God, man, or nature; but that with respect to the first, the members of the society would "meddle no otherwise with *Divine things,* than onely as the *Power,* and *Wisdom,* and *Goodness* of the Creator, is displayed in the admirable order, and workmanship of the Creatures." Science, in other words, would celebrate God's creations, not challenge theological doctrines.[28]

The institutionalization of modern science is also marked by new attempts to define the scientific personality. Sprat, in his *History of the Royal Society,* argued that the calm of the natural philosopher was to be contrast with the reformist zeal of the political or religious enthusiast. The ancients, Sprat claimed, were men of hot and hasty minds, preoccupied with war and strife. The Romans only conquered, the Christians only quarreled. But the men of the Royal Society, by contrast, had labored "to separate the knowledge of *Nature,* from the colours of *Rhetorick,* the devices of Fancy . . . ; to enlarge it from the service of private interest." Natural philosophers had pledged to maintain that "calmness and unpassionate evenness of the true Philosophical Spirit," a spirit lacking among the proponents of scholastic dogma. Dogma, Sprat wrote, that "most pernicious of tempers," leads men to undervalue other men's labors, even to fight against them. Natural philosophers reject dogma, along with all other moral disputes: "the Natural Philosopher is to begin, where the moral ends."[29]

On the Continent, too, scientific academies proclaimed the "neutralization" of science in the course of its political incorporation. The Accademia del Cimento announced its intent to avoid "Controversie with any, or engage in any Nice Disputation, or heat of Contradiction." And Christiaan Huygens, describing early plans for the French Académie Royale des Sciences in his 1663 "Projet de la Compagnie des Sciences et des Arts," suggested the proviso that "in sittings of the Académie neither the secrets of religion nor the affairs of the state shall be entered upon; and if, from time to time, discussion should turn to metaphysics, or morals, or history or grammar, this shall only be in passing and insofar as they pertain to physics or to exchanges among men."[30]

George P. Murdock argues in his *Culture and Society* that professionalization always involves the loss of something, through restrictions on the form and content of a discipline.[31] Professions are exclusive. This appears to be true, not just for recent professional associations but for the formation of any school or discipline. Above the door of Plato's Academy were inscribed the words, "Enter only those who know geometry." Thomas Hobbes excluded God and the angels from philosophy, on the grounds that philosophy could study only that which shows evidence of growth or decay—which supernatural beings do not. Hobbes also excluded history, both natural and political, on the grounds that historical knowledge was "but experience, or authority, and not ratiocination"; astrology, on the grounds that it was not well-founded; and doctrines centered on the worship of God, on the grounds that these were based on faith, not knowledge.[32] William Petty in his *Political Anatomy of Ireland* and his *Political Arithmetick* declared that he "professes no politicks" and that he intended to express himself only "in Terms of *Number, Weight,* or *Measure;* to use only Arguments of Sense, and to consider only such Causes as have visible Foundations in Nature; leaving those that depend upon the mutable Minds, Opinions, Appetites and Passions of particular Men, to the consideration of Others."[33]

Petty's oft-quoted maxim demonstrates that restrictions in modern science extend to questions of form as well as content. The language scientists should use becomes of particular importance; indeed, *how* something is said is often as important as *what* one is saying—form is often a key to content. Bacon and his seventeenth-century followers insisted that the works of science be presented in plain and ordinary language, appearing "naked and open" for all the world to see and judge. Descartes asked that philosophical ideas be "clear and distinct," and hence amenable to reasoning and empirical test. John Wilkins's 1668 *Essay Towards a Real Character and a Philosophical Language* championed "the distinct expression of all things and notions that fall under discourse" in a language in which things—not words—would be primary and well-chosen words would reflect the true nature of things.[34] The achievement of clarity was to be a historic mission: in regaining the original "nakedness" and simplicity of speech, the confusions of the scholastics might thereby be overcome and the understanding men shared before the Tower of Babel might once again be restored. "Ornaments of speech" were to be avoided, as philosophy would move from words to things, from the artificial languages of the various nations to the natural language of the Garden.

Incredible as it seems today, much early discussion of the question of style in early modern science was associated with this effort to restore the language and philosophy of Adam. Adam, in Bacon's view, had spoken before the Fall a "pure language," recognizing the true names of things with a "nakedness of the mind" that was "as nakedness of the body once was, the companion of innocence and simplicity." The goal of philosophical language was to restore the nakedness of speech men shared before the Fall. This was a goal of the Royal Society of London, as well. Both Wilkins and Robert Boyle proposed remedies: Boyle in 1647 expressed his hopes that a clarification of language "will in good part make amends to mankind for what their pride lost them at the tower of Babel." Henry Stubbe, the Galenic physician, found the Baconian attempt to link the mission of science with the return to grace a curious one—so curious, in fact, that he chided his fellow virtuoso Thomas Sprat for presenting the matter "as if natural and experimental philosophy, not natural theology, had been the religion of paradise."[35]

Efforts to replace the scholarly language of Latin by local or national dialects can also be understood in this context. Latin came under attack in the early decades of the century, as philosophers sought a broader audience for their works outside the universities. Galileo published his *Dialogue Concerning the Two Chief World Systems* in Italian, having it translated into Latin shortly thereafter. Bacon wrote a number of his works in English. Descartes, whose own *Discourse on Method* (1637) was published in French, once claimed that "there is no more sense in studying Latin and Greek than old Breton or Swiss German."[36] By the end of the seventeenth century the use of Latin in scholarly publications was in rapid decline. Newton's *Principia* was published in Latin in 1687, his *Optics* in English in 1704. While Latin remained in use in certain fields (medical and scientific nomenclature, for example), by the middle of the eighteenth century it had become anachronistic for English or French scholars to publish in Latin. In

Germany, the medieval scholastic tradition lasted a few decades longer. Kant's inaugural dissertation of 1770, for example, was published in Latin. Even in Germany, however, publication in Latin after 1800 was virtually guaranteed to relegate a work to obscurity.

Opposition to Latin was, among other things, political. Gutenberg's invention of printing by moveable type (in 1453) made possible the mass distribution of written materials; scholars who continued to write in a language foreign to their countrymen could be criticized as elitist. Nicholas Culpepper in 1648 accused physicians of using Latin to hide the truth from their patients; English Puritans identified Latin with the despised Catholic Church. John Webster, an early advocate of introducing Baconian methods into British universities, ridiculed the learning of Latin as "toilsome, almost lost labor."

At the same time that Latin began to decline as the *lingua franca* of Western scholarly discourse, others sought to devise some "artificial character" to replace Latin as a common medium of communication. Chinese and hieroglyphics were held up (wrongly) as languages capturing reality not by words but by pictures; the Lunarians of Francis Godwin's 1638 *Man in the Moon* (the first English work of science fiction) were given to speak a universal language clearly modeled on Chinese (consisting "much of Tunes"). Descartes, Wilkins, and Leibniz each sought to create a universal character—an effort that for Leibniz resulted in his invention of the "base two" number system used in computer assembly languages today. Wilkins hoped that the construction of a common language might heal the religious conflicts ravaging England.[37]

One important consequence of efforts to eliminate ornaments of speech from scientific discourse was a dramatic narrowing of the acceptable bounds of scientific style. Metaphor and analogy both came under attack as either useless or deceptive. Hobbes complained that "men use words metaphorically; that is, in other sense than that they are ordained for; and thereby deceive others." Sprat urged that eloquence "be banished out of all civil societies, as a thing fatal to peace and good manners." Samuel Parker, also of the Royal Society, advocated an act of parliament to stop preachers from using metaphors.[38] A century later, it was in revolt against such attitudes that romantic philosophers such as Rousseau accused Descartes of having tried "to cut the throat of poetry."

Wolf Lepenies has described what he calls the *Entliterarisierung* (deliterarization) of the sciences—the gradual acceptance by scientists of the belief that the presence of poetry or other literary frills in a scholarly text was grounds for suspecting its scientific substance. In the eighteenth century, poetic expression is increasingly contrasted with the clarity and honesty of scientific style. Georges Louis Leclerc, comte de Buffon, one of the most popular writers of the century (his *Histoire Naturelle* went through

250 popular editions), was accused of having sacrificed substance for style, and his career eventually suffered. In one instance he was derided for introducing a new member to the Académie Française as *ce jeune homme de trente ans,* when in fact the man in question was *vingt-sept ans,* which Buffon apparently felt to have an insufficiently poetic ring. His early works were written at a time when literature was not yet divorced from science; by the end of his life, and especially in the wake of the French Revolution, his polished "royal" prose had fallen from scientific favor. Erasmus Darwin's 1798 *Loves of the Plants* may be the last example of an original contribution to science written entirely as a poem.[39]

The scientific attack on superfluous language or flowery rhetoric was part of a larger attempt to replace the elaborate prose of the aristocratic man or woman of letters with a more concise, "business-like" style suited to the practical mentality of the emerging bourgeoisie. In England, it was also an attack on other aspects of society—notably women, the French, and the ancients. One should recall the class and gender character of early modern science. Baconian natural philosophy was championed by men who felt themselves more comfortable in the shops of merchants than the halls of academe or the courts of nobility; the flowery prose of the aristocrat was a luxury the men of science did not want. In Germany and in France, women were often leaders in the aristocratic salons where philosophers met for informal discussion or experimentation. When philosophers criticized ornamental speech as feminine or aristocratic (or French, as the British sometimes argued), it was often the woman-dominated, salon-style of inquiry they had in mind. Rousseau attributed the decay of French intellectual life to the power of the salons (where women gathered together a "harem" of effeminate men); Pierre Bayle claimed that the marriage of a man of science was a waste of national resources. When the salons collapsed during the French Revolution, so did one of the institutions where women had been able to pursue philosophical inquiry.[40]

Women were excluded from the leading academies of Europe—the Royal Society, the Académie des Sciences, the Berlin Akademie der Wissenschaften—until late in the twentieth century, despite the fact that women often engaged privately in science. But women were not the only group excluded. In 1784 the Royal Society of London announced that shortly would be published *A History of certain late Exclusions from the Royal Society,* "In which it is shewn, that neither Mathematicians, nor Country Physicians, nor London Physicians, nor Army Agents, nor Practical Astronomers, nor Authors, nor Scholars by Profession of any description, are proper persons to be made Fellows of the Royal Society of London, instituted for the Promotion of Natural Knowledge."[41]

The exclusion of "Moralls, Politics, and Rhetorick" was therefore only part of a larger series of exclusions touching upon the form, content, and personnel of science in seventeenth- and eighteenth-century Europe. Part of this, as I have already suggested, concerned certain notions of what was required to secure reliable knowledge. But there was also a political element. Moral knowledge was dangerous knowledge, and to secure institutional legitimacy, the founders of the first scientific societies promised to ignore moral concerns.

The conformity of science to intellectual fashion or political expediency has a long history, and will no doubt have a long future. J. L. McIntyre, in his book on Giordano Bruno, recalls how, according to the statutes of Oxford prior to the seventeenth century, "Bachelors and Masters who did not follow Aristotle faithfully were liable to a fine of five shillings for every point of divergence, and for every fault committed against the Logic of the Organon."[42] In the institutionalization of modern science, it was not just a case of conformity but of exchange: science needed funding, and the Crown needed gunpowder.[43] In the bargain struck, scientists received funds and social space to work (especially freedom from censorship), but traded away all rights to moral disputation and political engagement. Self-censorship replaced state censorship in the bargain that was struck. The institutionalization of science, in other words, was bought at a certain price: science would not interfere in matters of state, if the state would not interfere in matters of science.

3

The Devalorization of Being

The emphasis upon scientific neutrality in Restoration England was part of a broader shift in the place of value in the world—a shift that sees the rise of a new and quintessentially modern ethical ontology. Value for the ancients lay in the world itself—a hierarchy of perfection rose from the fires of hell to the spheres of heaven, from earthly *potentia* to celestial *actualia*. For the ancients, the natural order was an ethical order—there was *telos* in all *physis*—the order of nature was not separate from the order of the good. The ethical order was a natural one: morality was rewarded with good crops and good fortune; crimes were punished by plagues or natural disasters.

For the moderns, things are different. Alexandre Koyré in his *From the Closed World to the Infinite Universe* has argued that the key transformation in the rise of modern science was not, as some have argued, the change from *theoria* to *praxis,* or from *scientia contemplativa* to *scientia activa,* but rather the destruction of the medieval cosmos and the "geometrization" of space. The Aristotelian universe was an ethical universe: all things moved toward their "natural ends" in concert with that perfect harmony that ruled the universe. The cosmos was also hierarchical, ranked from the meanness of earthly things below to the perfection of the starry sky above. On the earth, the four elements rose from one to the other in ascending layers—first earth, then water, air, and fire. Above these four terrestrial elements (or essences) was the fifth and perfect essence (the so-called *ether* or *quint-essence*), constituting the realm of the seven planets (including the sun and the moon) and the fixed stars, all perfect and unchanging.

For the ancients, each of these spheres—the terrestrial and celestial—had its own distinctive laws. The "ethereal" celestial sphere circumnavigated the earth with a motion described by that simplest of all curves, the circle (or circles upon circles if observation so demanded, as in the perennial problem of the retrograde motion of the planets). Objects in the terrestrial sphere—including everything below the sphere of the moon—moved according to a quite different physics. Earthly objects possessed either gravity or levity, causing them to want to move either down or up. That which was heavy tended to fall toward the earth; that which was light tended to rise toward the heavens. Observation accorded with this hierarchy—air forced into water rose as bubbles, earth placed in water sank to the bottom. All things in this world occurred for a purpose: the arrow shot into the sky returned to the earth, its natural place, and in doing so fulfilled its natural end.

With the rise of modern science, however, the closed and hierarchically ordered world of the ancients is replaced by a geometrized, Euclidean conception of the universe as infinite in extent and everywhere the same. The turning point in the rise of modern science, for Koyré, is the rupture of the natural from the moral order: the world of value is finally torn from the world of fact. Nicholas of Cusa in the fifteenth century was among the first to challenge the doctrine of two entirely separate spheres—one perfect and heavenly, the other corruptible and earthly. The inhabitants of the moon and the sun were still, in Cusa's view, more perfect than those of the earth. But there was now only one universal world, within which everything influenced everything else; hence, one had no reason to believe that change and decay occurred only on the earth and not everywhere in the universe. Corruption and decay was everywhere the same, for the spirit of God was in all things and in all things the same.[1]

Galileo was to continue the process of recasting the universe. With his telescope, he is the first to find (in 1609–1610) that the sun has spots, the moon has mountains, Saturn has rings, Venus has phases, and Jupiter has moons. The heavens, in other words, are not perfect. But neither is the earth ignoble. In his *Dialogue Concerning the Two Chief World Systems,* Galileo has Simplicius ridicule those who value "precious" jewels over "base" earth and soil. It is only scarcity or plenty that makes things precious or worthless; indeed, he says, if soil were something difficult to find, princes would exchange a cartload of gold for soil to plant a jasmine in a pot or to sow an orange seed and watch it grow. Galileo criticized those who found beauty only in the timeless or immutable—it was fear of death, he argued, that made them loathe the transient and earthly, numb to the beauties of a fragile earth. With others of his time, Galileo articulated a different view—that the earth is nobler the way it is, alterable and in flux, than if it were a mass of stone, "even a solid diamond," hard and invariant.[2]

Cusa's and Galileo's are new views of nature—views that challenged the classical ideal of a corrupt earth beneath the perfect heavens. But more was at stake than the question of the nobility of the earth or the corruptibility of the heavens. In Charles Gillispie's words, Galileo "stripped from the skeleton of the cosmos the obscuring layers of sentience and pious moral and edifying lesson," leaving us instead "the hard, straight bones of Euclidean dimension, Platonism bleached bare, sterilized of its mystical nonsense in the Tuscan sun."[3] The scientific revolution is a revolution in our views of *value,* not just our views of *nature.* Value in the modern world is a human creation—the product of human arts and labors. The natural order is no longer a moral one; the cosmos is indifferent to the plight of humans. Things may be good or bad, but only in their relation to humans or their actions. Something is good if it pleases us or may be put to use. Value is no longer etched in the nature of things: "Being," in the expression of Koyré, has been "devalorized."

The changes that emerge with this "devalorization of Being" are dramatic. Consider the transformation in the idea of *cause.* Aristotle in his *Physics* had argued that in order to understand a thing one must consider four causes: material, formal, efficient (or moving), and final (there was also the "first cause" or unmoved mover—God—responsible for setting everything in motion in the first place). A house, for example, is a product of each of these four causes. Its material cause is *wood,* the substance of which it is made. Its formal cause is the plan or *design* according to which it is made. The moving cause is the *carpenter* who puts the pieces together; and the final cause (or goal) is the house itself, providing *shelter* for those who live in it.

All that remains today of Aristotle's four causes is the "moving" or efficient cause. The shape or substance of a thing we no longer consider part of its cause; nor are we satisfied with assurances that it rains "to make the crops grow." The ancient world view was *teleological* (from *telos* = goal or purpose)—to know a thing was to know its purpose, not just its form or substance or the moving force bringing it into being.

The teleological world view comes under attack in early modern science. Aquinas had already restricted final cause to the end-product itself (an olive tree—not the human use of its oil—is the "final cause" of an olive seed). And by the turn of the seventeenth century the notion of final cause has been largely expunged from physics. Francis Bacon restricts the object of physics to the study of "the efficient cause, and of matter."[4] Galileo, in his pioneering treatise on *Two New Sciences* (statics and kinematics), abandons the entire notion of causes. He still distinguishes between "violent" and "natural" motion (the upward and downward motion of a projectile, for example), but these are no longer entirely different qualities but rather conveniences in describing the motion of a body—first decelerating upward then accelerating downward, the difference being only in the mathematical sign of the movement. Galileo was not interested in the cause or quality of motion at all, but rather in what he called the axioms of motion—mathematical principles which could be used to predict the movement of an object in a general fashion. (The language of purpose, goal, and design continues in fields outside physics, as we shall see in a moment. Robert Boyle reported that William Harvey discovered the circulation of the blood after his realization that "so Provident a Cause as Nature" would not have placed valves in the veins unless there was some design or purpose for them, namely, to keep the venous blood moving toward the heart and the arterial blood away.)[5]

The fate of magic is also tied to these changes. Magic begins to decline, as links are broken between words and things, natural forces and moral fortunes. Power in the secular order is no longer vested in words or coincidence or similitude, but only in things. Paracelsus had thought that ferns could be used as medicine for stab wounds, because the leaves look like stitches. Palmists and physiognomists believed that the fate of a man could be read in the lines of his hands or the bumps on his head. For the magi, an intricate web of correspondences joined macrocosm and microcosm; it was possible to believe that music made from the strings of wolf gut and sheep gut would not harmonize, because of the struggle that must always exist between these animals. For the moderns, these links are broken. Diamonds no longer destroy magnets, garlic no longer detects unfaithful women.

The relations of science and magic are now thought to be more complex than when Bronislaw Malinowski contrasted science, "founded on the conviction that experience, effort, and reason are valid," with magic, based on the belief that "hope cannot fail nor desire deceive." Historians have shown that many of the earth-shaking achievements of the scientific revolution—Copernicus's model of the heliocentric universe; Harvey's discovery of the circulation of the blood; Bruno's postulate of the infinity of worlds—were inspired by magical (and specifically Hermetical) ideas. Furthermore, certain goals of magic—the achievement of some practical effect through manipulation of the forces of nature, for example—were close to those of early modern science. The magi wanted to transmute base metal into gold, heal wounds, or otherwise increase one's fortune by harnessing occult or hidden forces. Magic was like technology, except that it appealed to esoteric forces rather than those commonly known. In fact, as Bert Hansen points out, what separated magic from other forms of technical skill was often nothing more than this esoteric character: treatment with a rare medicinal herb, for example, might be considered magical in one context but quite unmagical in another, once the procedure became generally known. (Whether the procedure was effective was not necessarily relevant for its status as magic.)[6]

Early modern philosophers rejected the secretive character of magic. Techniques that could be made public were to be accepted; those that were to remain secret were not. Like magic, modern science was practical; but unlike magic, it was public. Philosophers also came to reject the *kinds of forces* appealed to in magical manipulations. Magicians had long appealed to nonmaterial forces—principles of affinity, contagion, and correspondence—to produce some desired effect. According to the "doctrine of signatures," for example, every herb bears some visible sign of its medicinal value (herbs used to treat liver ailments, for example, were likely to be liver-colored). According to the doctrine of correspondences, various parts of the world answer to or resonate with one another, independent of any physical contact. The Rosicrucian physician Robert Fludd (1574–1637) believed not only that magnetism could be harnessed to cure certain ills, such as a knife wound, but that one should apply the salve *to the weapon* that inflicted the wound—in this case, the bloodied knife. The logic here was that the wound and the wounding instrument are connected by a series of correspondences and that by appropriate means it is possible to draw lost spirits back into the body. Magical manipulations generally assumed both that "like effects like" and "like affects like."[7]

Early modern scientists rejected the principle of affinity, supposing instead (with Aristotle) that effects must be proximate to their causes. They rejected the power of words and proposed a science that would focus instead on things. They rejected the notion of a correspondence between micro- and macrocosm in favor of the view that there are universal laws that control both the large and the small. And they rejected the secrecy enjoined by magicians and tradesmen alike, protesting that science advances more easily in the light than in the dark.

It is true that, even into the latter part of seventeenth century, many natural philosophers entertained the possibility of one or another form of natural magic. Both Galileo and Newton still dabbled in alchemy (Newton also wrote a book on the Apocalypse); Bacon considered magic (as part of metaphysics) to allow an even greater command over nature than could be gained through physics. But such views were increasingly in the minority—and Newton, one should note, kept his alchemical work secret. John Ray rejected the idea of signatures in the classification of plants; Descartes disparaged the "disreputable doctrines" of alchemy and astrology in his *Discourse on Method*. John Harris's 1704 *Universal Dictionary of the Arts and Sciences* dismissed astrology as "a ridiculous piece of foolery" and alchemy as "an art which begins with lying [and] . . . ends with beggary."[8] The secrecy in which magical principles had been shrouded also began to be lifted. Magnetism and electricity, previously cloaked in magic, received a purely naturalistic, even mechanical explanation. Much of alchemy was incorporated into chemistry, much of medical magic into medicine. By the nineteenth century it becomes difficult for many even to imagine that philosophical giants like Newton and Galileo had concerned themselves with such affairs.⁹

Secrecy, emphasis upon the power of words, the principles of affinity and correspondence, the doctrine of signatures, the view that microcosm reflects macrocosm—these were all eventually rejected in one form or another in early modern science. But more even than the form of discourse or the character of physical force was involved in this rejection. Ultimately, it was the *ethical ontology* of the magical world view that put it at odds with modern science. In the magical view, certain numbers were inherently good or evil; events of the heavens spelled doom or fortune for people on earth. The magical world presupposed that good and evil are *cosmological* concepts—and that with proper images or incantations one could harness the forces of the cosmos and turn them to one's advantage.

For materialists, such ideas were anathema. Thomas Hobbes, for example, recognized that humans tend to project human qualities onto inanimate objects; in his great *Leviathan,* the English materialist noted that people tend to measure other things, as well as other men, against their own feelings and desires. People assume that things in motion tend toward rest, like men who, growing weary, seek repose. Whereas Aristotelian philosophers claimed that bodies fall out of an appetite to rest, moving to occupy their natural and proper place, Hobbes says that in reality it is only man who has an appetite, only man who has passion and reason. *Man* and *rational* are terms of equal extent.

Value in Hobbes's view lies not in the structure of the cosmos but in the productions of humans. Value is made by humans and hence will be estimated differently among them. All men feel passions; all desire and fear, hope and love. But the objects of passions will depend upon one's education and constitution. Desires are at base material, "motions" toward things that have excited our appetites. And judgment that a thing is good is nothing more than an expression of our feelings of desire—our tendency of motion—toward that thing. Good and evil are only words used with respect to a certain person's desires or aversions. There is thus no moral absolute, no "common rule of good and evil to be taken from the nature of the objects themselves"; these are taken from the judgment of man, or, when in commonwealth, from an arbiter or judge acting as representative of the people. Pleasure is the appearance or sense of good; displeasure the sense of evil. But good or evil do not exist in the things themselves: there is no ultimate end or greatest good *(finis ultimus, summum bonum).*¹⁰

Critiques of cosmology, causality, and magic were all elements in the collapse of value in the universe, the "devalorization of Being." Consider one further example: the curious case of trials and punishment of animals in early modern Europe. Odd as it seems today, from the thirteenth through the eighteenth century animals were brought before European courts of law and prosecuted for various crimes. Pigs, horses, birds, and even insects could be tried, convicted, and punished for crimes such as homicide, theft, and the destruction of property.[11] In 1386, for example, an infanticidal sow was executed in the Norman city of Falaise, a scene which was depicted in the west wall of the Church of Holy Trinity in that city. In 1516, officials of Troyes pronounced a sentence on certain insects which had destroyed the local vines. And as late as 1685, a wolf supposed to be the reincarnation of a deceased mayor of Ansbach was captured after having devoured several women and children; the wolf was tried for murder and, with its carcass dressed as a man with a chestnut wig, mask, and long white beard, was finally condemned to be hung by order of the local court. Such prosecutions continued in sporadic forms, even into the eighteenth century.[12]

Many early trials of animals were connected with church practices. The Catholic Church would typically excommunicate animals such as rats, mice, snails, snakes, or frogs for creating a pestilence or other nuisance. Such acts could be preventive as well as vindictive. In 1710 the bishop of Grignon, near the Côte d'Or, excommunicated a group of insects for destroying trees and thus robbing the local people of fruit.[13]

What these examples illustrate is that until well into the eighteenth century, nature commonly carried moral weight and implication. With the growing separation between natural and positive justice, however, as between secular (material) and spiritual (nonmaterial) life, human moral and nonhuman natural spheres are increasingly isolated from one another. Moral lessons begin to disappear from books on nature; the cosmos is conceived less and less to "care" about the fate of humans. In 1668, Jean Racine mocked the prosecution of a dog for a stealing a capon in his *Les Plaideurs*. Two decades later, Pierre Bayle could scoff at the idea that the comet of 1680 spelled doom for the peoples of Europe. By the beginning of the nineteenth century, most of Europe's leading philosophers were united in arguing that nature cannot be seen as the source of either moral messages or proofs of the existence of God.[14]

Most, but not all. Especially in Britain, the early nineteenth century witnessed the flowering of the movement known as *natural theology,* a body of doctrine that was to become the object of Darwin's ridicule but also his respect. With its strong emphasis upon the presence of God's hand in nature, natural theology represented a countervailing trend in the move to separate religion from science, an argument against the view, attributed to Cardinal Caesar Baronius in the sixteenth century, that the purpose of the Bible was to teach us "not how the heavens go, but how to go to heaven."

We are of course familiar with stories of religious persecution of scientific innovators. Giordano Bruno was burned at the stake in 1600 for upholding the doctrine of the plurality of worlds; Galileo was imprisoned for having taught that it is the earth, not the sun, that moves. And there are many other examples: Linnaeus was denounced for having classified plants by their sexuality; Buffon was forced to recant his theory of the earth as "contrary to the narrative of Moses."

The boundary between science and religion, however, is a shifting one. Prior to the second half of the nineteenth century science and religion were often seen as compatible or even inseparable. Many scientists were trained for the clergy: Pierre Gassendi (1592–1655), for example, was an ordained priest who attempted to reconcile mechanistic atomism with Christian dogma; Robert Boyle (1627–1691) bequeathed in his estate monies to found a lecture series devoted to the defense of Christianity against unbelievers. John Wilkins was a bishop in the Anglican Church, as were Thomas Sprat and Samuel Parker. Most natural philosophers were either rational supernaturalists or defenders of some sort of "natural religion." (Darwin himself began his education training for the ministry.) Scientists commonly studied nature to celebrate the power and wisdom of God's creation: this was the natural-theological vision, the view that (in the Dutch naturalist Jan Swammerdam's words) if one were astute, one could find proof of "the glory of God in the anatomy of the louse."

Natural philosophers held sharply differing views on the question of how or even whether God intervened in the workings of the world. Newton believed God to be continually acting in the world; indeed, gravity itself was made possible by the sensorium of God in the world. God made possible action at a distance; God intervened periodically to re-adjust the planets in their orbits. For Descartes and his followers, this idea was repugnant. God did not intervene in the natural course of events; instead, He created natural laws when He brought the world into being, thereafter retiring to the sidelines. (Laplace in the eighteenth century proved that the periodic "adjustments" postulated by Newton were not necessary for the solar system to remain stable.) Alexandre Koyré distinguished these two kinds of views by speaking of the "God of the workday" and the "God of the Sabbath."

For those who celebrated the hand of God in nature there was one perennial problem: if nature is the product of intelligent design, why is there so much *evil* in the world? Why has God created useless or pernicious creatures—such as tigers and serpents, thorns and thistles? One common theory was that prior to Adam's temptation and the Fall of man there had been no conflict among animals: no poison, no ferocity. Augustine, for example, claimed that the vegetable and animal kingdoms were both cursed as a result of Adam's sin. The Venerable Bede expressed the view that onerous creatures had been created for a specific purpose: "fierce and poisonous animals were created for terrifying man, in order that he might be made aware of the final punishment of hell." This was a common view, even in the eighteenth and much of the nineteenth century. John Wesley, the Methodist cleric, maintained that before the Fall no animal "attempted to devour or in any wise hurt" another; the spider was "as harmless as the fly, and did not lie in wait for blood." Serpents were said to have originally stood erect, walked, and talked; only later were they forced to go on their bellies. According to the educator and diplomat Andrew White, author of the widely read *History of the Warfare of Science with Theology in Christendom*, it was not until the nineteenth century, when geologists discovered "vast multitudes of carnivorous creatures, many of them with half-digested remains of other animals in their stomachs," that people were forced to realize that beasts had been vicious long before the appearance of humans on the planet.[15]

The problem of evil or useless creatures was a vexatious one. Augustine of Hippo confessed that it was unknown to him why "mice and frogs were created, or flies and worms." All creatures, in his view, were either useful, hurtful, or superfluous. It was clear why God had made the useful ones—but what about the others? Superfluous animals were of no use to humans, yet their creation guaranteed that "the whole design of the universe is thereby completed and finished." And as for the hurtful creatures, "we are either punished or disciplined or terrified by them, so that we may not cherish and love this life." Augustine was reluctant to invoke demons, but Martin Luther was less hesitant: in his view, flies were not merely superfluous—they were sent by the devil to vex a man while he is reading.[16]

Such was the world in the human-centered cosmos: Adam gave names to animals—the task of science was to recover these names. Plants were classified by use, animals by aesthetic qualities (symmetry, elevation above the earth) or their resemblance to humans. In medieval "bestiaries," parallels were drawn between human morals and the lives of the beasts: the phoenix rising from the ashes proved the doctrine of resurrection; the mischief of monkeys proved the existence of demons. And so forth and so on. Nature was not neutral: there were good animals and bad ones. This was the tradition Francis Bacon drew upon for his comparison of scholars, scientists, and empirics with spiders, bees, and ants; Mandeville's *Fable of the Bees* also played upon such metaphors.

In England, William Paley was the leading figure in this tradition in the early years of the nineteenth century. In 1802 he published his *Natural Theology; Or, Evidences of the Existence and Attributes of the Deity*, a book that was to become required reading in British universities for several decades. Paley begins by asking his reader to imagine finding a pocket watch on the path while out for a walk. Would we ever seriously imagine that such a device could have been produced by accident? Certainly not—

we recognize instantly that such a device, so carefully designed, is the product not of chance but of intelligence. Imagine further, he asks, that the watch is not an ordinary watch but has built within it a tiny system of lathes and hammers, all of which acting together have the marvelous capacity to produce other watches, similar to the first. What would we think of such a watch? Would we not be even more convinced that an intelligence had created it, and would this not make us wonder even more at the greatness and intelligence of its creator? Such is the case, Paley argues, in the natural world. The evidence of creation is all around us: look at the membrane protecting a fish's eye, or at how the lens is perfectly adapted to the refractive index of water. Look at the armored tail of the lobster, and how perfectly it protects the creature from the attacks of voracious fish. How fortunate that the world is created the way it is! How fortunate for spiders that they make such excellent webs, given that they cannot fly and hence cannot procure their food by any other means.

This was the purpose of natural theology: to examine the mechanical contrivances in the world around us for answers to the question, *Why?* Why do young deer have no horns? Because they otherwise would spike their mothers' udders. Why did ancient fish have such strong protective plates? To protect them from the high temperatures of early oceans. There is fortune even in the physical world: How fortunate it is that particles of light exert no force upon us when the sun shines! How fortunate it is that gravity falls off as the *square* of the distance (and not, say, as the cube or some other proportion), for only such a force allows our planet to have a stable orbit. And how fortunate it is that there is neither more nor less water upon the earth: with only a little more, all the land would be covered; with only a little less, life would be impossible. It is a happy world, after all: "What a sum, collectively, of gratification and pleasure have we here before our view! . . . At this moment, in every given movement of time, how many myriads of animals are eating their food, gratifying their appetites, ruminating in their holes, accomplishing their wishes, pursuing their pleasures, taking their pastimes?"[17] God is working all around us: as the watch implies a watchmaker, natural design implies a natural designer. Paley advises that an examination of structures like the eye might well provide "a cure for atheism."

Paley also has a solution to the problem of evil. Why are there venomous bites and stings or bothersome gnats and mice? Paley warns that we must not think only of ourselves when considering such matters. For "what we call blights, are, oftentimes, legions of animated beings, claiming their portion in the bounty of nature." What we may find cruel is often just another creature's way of making a living. Given that snakes must eat, perhaps it is better that they first stun their victims with the poisonous fang; indeed, frogs and mice might otherwise "be swallowed alive without it."

Evil and suffering certainly do exist, but (by contrast with many human arts) these are never the purpose for which they are designed: "Teeth are contrived to eat, not to ache; their aching now and then is incidental to the contrivance." Moreover,

> you would hardly say of the sickle, that it is made to cut the reaper's hand; though, from the construction of the instrument, and the manner of using it, this mischief often follows. But if you had occasion to describe instruments of torture or execution; this engine you would say, is to extend the sinews; this to dislocate the joints; this to break the bones; this to scorch the soles of the feet. Here, pain and misery are the very objects of the contrivance.

Nothing of this sort exists in nature. "No anatomist ever discovered a system of organization calculated to produce pain and disease; or, in explaining the parts of the human body, ever said, this is to irritate; this to inflame; this duct is to convey the gravel to the kidneys; this gland to secrete the humour which forms the gout."[18] There is certainly illness, but health is the more usual state. There is death, but how else could there be sex, or parental relations, or anything else conducive of happiness? There are predators and prey, but how else are animals to eat and the number of creatures kept in moderation?

The natural theology tradition culminates in a series of eight books, published in the 1830s, commissioned in accordance with the last will and testament of the eighth earl of Bridgewater to illustrate "the Power, Wisdom, and Goodness of God, as Manifested in the Creation." The purpose of these works, collectively known as the *Bridgewater Treatises,* was to demonstrate God's design "by all reasonable arguments, as for instance the variety and formation of God's creatures in the animal, vegetable, and mineral kingdoms; the effect of digestion, and thereby of conversion; the construction of the hand of man, and an infinite variety of other arguments." Authors were paid 1,000 pounds sterling; several of the works are written by leading scientists of period, including William Whewell (who coined the term *scientist*) and William Buckland (one of the period's foremost geologists).[19]

The authors of the *Bridgewater Treatises* followed Paley in searching for proof of providence in works of nature. William Buckland marveled at the fact that the earth's soil is so fit for agriculture, its metals for mining, its animals and plants for eating. John Kidd argued that the fact that night followed day was a "gift of heaven" (otherwise "all our faculties would soon be exhausted"). It was a merciful provision of nature that the earth was neither too hot nor too cold, too light nor too dark; happy, too, was the fact that the human body is equipped to sweat when hot and shiver when cold. Kidd thanked God for the liquidity of water, the solidity of the earth, and the transparency of glass (otherwise, how could we ever have built microscopes or telescopes?).[20] The natural universe was a stage for the drama of human sin and redemption; nature was adapted to man, and not vice versa.

Criticism of such views emerged from various quarters. Goethe, with tongue planted firmly in cheek, praised God for growing cork trees to allow men to cork their wine. Voltaire's *Candide* satirized the optimistic Pangloss, who, in the face of endless adversity, remained firm in his belief that ours was "the best of all possible worlds." Others criticized the view that man is the center of all things: Abraham Tucker's *Light of Nature* ridiculed the conceit of those who supposed that magnetic effluvia traversed land and sea only to move the mariner's compass, or that the stars were put in the sky to twinkle in our eyes at night: "Surely he must have an overweening conceit of man's importance, who can imagine this stupendous frame of the universe made for him alone."[21]

It is important to realize, though, that much of the natural-theological research program consisted in an analysis of the mechanisms of organic life—and in this sense, much of the work behind these texts was empirical. In their search for evidence of design, natural theologians sought to prove that there is no structure without a function, that the mechanisms of organic "artifice" are complex and worthy of study. Natural theologians urged an exhaustive investigation of organic adaptation, confident in their belief that nature is prodigious in this department.

These were arguments Darwin respected to the highest degree. In his autobiography, Darwin claimed that reading Paley had been the only part of his university training that was "of the least use to me in the education of my mind." (Paley's *Natural Theology* gave him "as much delight as did Euclid": "I was charmed and convinced by the long line of argumentation.")[22] Paley's was one of the few books Darwin took with him on his five-year voyage aboard the *Beagle;* the explorer-evolutionist recalled later in life that Paley's work had exercised a profound influence on the early development of his thought. Here was a catalogue of countless wonderful adaptations—the downy feathers on ducks, beautifully adapted for warmth; the long tongues of woodpeckers, perfectly designed for catching grubs in trees. Darwin did not deny these facts, but simply interpreted them through a different lens. The mechanisms discussed by Paley and the others were evidence not of design but of evolutionary adaptation. Darwin once mused that while mistaken observations were usually sterilizing, mistaken theories could be stimulating. This was the case with Paley: he had made many good observations, but he simply got the theory wrong.

Darwin was impressed with Paley's argument but could not follow him in his belief that the organic world was either perfect or prefigured. In a letter of 1860 to Charles Lyell, Darwin pointed out that we no longer think the planets are divinely ordained to follow their particular paths—why should we imagine this to be true for living beings?[23] The wonder of nature was that there are *laws* that produce the marvelous adaptations we find. Over thousands of generations, modifications preserved by natural selection could slowly accumulate, creating the beauty and wondrous adaptations we see around us. It was not a negative thing that nature's laws were able to produce "design without a designer"; indeed how much more wonderful it was that the Creator did *not* have to intervene at every point, but rather had designed certain natural laws (like natural selection) that, acting alone, could produce the wondrous diversity we find in nature. Is the variety of nature preordained? Who ordained the variety of pigeons that breeders have created? And if ours is such a perfect world, then why must creatures produce far more offspring than can possibly survive? Why does variation proceed at random, "in all directions," and not directly toward that which is fitter? Why is nature at war and not at peace? For Darwin, nature was not the world of perfect harmony imagined by natural theologians. Nature was cruel and capricious; nature was red in tooth and claw.

Darwin did not deny the wondrous beauty of natural adaptations. This was apparent to anyone who bothered to look. Darwin differed from others, however, in his views on the *origins* of this beauty. The origin of apparent design was the adaptation to changing environments resulting from competition and the struggle for existence. It was an irony of natural history that the struggle that appeared so cruel in nature was, in fact, the very force responsible for the appearance of design. In the concluding lines of the first edition of the *Origin*, Darwin waxed poetic in his celebration of this contradiction:

> from the war of nature, from famine and death, the most exalted object which we are capable of conceiving, namely, the production of the higher animals, directly follows. There is grandeur in this view of life, with its several powers, having been originally breathed ["by the Creator" added in later editions—R.P.] into a few forms or into one; and that . . . from so simple a beginning endless forms most beautiful and most wonderful have been, and are being evolved.

I began this chapter with Koyré's thesis that the key event in the rise of modern science was the devalorization of Being—the radical separation of the world of value from the world of fact. The triumph of evolutionary theory in the nineteenth century fulfilled part of this mission, by reinterpreting the design that had commonly been attributed to the realm of living beings. But it would be wrong to think that the triumph of evolution meant the end of moralizing naturalism, or naturalistic morality. Despite the protests of many of Darwin's apostles (Huxley, for example), evolution itself came, for a time, to serve as a substitute ethic for the religious ontology it had displaced. We shall return to this story in a subsequent chapter; but now let us look more closely at how values came to be stigmatized as "secondary qualities" in early modern science and philosophy.

4

Secondary Qualities and Subjective Value

Bacon's warning that morality served to distort and hinder the progress of the sciences (the "moral taint"), and the agreement on the part of early natural philosophers to avoid meddling with matters of morals and religion (the "royalist compromise"), are two early and distinct aspects of the exclusion of morality from modern science. The new ethical ontology associated with the secularization of the study of nature (Koyré's "devalorization of Being") provides a third strand of my argument thus far. Alongside these, however, it is also important to appreciate the new dialectic of subjectivity and objectivity that arises with modern science.

The exclusion of values from science represents, at least at one level, the consequence of a new and larger recognition of the *subjectivity* of human knowledge—a recognition of the agency of the human subject in all constructions or conceptions of nature. Consider the case of pictorial representation. In the fifteenth and sixteenth centuries, European artists and engineers began to search for new ways of drawing that would be true to life. Realism, however, was only one of the goals of Renaissance art. In Italy and in Germany, artist-engineers developed techniques to give a precise meaning to drawing from a given perspective. The development of perspective drawing signaled not only a new realism but also a broader appreciation of the fact that what one sees depends upon where one stands.

A parallel phenomenon can be seen in the new ideal of experimentation, according to which nature reveals her secrets only when "constrained and vexed" by human experiments. The purpose of this constraint was to give the experimenter a measure of control, which could then be used to identify and eliminate the various kinds of bias that spoil what we see, think, or do. Bacon's attempt to rescue thought from the various "idols" of the tribe, cave, market, and theater was designed to reduce the impediments to learning that spring from human subjectivity. In his *Advancement of Learning,* Bacon produced his famous observation that the mind of man "is far from the nature of a clean and equal glass, wherein the beams of things should reflect according to their true incidence; nay it is rather like an enchanted glass, full of superstition and imposture, if it not be delivered and reduced."[1] To "deliver and reduce" these distortions of the human mind was the task of experimental philosophy. The establishment of reliable knowledge required that one recognize the character of this "enchanted glass" and eliminate or minimize its distortions through proper methods. This would have consequences for the relation of science and ethics.

The subjectivity that arises in the science of the seventeenth century is associated with broader social movements sweeping across Europe at the time. With the rapid growth of towns and town-based trade came the rise of institutions sanctifying the economic independence of the European bourgeois individual. Individualism brought with it a distrust of authority and an eagerness to get at nature directly, without the mediation of books or priests or the worn-out wisdom of the ancients. New religious movements embraced the new individualism. As the northern states of Europe began to cast off the religious hegemony of Rome, Martin Luther demanded that every man read the Bible for himself, free of the intervention of priests. The "puritan ethos" emerging in England provided a further spur to science, as did voluntarist theology and secular efforts to trust in nothing but what Bacon called "the faith of eyes."[2]

When Bacon argued for the need to separate the original of nature from the additives of human perception, his goal was to capture the true essence of things. Most early modern natural philosophers reaffirmed this goal by distinguishing between "primary" and "secondary" qualities. Johannes Kepler distinguished secondary qualities of the mind from primary qualities which inhere in nature. Galileo distinguished between qualities absolute and fixed, which form the objects of mathematical analysis, and qualities subjective and in flux, which derive from the constitution of the observer.

Galileo's distinction can be taken as typical. For Galileo, the primary reality of the universe was a mathematical one—human experience could only approximate that reality, because human sense is fallible. The qualities produced by human sense—taste and color, odor and sound—are subject to the distortions of the senses, the peculiarities of persons, and the changing climate of opinion. The primary qualities of nature alone are real. The fact that an object has a certain size, or shape, or position; whether it is in motion or at rest; one or many; touching another body or not—these are the *primary* qualities of a body, qualities which no amount of imagination can separate from that body. By contrast, whether it is "white or red, bitter or sweet, noisy or silent, and of a sweet or foul odor" is of an entirely different nature: tastes, odors, colors, and the like differ from the primary qualities of a thing insofar as they reside only in the consciousness of the person receiving these sensations. Indeed "if the living creature were removed, all these qualities would be wiped away and annihilated."[3]

According to Galileo, it is simply good experimental method that locates the secondary qualities in us and primary qualities in the object:

> A piece of paper or feather drawn lightly over any part of our bodies performs intrinsically the same operations of moving and touching, but by touching the eye, the nose, or the upper lip it excites in us an almost intolerable titillation, even though elsewhere it is scarcely felt. This titillation belongs entirely to us and not to the feather; if the live and sensitive body were removed it would remain no more than a mere word . . . if the ears, the tongues, and noses were removed, shapes and numbers and motions would remain, but not odors or tastes or sounds. The latter, I believe, are nothing more than names when separated from living beings, just as tickling and titillation are nothing but names in the absence of such things as noses and armpits.[4]

Again, Galileo wants to get to the bottom of things—to distill what is necessary and essential from what is spurious and accidental. The point in distinguishing primary and secondary qualities is to allow us to recognize that appearances can deceive—that even if the earth appears flat, our reason and our experience tell us otherwise. We must distinguish essence from appearance, leaving aside our subjective impressions to capture some sense of the essence of things.

It is true of course that the ancients had also distinguished essence from appearance, and for somewhat the same reasons—that appearances can deceive, that one must look beyond the surface and into the heart of the matter, and so forth. Underlying mathematical principles are the goal in each case, and Galileo for this reason is commonly considered a neo-Platonist. But the means by which the moderns achieve this distinction are quite different, on several counts. For one thing, the modern notion of experiment marries the natural and the artificial in ways entirely foreign to the ancients. In the modern experiment nature is abstracted but also manipulated—in Bacon's words, "under constraint and vexed; that is to say, when by art and the hand of man she is forced out of her natural state, and squeezed and molded."[5] Galileo, too, believed that the world must be explored not in its pristine condition, nor as the shadows of eternal forms, but in artificial circumstances created by the experimenter. For the ancients, by contrast, there is little sense that nature must be manipulated to be known, that a special kind of artifice—the experiment—is required to grasp reality. In its emphasis on artifice as the key to essence, modern experimental science represents a challenge to both Platonic formalism and the common-sense world of the Aristotelians.

New also is the notion that matter, shape, and motion (primary qualities) are more real than color, taste, and feel (secondary qualities). Few among the ancients would ever have said that the color, taste, or smell of a thing was incidental to its nature, or that such qualities were "in us" but not in the things themselves. For the moderns, however, the entire realm of human sensory experience is relegated to a secondary and nonessential status. Matter (or "extension," in Descartes's vocabulary) is identified with essence, human sense with appearance. As E. A. Burtt describes it, the external world is exalted over the world of human experience: "The features of the world now classed as secondary, unreal, ignoble, and regarded as dependent on the deceitfulness of sense, are just those features which are most intense to man in all but his purely theoretic activity."[6] Experimental method requires that the human be excised from the situation to get at underlying principles.

Especially important for our purposes, however, is that all purpose, plan, and value are removed from the physical world and lodged instead in the private, inner space of the human mind. Social philosophers, eager to apply the New Learning to the realms of human social life, were among the first to articulate these conclusions. The most comprehensive formulation of the ideal of subjective value—of value as a secondary and purely personal quality—can be found in the philosophies of Hobbes, Locke, and (especially) Hume.

Thomas Hobbes was the first to explore systematically the import of the new mechanical philosophy for the science of human affairs. Drawing from both Bacon and Galileo, Hobbes is an empiricist: the origin of all thought is sense; there is no thought or knowledge not originally derived from human touch, sight, smell, taste, or hearing. He is also a materialist—for the real qualities of things are matter and motion. Aristotelian philosophers had taught that objects send forth audible or visible forms ("intelligible essences") that, upon entering the eye or ear, make sound or sight. For Hobbes, the idea of intelligible essences was absurd. The qualities of things are "but so many several motions of the matter, by which it presseth our organs diversely." Sensations we receive from the world are the products of matter in motion—as pressing the eye produces light or rubbing the ear noise. It is only *in us* that we fancy these motions to appear as sound or color. It is we who make sound and light into something more than matter in motion. And we know this by Galileo's method of variations: "For if these colours and sounds were in the bodies, or objects that cause them, they could not be severed from them, as by glasses, and in echoes by reflection, we see they are."[7] We must distinguish between an object and its appearance. For though we know a thing to be in one place, its appearance can be in another, as when lenses distort an image or walls echo a sound.

And so it is with the passions, our moral sense of what is right and wrong. Hobbes is one of the first to suggest that "the good" is entirely the product of human desire. Good and evil are simply words used to describe human attraction or aversion—a person's motion toward or away from a thing. Moral right and wrong are not in the nature of things: there is no "common rule of good and evil, to be taken from the nature of the objects themselves." Nor is there morality apart from civil society; in a state of nature men would be in a perpetual state of war—a state in which right and wrong have no meaning, where "force and fraud" become the two cardinal virtues. Ideas of vice and virtue spring instead from the judgments of men or, when in commonwealth, from an arbiter or judge acting as representative of the people.[8]

John Locke elaborated on this argument of Hobbes in his 1689 *Essay Concerning Human Understanding,* one of the two or three most influential texts in all of British philosophy. Like Galileo and others before him, Locke distinguished primary from secondary qualities in the course of his search for the "original, certainty, and extent of human knowledge, together with the grounds and degrees of belief, opinion, and assent." The task of the student of human understanding was to discover how these two sets of qualities relate to each other. Primary qualities (size and shape, motion or rest, number and solidity), he says, cannot be separated from objects: they are "essential" to objects, the proof of this being that we cannot imagine an object without them. Secondary qualities (smell, sound, taste, color, heat or cold) by contrast are epiphenomenal—not really qualities of matter at all, but rather the product of certain powers of primary qualities to impact on our senses. Fire produces both warmth and pain—yet certainly the "pain" is not in the fire—why, then, the warmth? In this, Locke follows the lead of Galileo, Boyle, and Hobbes.[9]

New in Locke, however, is the extension of the idea of secondary qualities to the realm of moral principles. No ideas, he writes, are innate—this is as true for morals as it is for other ideas about nature, man, or God. There is no moral principle invariably accepted by all men. The Megrelians bury their children alive; the Caribs castrate their children, then fatten and eat them. In parts of Asia, Locke states, the elderly are put out to starve and freeze. Entire nations reject certain moral rules, and the very fact that moral rules are in need of proof shows that they are not innate in the human mind, but rather impressed upon it through sensation and reflection. Virtue we approve not because it is innate, but simply because it is profitable.

Moral principles are based on moral ideas, but what are those ideas? As for Hobbes, good and evil for Locke are ultimately based on sensations of pleasure and pain:

> That we call *Good,* which *is apt to cause or increase Pleasure, or diminish Pain in Us; or else to procure, or preserve us the possession of any other Good, or absence of any Evil.* And on the contrary we name that *Evil,* which *is apt to produce or increase any Pain, or diminish any Pleasure in us; or else to procure us any Evil, or deprive us of any Good.*[10]

Pain and pleasure, like all secondary qualities, are simple ideas in us, not in the world. Morals ideas are "creatures of the understanding rather than the works of nature."

John Herman Randall in his *Career of Philosophy* distinguishes two traditions in the history of moral philosophy: one that champions a morality of commitment and one that stresses the morality of the spectator. The one is designed to teach us right and wrong, the other to explore the springs of moral action—from a distance, as it were. Locke is clearly in the latter tradition. Moral truths are capable of demonstration (that "no government allows absolute liberty," for example, or that "where there is no property there is no injustice"), but only if we apply ourselves with "indifferency" to such problems. Locke chooses to describe the psychology of moral action, not "what ought to be done" or the features of the good life. Locke is not alone in this. The majority of subsequent British moralists—from Hutcheson and Shaftesbury through Hume to Moore, concern themselves more with the psychological question—Why do men act morally?—rather than the commitment question—What are the moral ways to act? As Randall puts it, "The limitation of the British moralists lies in their consistent refusal to ask, even to raise, the central questions of ethics."[11]

David Hume is probably the leading culprit in this regard. Hume is famous for having recognized (and denounced) the "naturalistic fallacy"— the illegitimate derivation of "ought" from "is," ethical imperatives from facts about the world. The classical formulation of the fallacy appears in his *Treatise of Human Nature* (1739–1740), commonly considered the most important philosophical treatise in all of British history.

Hume begins his *Treatise* by noting that there are those for whom virtue is nothing but conformity to reason and for whom the ethically good must appear the same to every rational being. Such a view holds that morality, like truth, is discerned merely by ideas, by their juxtaposition and comparison. But is it in fact possible from reason alone "to distinguish betwixt moral good and evil"? Or must there be some other principles that enable us to make that distinction? Morals foster passions and stimulate actions. Reason by contrast "is perfectly inert"—it can neither excite passions nor produce actions. But if reason cannot influence passion, then morals cannot be derived from reason. For "as long as it is allow'd, that reason has no influence on our passions and action, 'tis in vain to pretend, that morality is discovered only by a deduction of reason. An active principle can never be founded on an inactive; and if reason be inactive in itself, it must remain so in all its shapes and appearances."[12]

The object of reason, Hume argues, is the discovery of truth or falsehood; only that which is capable of being true or false can be the object of our reason. Yet actions and morals, and the passions upon which they are founded, are neither true nor false, and hence not the proper object of reason. "Actions may be laudable or blamable; but they cannot be reasonable or unreasonable." Reason is to the passions as moved is to mover: "Reason is and ought to be the slave of the passions, and can never pretend to any other office than to serve and obey them." Reason is the impotent servant of desire, desire the blind motor of reason.

The ethical impotence of reason in Hume's philosophical scheme is ultimately based on what he calls "the fundamental principle" of modern philosophy: the principle of primary and secondary qualities. Qualities perceived by the senses (color, taste, smell, heat and cold) are nothing but "impressions of the mind, derived from the operation of external objects, and without any resemblance to the qualities of the objects." In harmony with Galileo and Locke, Hume finds justification for this principle in the fact that whereas some qualities of an object remain fixed (extension and solidity, gravity and shape), those relating to our perceptions may change, depending upon our health or mood or other circumstances. Meat to a healthy man tastes different than it does to a sick man; what seems bitter to one may seem sweet to another, and so forth. There are, in other words, qualities of things that are primary and do not change in things themselves; and there are qualities of things that are secondary and may vary according to circumstances affecting our perception.

Hume admits that there are problems in a strict separation of primary and secondary qualities, not the least of these being the fact that ultimately, our knowledge even of "primary" qualities must be traced back to human sense. He also concedes that it is impossible to conceive of bodies in motion—or even matter itself—without having color or being composed of parts. Yet despite these objections, Hume is not averse to using the distinction in the argument crucial for his attack on the "naturalistic fallacy." Praise and blame, he writes, we bestow upon an action or object according to how we feel toward that action or object. And therefore "when you pronounce any action or character to be vicious, you mean nothing, but that from the constitution of your nature you have a feeling or sentiment of blame from the contemplation of it. Vice and virtue, therefore, may be compar'd to sounds, colours, heat and cold, which, according to modern philosophy, are not qualities in objects, but perceptions in the mind." Vice and virtue have nothing to do with the world at large, nothing that might form the object of reason or science. It is the essence of virtue to produce pleasure, the essence of vice to produce pain. If, then, there is any principle we can learn from philosophy it is this, that "there is nothing, in itself, valuable or despicable, desirable or hateful, beautiful or deformed"; these instead are simply attributes that arise from "the particular constitution and fabric of human sentiment and affection." Moral qualities are secondary qualities: they lie *in us,* and not in the objects to which we refer. It is directly following this discussion that Hume introduces his famous principle that "ought" cannot derive from "is":

> In every system of morality, which I have hitherto met with, I have always remark'd, that the author proceeds for some time in the ordinary way of reasoning, and establishes the being of a God, or makes observations concerning human affairs; when of a sudden I am surpriz'd to find, that instead of the usual copulations of propositions, *is,* and *is not,* I meet with no proposition that is not connected with an *ought,* or an *ought not.* This change is imperceptible; but is, however, of the last consequence. For as this *ought,* or *ought not,* expresses some new relation or affirmation, 'tis necessary that it shou'd be observ'd and explain'd; and at the same time that a reason should be given, for what seems altogether inconceivable, how this new relation can be a deduction from others, which are entirely different from it. But as authors do not commonly use this precaution, I shall presume to recommend it to the readers; and am persuaded, that this small attention wou'd subvert all the vulgar systems of morality, and let us see, that the distinction of vice and virtue is not founded merely on the relations of objects, nor is perceiv'd by reason.[13]

Hume's approach represents a radical departure from classical morality, according to which moral claims are either true or false. As Alasdair MacIntyre describes it, Hume's moral theory postulates a new and restricted conception of reason, one that emerges (ironically, given Hume's atheism) from heterodox Protestant and Jansenist Catholic theologies. According to these theologies, reason is powerless to grasp human moral ends; the Fall of man in the Garden destroyed this power once and for all. Reason for Hume as for Pascal is purely calculative; reason deals with means, not with ends.[14]

Hume's radical separation of reason and morality emerges in the context of the broader Enlightenment critique of natural theology and rationalistic ethics. According to Hume's empiricist epistemology, something can be known only if it is derived from either perceptual experience or relations among ideas (logic), which together and in combination exhaust the possible sources of truths about the world. Something is true either by logical necessity or experience. Whatever is true by logic must be true either by definition or by tautology. And for anything that is true by experience, there must be some kind of corresponding evidence. The implications of this line of reasoning for religious and dogmatic philosophy were not lost to the times. God, it could now be argued, existed neither by definition nor by tautology and could only be known to exist (if at all!) by evidence. Similar doubts could be raised with respect to the problem of evil or, stated in reverse, the problem of theodicy. For even if God did exist, it would not follow that it was moral to obey him. One cannot derive ethical conclusions from empirical facts.

Hume never intended that his "ought-is" paragraph be used as a defense of "pure science" or any particular political vision of the task of science. In fact, the celebrated division of fact and value for which he is known was repeatedly violated by the master himself. Hume's writings are full of moral judgments and their justifications from facts and consequences. Society was to be ruled by just laws (ensuring protection of property, for example); indeed, the root of all moral sentiments was the "fact" of sympathy. Sympathy was the basis of all the indirect passions, all social virtues. Sympathy explained "why utility pleases, and how social organization is possible"; sympathy accounted for national character and "the contagion of manners." Sympathy was in fact "the chief source of all our moral distinctions," a necessary precondition for both science and morality. Hume's Scottish friend and colleague, Adam Smith, agreed: "Though our brother is upon the rack, as long as we ourselves are at ease, our senses will never inform us of what he suffers." Moral science presupposed sympathy, as did moral life more generally.[15]

In the twentieth century, Hume has commonly been regarded as the architect of the separation of ought and is, fact and value. Moral philosophers often credit him as being the first to have discovered the "naturalistic fallacy" of deriving values from facts, moral statements from scientific truths. In fact, as Stuart Hampshire, Alasdair MacIntyre, and others have shown, Hume himself intended nothing of the sort. Hume never denied that moral judgments could be derived from matters of fact; he simply showed that they (like all nondeductive derivations) were not logically compelling.[16]

But ideas often live a life apart from the intent of their authors. In the midtwentieth century, Hume's call for a separation between "ought and is" became a rallying cry for scientists and philosophers defending the neutrality of science. Hume was used to widen the rift between the empirical world and moral obligation, one that could be used in a liberating sense to criticize church and state domination but also (as we shall see) in a more conservative sense, in defense of the immunity of science from moral and political critique.

Baconian science announced its Enlightenment mission of learning that was to be both political and practical, serving both the glory of God and the relief of man's estate. Bacon qualified his vision, however, advising caution in the rush to apply the principles gained by science, in order not to "mow the moss or reap the green corn." Knowledge was power. But Bacon also advised caution in dealing with matters of morals to avoid the dangers that come with "the proud knowledge of good and evil." With the institutionalization of science in late-seventeenth-century England, the reformist associations of Baconian science were compromised: science, under the auspices of the crown, was to avoid the explosive issues of morals and politics. Science was to be practical but not political. Science was to serve two masters: the cause of truth, but also the practical needs of those who pay.

The rise of the practical ideal of science also coincides with the rise of a new ethical ontology, one that found the origin of value not in nature but in labor and its products, a view that was to merge with a conception of the neutrality of natural (and ultimately also of social) science. The exclusion of morals from science in the seventeenth and eighteenth century was defended on the basis of a separation of primary and secondary qualities: desires and passions were subjective, human, additives to the original of nature, distorting our understanding of the true nature of things.

In this way, the rise of ethical individualism and subjectivism are associated with the decline of the doctrine of absolute good and evil in nature apart from man. The world becomes matter in motion; the good becomes purely epiphenomenal, squeezed into the private spaces of the individual mind and reflected in the price of goods in a market. Value, divorced from the world, is also forced from the science that studies that world. This was not something poets failed to denounce, nor was it a struggle won without a fight. But it was a change that has marked, and continues to mark, our time.

Part Two

The Politics of Neutrality in German Social Theory

Science is a goddess, not a milk cow. —FRIEDRICH SCHILLER

As sociologists we are neither for nor against socialism, neither for nor against the expansion of women's rights, neither for nor against the mixing of the races. —FERDINAND TÖNNIES, 1910

Now that the "disinterested" are praised so widely, one has, perhaps not without some danger, to become conscious of what it is these people are really interested in.
—FRIEDRICH NIETZSCHE, 1886

No one today would speak of a "liberal physics" or a "conservative chemistry." But should this be any different for the science of human history and institutions? I, for one, say no, and cannot consider a liberal, conservative, or socialist social science anything but a contradiction in terms.
—EDUARD BERNSTEIN, 1901

5

The German University and the Research Ideal

Contrary to popular philosophical legends, David Hume was not the first to distinguish *ought* from *is*. Philosophers had long distinguished between fair observation and wishful thinking, between the world as it is and the world as we would like it to be. Machiavelli in *The Prince* calls for a study of man "as he is, and not just as we might want him to be"; Montaigne in his *Essais* declares that while others try to shape man, he contented himself with description ("Les autres forment l'homme; je le récite"). Bernard Mandeville in his 1732 *Fable of the Bees* complains that "One of the greatest Reasons why so few people understand themselves, is, that most Writers are always teaching Men what they should be, and hardly ever trouble their Heads with telling them what they really are."[1]

It is not until the nineteenth century, however, that value-neutrality becomes a widespread and dominant ideology in social thought. It is not until then that social science *qua* science emerges, and neutrality is advocated as a conscious program for the sciences. Here we explore the conceptions of value-neutrality that emerged in the late nineteenth and early twentieth century with the rise of continental, and especially German, social science. The argument may be sketched as follows. Neutrality was not a single notion but rather a collection of loosely associated ideals that emerged at different times to serve different functions. The ideals of pure science, of science for its own sake, and of value-free science which emerged at this time drew upon earlier traditions. The classical ideal of the separation between theory and practice, the modernist conception of "devalorized Being," and the political imperatives involved in the royalist compromise were all drawn upon in late-nineteenth-century social thought. There were new elements as well.

Social scientists proclaimed their work neutral as part of an attempt to make that work scientific. But the exclusion of values from social science was also associated with the division of scholarship into increasingly narrow specialties and subspecialties. The nineteenth century sees the emergence of the separate disciplines of economics and sociology (out of classical political economy) and the modern sciences of biology, psychology, and statistics, all as part of the specialization of academic labor. Indeed the century sees the coining of more new words to designate new sciences than any previous time. English words coined in the nineteenth century include: *hematology* (1811), *phrenology* (1815), *biology* (1819 in the modern sense), *paleontology* (1838), *toxicology* (1839), *ethnology* (1842), *ophthalmology* (1842), *climatology* (1843), *gynecology* (1847), *epistemology* (1856), *embryology* (1859), *egyptology* (1859), *epidemiology* (1873), *topology* (1883 in the modern sense), *bacteriology* (1884), *criminology* (1890), *limnology* (1895), and many others either less well known or now extinct. *Methodology* (1800), *morphology* (1830), and *terminology* (1801) are all new to the century.[2]

The growth of empirical science in the nineteenth century is associated with two conflicting tensions. On the one hand, scientists experiment with the extrapolation of scientific methods and models from one field to another. At the same time, scientists feel increasing pressure to establish and maintain certain "boundaries" for their fields—implicit rules and regulations concerning what might and might not be said. The consequence of this is the rise of a new and more specialized style of science. Scientists begin to abandon grand "systems" in favor of narrow specialties and to cultivate the ideals of radical modesty and trans-disciplinary ignorance as honorable norms of science. Scientists begin to devote the early paragraphs of their works to describing what they will *not* do, as well as what they will do. The expressions *dilettante* and *amateur* come to have their modern, pejorative meanings.

The end of the nineteenth century thus sees a growing hostility to grand, overarching systems of philosophy. "Systems" of philosophy for the turn of the century critic included Auguste Comte's positivism, Spencer's *Principles,* Mill's *Logic,* Hegel's *Phenomenology,* and Schelling's *Naturphilosophie* (the Marxian corpus was also sometimes included among the systems). Wilhelm Wundt was one who fell back into the fold with his 1889 *System der Philosophie;* his reward has been the obscurity of the work. The attack on systems continues into the twentieth century, as systems are replaced by "introductions" and "remarks," "notes," "letters," "comments," and ultimately "rapid communications" (as, for example, in the *Physical Review* beginning in the second half of the twentieth century).

The description of these changes in terms of the rise of empiricism or positivism or even naturalism captures part of this trend but not the whole. Across all fields of knowledge, scientists become concerned to eliminate all residual traces of the metaphysical or theological past, especially insofar as these could be seen as impeding the progress of science. But changes in the practice of science cannot be reduced to changes in the *sources* of knowledge (experience vs. reason or revelation, for example), as is commonly the focus in philosophical literature. Science also becomes important for national power and prestige. The nineteenth century sees the rise of the industrial research laboratory, the science museum, the scientific exhibition, and many of the scientific associations that persist today (the British and American Associations for the Advancement of Science, for example). Science claims a significant share of university curricula; technical schools are established to promote engineering and the applied sciences.

As we shall see, each of these transformations is important in the rise of the modern ideals of pure and value-free science. Also important are changes in the status of the physical sciences—changes that set the standard for transformations in the social sciences. The military, industrial, and political power provided by science-based technologies brings prestige to the sciences, fostering confidence that science is capable of solving social problems. Impressed with the apparent link between science, power, and prosperity, many social theorists turn to the physical sciences for models or methods for their own disciplines. Moral and political engagement are equated with the metaphysical and theological baggage of an earlier and unscientific age. The elimination of values from science becomes part of a broader program to purify theory from the ill effects of practice; social theory is transformed from a practical art (or speculative philosophy) into a positive science.

It is important to distinguish reasons from causes: the explanations provided for neutrality by its advocates are not sufficient to account for its origins. In their own terms, sociologists and political economists often saw the adoption of neutral methods as a natural consequence of their discipline becoming "scientific." Value-neutrality was associated with precision and rigor, eventually with prediction and control, and ultimately with an eye to the prestige won by physicists and chemists in their mastery over nature. There were, of course, national differences in this regard: English and French economists tended to defend the rise of neutral science as part of a transition from art to science; German and Austrian sociologists tended to focus more on the elimination of speculative metaphysics.

But it would be wrong to associate the rise of the modern ideal of neutral science with the simple incorporation of physical science methods into the social sciences. For one thing, value-neutrality was no longer central to the rhetoric of physical science by the end of the nineteenth century, the period when objectivity in the historical sciences and *Wertfreiheit* in the social sciences become popular slogans. The ideological battles for a morally neutral physics had largely been won in the seventeenth century, and the rhetoric of warnings against the dangers and distractions of morals had largely been supplanted by rhetorics either of the value of science "for its own sake" or of science as a source of industrial and economic power. Few social theorists had any illusions about the value of social science for military or industrial strength. More to the point, many social scientists proclaimed science value-free not in order to imitate the natural sciences but to do precisely the opposite: to avoid what they saw as the illegitimate intrusion of science into the free realm of values and politics. Value-freedom is an ideology of science under siege—a defensive reaction to threats to the autonomy of science from political tyrants, religious zealots, secular moralists, government bureaucrats, methodological imperialists, or industrial pragmatists asking that science be servile or righteous or politically correct or practical or profitable.

This may help to explain why it is that the ideal of pure science arises, curiously enough, at the very time that science is perceived as vital for the realization of political, military, and industrial objectives. The connection is not a spurious one. The movement to preserve science as "pure" emerges only with the rise of "applied science"—science applied to the new chemical and pharmaceutical industries, science in the style of Bayer and Höchst or as taught in France's Ecole Polytechnique and Germany's Technische Hochschulen. The modern ideal of purity emerges with the development of new relations between science and society. Before the mid-nineteenth century, science is largely an amateur affair, practiced outside the universities by the well-to-do in private cabinets, salons, or out on the open fields. In the course of the century, however, science enters the universities and the laboratories of industrial concerns. Science becomes a profession, a salaried occupation; scientific knowledge becomes a source of national pride and industrial power.

The distinction between science "pure" and "applied" is of recent origins. Scientists did not commonly distinguish pure from applied sciences until the second half of the nineteenth century. Indeed most earlier science has a clearly practical cast. In the seventeenth century, Royal Society fellows experimented with ways to improve the making of wine, gunpowder, and artillery; scientists in the early German Academy of Sciences were granted monopolies for constructing calendars and making silk. In 1835, in Robert Thompson's *Records of General Science* (one of only three general science journals in early-nineteenth-century Britain) we find, alongside articles on "The Magnetic Intensity of the Earth" and "Researches into the Number of Suicides and Murders committed in Russia in 1821–22," articles on "Calico Printing," "A journey to Spain," and "A Method of destroying Mice in their Lurking Places." Today's sharp distinction between science and technology (or science pure and applied) was simply not an issue.

This is not to say that the distinction is not to be found much earlier, though admittedly in different forms. Aristotle in his *Eudemian Ethics* distinguished "theoretical" and "productive" sciences—sciences concerned with the contemplation of nature or the properties of form (such as astronomy and geometry), and sciences whose goal was something different from knowledge (medicine, for example, the goal of which is human health, not just medical knowledge). Seventeenth-century philosophers raised a different set of concerns. In his *Advancement of Learning,* Francis Bacon distinguished between mathematics "pure" and "mixed." Pure maths were those sciences "which handle quantity determinate, merely severed from any axioms of natural philosophy," that is, geometry and arithmetic. Mixed mathematics considered "axioms or parts of natural philosophy" employed in association with music, astronomy, architecture, and engineering. In 1648, John Wilkins spoke similarly of "pure mathematics," meaning by this mathematics intended to handle "only the abstract quantity," while that which is mixed "doth consider the quantity of some determinate subject." Samuel Johnson in the 1750 issue of his literary magazine, *The Rambler,* distinguished between "pure science, which has to do only with ideas," and the practical application of science "to the use of daily life."[3]

According to the *Oxford English Dictionary,* Charles Babbage was the first to use the expression "applied science." In his 1832 book *On the Economy of Machines and Manufactures,* Babbage declared that "the applied sciences derive their facts from experiment; but the reasonings, on which their chief utility depends, come more properly within the province of what is called abstract science." In Germany, Justus Liebig's 1840 book on the application of organic chemistry to agriculture represents one of the earliest explicit statements of how a pure science might be applied to practical problems. By the 1870s Louis Pasteur could declare his opposition to any notion of a separate "applied science"; science and its applications were "bound together as the fruit to the tree which bears it."[4]

Scientists first distinguished the value of pure science apart from its applications when (a) the long-term practical utility of science began to be widely appreciated; (b) separate schools were established for teaching of science with a eye for its application; and (c) fears arose on the part of intellectuals that the success of science in solving practical problems would result in efforts to subordinate all other knowledge to practical pursuits.

For social scientists, however, rejection of practical applications was only one of several reasons offered for neutrality. For many social theorists at the end of the nineteenth century, the imperative of avoiding pronouncements of value derived from the supposed "subjectivity" of values—from the impossibility of their resolution (when in conflict) or demonstration. The neutrality of science was justified as part of a subtle dialectic of fact and value, ought and is. At question were the limits of knowledge and the relations of knowledge to ethics. This in turn entailed a vision of the organization of society and the proper political order. Neutrality was a liberal response to pressures by governments for censorship, industry for practical results, and social movements for relevance. Neutrality was a political statement as well as an ontological position, part of a more general, liberal, vision of the relations of knowledge and power.

Among historians, explanations for the origins of the ideal of *Wertfreie Wissenschaft* (value-free science) have been varied. Liberal historians have tended to stress the continuity of scientific sociology with traditional German academic ideals of *Kultur* and *Bildung*, deriving from German pietism and idealism.[5] Explanations of the separation of the Deutsche Gesellschaft für Soziologie from the older Verein für Sozialpolitik, often written from the perspective of the victorious neutralist camp, have tended to focus exclusively upon debates internal to the academic polity (the *Werturteilsstreit*, the *Methodenstreit*), where questions concerning the distinctive nature of social knowledge or the abstract place of "values" in the sciences come to a head.[6]

Without disputing the importance of these trends, I would like to suggest two modifications of this line of reasoning. First, I want to suggest that at least in the case of Germany it was largely in reaction against movements *outside* the universities—socialism, the women's movement, and racialist nationalism—that the ideal of neutrality becomes important in German social theory. Neutrality emerges as a self-conscious ideology of science partly in reaction to political challenges to state power—Marxism, feminism, and social Darwinism—but also as an outgrowth of fears that practical, and specifically industrial, concerns were about to swamp the pursuit of science "for its own sake."

Second, I want to show that despite their protests, many of the greatest advocates of neutrality played an important role in the politics of their time. The German academic was neutral not in the abstract, but in opposition to certain ideas and social forces. Neutrality was calculated to have certain specific effects on life and society. Neutrality was political not only in the sense that it was devised to counter certain political movements, but also in the sense that it served to mask the broader political ideals of its advocates. But first, the German research ideal.

In Germany, the ideal of pure science emerges first with the reorganization of the universities in the early years of the nineteenth century. After Prussia's defeat in 1806 at the hands of Napoleon's armies, Wilhelm Friedrich III, king of Prussia, proclaimed that Germany would rise again through spiritual rather than military strength.[7] Wilhelm Friedrich intended this spiritual strength to be embodied in a new university ideal, one that stressed the purity of science and its freedom from political controls. His reign thus sees the rise of the great University of Berlin, an institution that would soon become famous for the devotion of its teachers to research; its academic freedom; its seminars and laboratories; its new and rigorous Doctorate of Philosophy. Nine thousand American students would study in Germany between 1820 and 1920, most of these in the period 1870-1900. German "graduate studies" became a model for scientific education throughout the world.

Wilhelm von Humboldt, setting forth plans for the University of Berlin in 1810, described in glowing terms this new ideal of intensive, independent research. Science was to be pursued "for its own sake," not for its applications. The task of the university was to transform the individual from an acquirer of knowledge to a creator of knowledge. Scientific institutions were to keep in mind the "pure idea of science and scholarship." Their goal was to combine "objective science" with "subjective, spiritual education" *(Bildung)* through "unforced and disinterested collaboration of scholars with one another" in an environment of solitude and freedom *(Einsamkeit und Freiheit)*—freedom, that is, from political or practical distractions.[8]

Political concerns in the new German university were to be subordinate to intellectual concerns. Only thirty years earlier, Johann Gottfried Herder had stressed the mutual benefits of science for government and government for science. But with building already under way for Berlin's new university, Humboldt announced that politics must not enter the university. The state, Humboldt wrote, "must remain conscious of the fact that it never has, and in principle never can, by its own action bring about the fruitfulness of intellectual activity." State authorities were to avoid intervening in the life of the mind, apart from providing financial resources. Selection of academy members or of university professors, for example, was to remain in the hands of scholars themselves. Academia was to remain as free as possible from state control. Friedrich Schleiermacher warned similarly that there was "nothing more disgusting—nothing that will more thoroughly erode mutual trust and friendly relations—than when a government takes a stand on philosophical questions."[9]

The new university was to differ from older models in form as well as content. As late as 1795, German universities still followed the medieval model whereby lectures consisted of an interpretive reading of some text or texts, which students would then copy down generally without ever having access to the books themselves (apart from those they could use in libraries). Students defended their work in *disputatio* before the entire faculty, generally in Latin or in French. Christian Wolff in the early eighteenth century was one of the first to break with tradition and write his philosophical works in German. In the new university all work was to be in the vernacular; Latin was to be studied as only one among several Romance languages. Leibniz had supported this policy at least in spirit: traveling to Paris in the late seventeenth century, he was impressed that scholars there wrote in their native French. Returning to Berlin, he advised his German colleagues also to write—in French![10]

It is possible to argue that the ideal of intensive and apolitical science that emerges in the early years of the nineteenth century represents a continuation of the medieval scholastic ideal that the scholar should remain isolated from the world of the flesh. The case can equally be made, however, that the new ideal represents a reaction against specific social and political conditions faced by German scholars at the end of the eighteenth century. In the years prior to the founding of the new university, student revolts in sympathy with the French Revolution had plagued German universities. Student riots in 1796 and 1797 brought Culture Minister von Massow to Halle to quell the rebellions, culminating in his 1798 "Order for the Punishment of Students Disturbing the Public Order in Royal Academies." Alte Dessauer, governor of Halle and head of the local Prussian army regiment, arrested the most active students and pressed them into service in his regiment. Fichte was concerned about student unrest when he assumed the position of first rector of the University of Berlin. His inaugural address, devoted to the question of "Threats to Academic Freedom," argued that retribution against student unrest posed the single greatest threat to academic freedom. Fichte advised the restoration of order and the restriction of student activities to "higher things."

It is also important to recall the precarious condition of German universities in the early years of the century. In 1810 Germany was just recovering from Napoleonic invasions that had wrecked the German economy. In 1792 there had been forty-two universities in the German-speaking world. By 1818, as a consequence of the impoverishment of the German state due to the ravages of war, more than half of these—including universities at Strasbourg, Stuttgart, Cologne, Mainz, Bonn, Trier, Ingolstadt, Fulda, Bamberg, Dillingen, Altdorf, Paderborn, Helmstedt, Rinteln, Salzburg, Innsbruck, Frankfurt an der Oder, Herborn, Erfurt, Wittenberg, Duisberg, and Münster—had disappeared. Proposals were put forward at this time that *all* German universities should be closed.[11]

In the first chapter of this study I suggested that the "purity" of science is generally defended in the degree that its autonomy is perceived to be under siege. In early-nineteenth-century Germany it was largely against what was seen as a French threat to science that purity was defended. Napoleon had conquered Prussia in 1806 and dissolved the university at Halle. The University of Berlin was to replace Halle, as part of the broader attempt on the part of Prussia to rebuild its cultural strength. This particular relationship of France and Germany helps explain the distrust many German scholars held toward French Enlightenment ideals.

The ideals of *Bildung durch Wissenschaft* (spiritual uplift through science) and of *Einsamkeit und Freiheit* (solitude and freedom) represented a rejection of the French and English utilitarian ideal of knowledge. Wilhelm Kahl points out that the founders of the new university raised a cry against the narrowly vocational studies *(Brot- und Butterstudium)* and "utilitarian sects" which they saw as "prostituting" the sciences. Friedrich Paulsen notes that the new university was organized "in direct opposition to the higher schools of the military dictator" (notably Napoleon's Ecole Polytechnique).[12] Germany's leading philosophers left no doubts about where they stood on this question. G. W. F. Hegel attacked the "crude materialism" of the French Enlightenment, generalizing his attack to include all empirical science with its mere "collection of facts and discovery of unusual examples."[13] Friedrich Schelling rejected that "empty-headedness that calls itself enlightenment," mocking those for whom the task of science was to perfect textiles or improve fermented juices. Schelling had little respect for those for whom geometry was valued because it teaches us "how to measure fields and build houses." It was not that such skills were not useful; astronomers were certainly needed to make calendars and doctors were usefully applying Galvanism to disease. But the philosopher was to draw from universals, not particulars. His was the spirit of unity and totality *(Geist der Ein- und Allheit)*: the task of *Bildung* (spiritual, holistic learning) was to allow the individual to transcend the realm of one-sidedness and enter the realm of absolute universality. Science was to be one with philosophy, that universal "science of all science" and the key to spiritual transcendence.[14]

Notwithstanding the prejudices of men such as Hegel and Schelling, one should not imagine that the utilitarian spirit was foreign to Germany at this time. The Enlightenment conception of the bond between knowing and doing was common enough in Germany at this time. One example can be seen in a rather surprisingly modern-sounding theory of the practical origins of science put forth a decade prior to the founding of the University of Berlin. In the very first (1799) issue of the *Annalen der Physik* Friedrich Gren, editor of the new journal, noted the importance of distinguishing the history of the invention of tools from the history of physics proper. Whereas the former was properly the realm of "mixed science," drawing its conclusions from experience, only the latter constitutes that "rational part" of science—including mathematics, astronomy, and chemistry—that achieves a higher independence of empirical truth. A true history of science, Gren suggests, must consist in a history of this rational part of science. But Gren also admits that the arts have been important for the sciences. It is certain, he writes, that "the origins of the arts and sciences lie in the needs of men." Indeed,

> It is to agriculture and the measurement of time that we owe astronomy, to the need for iron that we owe metallurgy. It is to the parceling up of land that we owe geometry; to trade that we owe arithmetic; and to the transport of goods and the building arts that we owe the science of mechanics. It is to wounds and sickness that we owe the sciences of botany, anatomy, and medicine.[15]

This was the kind of practical view that Schelling, Hegel, and the architects of the new research ideal opposed.

The ideal of pure and disinterested research that emerges with the University of Berlin represented a rejection of the Baconian vision of practical knowledge and the utilitarian ideals of the British and French Enlightenment. But science, in this new vision, was not supposed to be "value-free" in the modern sense. The new research ideal posed no fundamental challenge to Graf Eberhard von Württemberg's view, upon the founding of the University of Tübingen in 1477, that "nothing provides a shorter and better means of achieving purity and chastity than a scholarly education."[16] The German scholar was to cultivate moral as well as intellectual virtues, to combine objective scientific study with cultural development of the soul. As we shall see, this was to change in the course of the century. With the rise of science-based industry knowledge becomes a commodity; the ideal of the cultivated scholar is forced to accommodate to a new ideal: that of the specialized expert. And with this new ideal, the conception of knowledge as integral to moral life is progressively supplanted by a conception of knowledge as a neutral instrument in the service of goals external to science.

6

Empirical Science and Specialized Expertise

The German scholar of the late eighteenth and early nineteenth century was supposed to be a carrier of moral as well as intellectual culture. *Wissenschaft* (science) was equally the study of theology, medicine, law, and philosophy; the equivalent terms in French and English do not capture the breadth of the German term. Gleditsch's 1745 *Teutsch-Englisches Lexikon* translated *Wissenschaft* variously as "knowing," "knowledge," "notice," or "cognisance"; to possess *Wissenschaft* was to be privy to or practiced about, to have a knack or skill, craft or dexterity. A man *von grosser Wissenschaft* was "an artfull, able, learned, scientifical, skilled, skilfull, well-practiced expert, experienced, witty, ingenious or industrious, man; a great scholar; a man of great learning, scholarship, ability, erudition, doctrine, science, knowledge, skill or experience."[1] *Scholarship* or *study* better reflects the meaning of *Wissenschaft* than does the term *science*.

The "man of science" of the German Enlightenment was supposed to be a *virtuoso,* a *homo literatus,* a carrier of universal and encyclopedic learning. This broader sense of science, together with the Baconian vision of practical knowledge, was elaborated in Leibniz's plan for the Academy of Sciences at Berlin. The academy, according to Leibniz's letter of July 11, 1700, was established to promote the honor of God and the cultivation of all kinds of virtuous and useful practices for the common good. As an institution devoted to "the love and care of things German," the academy was to help make the mechanical sciences "practical and fruitful." The academy was to help maintain the German language in its purity, to help equip missions in heathen lands, to enrich the honor and glory of the German nation, and to "further the progress of true belief." Academicians were also to supervise the production and censorship of books and to sponsor lotteries. True to his encyclopedic vision, Leibniz insisted that science was to be valued not just for "its own sake" but for the glory of God, the benefits of human practice, and the honor of the German nation.[2]

This broader sense of the term *Wissenschaft* was maintained until sometime late in the nineteenth century, and indeed remains to some extent today. Yet by the middle of the nineteenth century this has begun to change. Universities, fueled by the promise of exact empirical science and science applied to industrial needs, begin to shift their priorities in the direction of laboratory-based research and teaching. The status of the philosophical faculty vis-à-vis the traditionally dominant faculties of theology, law, and medicine also undergoes a dramatic change.

Since late medieval times German universities had supported, alongside the faculties of theology, law, and medicine, a faculty of the liberal arts composed of the *trivium* of arts (grammar, logic, and rhetoric) and the *quadrivium* of the sciences (arithmetic, geometry, music, and astronomy). This "philosophical faculty," as it came to be called, was primarily responsible for preparing undergraduates for graduate study in the faculties of law, medicine, or theology, where one trained to become a lawyer, physician, or man of the cloth. Graduate training in science was generally not possible, apart from what might be learned in medical studies. This begins to change, however, in the seventeenth and eighteenth centuries. Indeed the history of the university from the fifteenth through the eighteenth centuries can be seen as a history of the gradual subordination of the Master of Arts *(magistri artium)* to the Doctorate of Philosophy *(doctor philosophiae)*; so that by the end of the eighteenth century the Master of Arts has begun to be regarded as an intermediate step on the path to the Ph.D.[3] As philosophy was the home for most of what we now call the sciences, the growth of the philosophical faculty provides a fairly good measure of the rising importance of science in the German university (see Table 1).

In 1800 only about 5 percent of all Prussian students were enrolled in the philosophical faculties of Prussian universities (what would eventually become our "arts and sciences"). By the early 1830s this figure had risen to nearly 15 percent, and by 1887 to nearly 30 percent. The most rapid growth in this regard occurred in the middle decades of the century, as the proportion of students studying science and the humanities as a fraction of the total population grew from 114 students per million in the 1860s, to 200 per million in the early 1880s, to 400 per million in 1906. This does not even count the students at Germany's eleven Technische Hochschulen (technical colleges), numbering 11,000 in 1900.[4]

The growing importance of the philosophical faculty in general,

Table 1 Students at Prussian universities

	Theology	Law	Medicine	Philosophy	Total
1799–1805	495	859	135	73	1,562
1832–1834	1,688	646	835	805	5,335
1887–1890	3,184	2,369	3,484	3,672	12,709

Note: Enrollments for 1799–1805 are calculated for the universities at Erlangen, Erfurt, Duisberg, Frankfurt an der Oder, Königsberg, and Halle; statistics for 1832–1834 are for Berlin, Breslau, Halle, Bonn, Königsberg, Münster, and Greifswald. Figures for 1887–1890 include these universities plus Marburg, Kiel, Braunsberg, and Göttingen, at this time in Prussia. See Wilhelm Dieterici, *Geschichte und statistische Nachrichtungen über die Universitäten im preussischen Staate* (Berlin, 1836), pp. 107 and 156–172; Wilhelm Lexis, *Die deutschen Universitäten* (Berlin, 1893), vol. 1, p. 144. The numbers in each case are annual averages for total student enrollment in all Prussian universities.

and the natural sciences in particular, reflected the Enlightenment ideal of the primacy of reason among the faculties of human understanding. Francis Bacon in his *Advancement of Learning* had distinguished three faculties of human understanding: reason, imagination, and memory, with which were associated the pursuits of philosophy, poetry, and history. When d'Alembert in his 1750 *Discours préliminaire* acknowledged his debt to Bacon, it was not simply to his method but to his division of the sciences in terms of the faculties of the mind according to which the tree of knowledge was to be constructed. Immanuel Kant, follower of both Bacon and d'Alembert, had introduced the Baconian ideal into German debates, arguing for the elevation of the philosophical faculty—the faculty of reason—to parity with the faculties of law, medicine, and theology.[5]

The growth of empirical science in nineteenth-century Germany reflects more than simply the triumph and spread of Enlightenment ideals. The value of science in producing military and industrial power was recognized, and it was believed that science would solve social problems. We shall return to these themes in subsequent chapters. Here, though, I want to explore one particular philosophical consequence of the rise of the empirical sciences. Specifically, I want to look at how the growth of empirical methods was accompanied by a narrowing of the conception of what science is and ought to be.

By the 1830s, students of biology, medicine, chemistry, and physics had begun to reject the mystical, encyclopedic visions of the *Naturphilosophen* (nature philosophers) in favor of more cautious and empirical procedures of laboratory science. In 1822 the Verein der Naturforscher und Ärzte was established to promote the exchange of ideas concerning the new experimental sciences; in 1824 Johannes Müller began his work in physiology at the University of Bonn. Two years later Justus Liebig opened his chemical laboratory in Giessen, marking the beginning of laboratory teaching in the German university.

Liebig was not alone by this time in complaining of his "delusion at the hands of the German nature philosophers"; discontent soon spread to an abandonment by many scientists of *all* philosophy. In 1844 Schelling delivered his last lectures as professor of philosophy at the University of Berlin. Hegel had already been dead more than a decade. Fewer and fewer scientists recognizing the power of laboratory techniques were satisfied with a conception of science according to which, as Schelling once asserted, every mineral was "a philological problem," or, as another *Naturphilosoph* put it, diamond was "quartz which has achieved self consciousness." Hermann von Helmholtz attacked Hegel's philosophy of identity for its pretensions to derive the structure of the world from the process of thought. Karl Ernst von Baer mocked those who had felt compelled to credit the Creator with every detail of nature, churning out books on "insect theology," "fish theology," and "rock theology."[6]

Mid-nineteenth-century critics no doubt exaggerated the hostility of early German idealism to experimental science. As recent historical scholarship has shown, the *Naturphilosoph*'s fascination with unity and polarity actually helped to inspire a number of creative scientific works, including Johann Ritter's discoveries of electroplating and ultraviolet radiation (1800–1801), Hans Christian Oersted's discovery of electromagnetism (1820), and Johannes Purkinje's discoveries of visual afterimages and the cerebellar nerve cells that bear his name (1837). Goethe wrote extensively in the areas of botany, optics, and mineralogy, and even Schelling was open to certain types of experimentation. Hegel was notoriously inept when it came to questions of mechanics or kinematics; but his historicist dialectics helped inspire both the biblical criticism of D. F. Strauss and the political economy of Marx.

By the mid-nineteenth century, however, the vogue was to contrast the new with the old, and few were willing to stand up for the Romantic idealism that had swayed earlier generations. The revolt against German idealism turned into a power struggle for control of the German university— between scientists advocating funding for laboratory research and philosophers advocating traditional ideals of humanistic scholarship. Philosophically minded scientists such as Helmholtz, Emil Du Bois-Reymond, A. W. von Hofmann, and Julius Lothar Meyer tried to steer a middle course, supporting an increased science curriculum while warning against the total disintegration of the universities into specialized technical schools. Helmholtz, despite restricting the task of philosophy to "criticism of the sources of cognition and to investigation of the functions of the intellect," urged scientists to explore connections of their specialties with the whole of science. Meyer advocated the integration of science and humanities curricula.

Yet pressures to establish separate specialized curricula were too great to stop. In 1863 Tübingen established the first faculty of natural sciences separate from the philosophical faculty; the University of Strasbourg followed suit shortly thereafter.[7]

Trade and technical schools, established in the seventeenth and eighteenth century to train students in the baking, engineering, and other practical trades, grew in the middle of the nineteenth century into the famous Technische Hochschulen, technical schools designed to train engineers in a course of study on a par with that of the traditional university. Until about 1880, the Technische Hochschulen were controlled by the Ministry of Trade and enjoyed considerably less status than the older universities. In the early 1880s, however, the schools were transferred to the Ministry of Education, signaling a move toward parity with the older universities. In 1892 Wilhelm II declared that professors at the technical schools should have equal status with university professors and could henceforth wear academic regalia; in 1898, the rectors of Prussia's three Technische Hochschulen were called to the upper house of the German parliament, a privilege formerly granted only to university rectors. In 1900, technical schools were granted rights to award doctoral degrees in the engineering and applied sciences.[8]

The triumph of science in German higher education was encouraged by the recognition of the value of science in producing military and industrial power. Science comes to have the cachet that philosophy as a whole once had. But what happens to philosophy? With the collapse of Romantic *Naturphilosophie,* scientists either abandoned philosophy altogether or cast about for alternatives to Hegel and Schelling. The materialism of the French Enlightenment (La Mettrie, Holbach, Helvetius) provided one source of inspiration; the "critical philosophy" of Kant provided another. Which of these alternatives one favored depended very much on one's politics. The more conservative choice was Kant—a choice that allowed one to argue against both the materialism associated with 1848 and the newly emerging Social Democratic movement; but also to articulate a justification for the autonomy of science in the face of attempts by business concerns to direct science onto paths useful to industry.

Kant's philosophy appealed to those wanting to isolate science and religion (or morality) from each other and to provide each with separate realms over which the other has no jurisdiction. Religious and practical truths for Kant lay in the sphere of the noumenal "thing in itself"; the truths of science embodied only the phenomenal "thing for us" as presented by human sensory experience. Knowledge of the two spheres—science and religion—could never conflict, for they deal with entirely separate realms of Being. Kant's doctrine appealed to the man of science who wanted to keep his science and his religion intact, without worrying about apparent contradictions between the two. Science played a crucial role in Kant's philosophy. He adopted as his method "that of the student of nature," which he described as "looking for the elements of pure reason in what admits of confirmation or refutation by experiment."[9] The *Critique of Pure Reason* is dedicated to Francis Bacon, and the English philosopher's presence is felt throughout the book.

Neo-Kantians in the second half of the nineteenth century used the distinction between noumena and phenomena to fix a boundary between science and religion. Kant, they argued, had defined the limits for knowledge based on experience, beyond which all knowledge must be based on faith. By radically separating knowledge of moral or religious questions from knowledge of the empirical world, Kant's philosophy, like that of Hume in England or Auguste Comte in France, served to liberate the pursuit of science from the constraints of philosophy and religion, allowing scientists to pursue empirical research without fear of violating religious or philosophical doctrines. Kantian dualism also served to insulate science from political critiques. In September 1854, in a debate over materialism at a scientific convention in Göttingen, Kant's doctrines were invoked to preserve a separate and unassailable realm for ideals of God, freedom, and immortality, ideals preserved in neo-Kantian doctrines as "convenient fictions" or "regulative ideals."[10] Divorced from religion and philosophy, science could now proclaim itself immune to religious or moral criticism.

In the second half of the century the neo-Kantian model of exact experimental science divorced from religious or speculative philosophy begins to penetrate every academic discipline. It is an age of empirical research. Hermann Lotze captured this sentiment when he lamented the fact that, at least in philosophy, "there no longer exists any merit in originality, but only in accuracy." Despairing of finding anything new in philosophy, many philosophers spent their time preparing elaborate commentaries on the masters, dissecting the philosophical works of Leibniz, Kant, or Hegel with a precision and attention to detail that rivaled biblical exegesis. Hans Vaihinger devoted 1,000 pages to his commentary on the first 70 pages of Kant's *Critique of Pure Reason*.[11] Others turned to psychology or to social or political theory, attempting to salvage some sort of synthesis in the nonexperimental sciences.

In the program of Eduard Zeller, Eugen Dühring, and Richard Avenarius, philosophy itself was to become a science, and the philosophy of science the heart of philosophy. In 1862, in his Heidelberg address on "The Meaning and Scope of *Erkenntnistheorie*," Zeller issued a call for a new "theory of knowledge" *(Erkenntnistheorie, épistémologie)* signaling the beginning of the predominance of epistemology over other kinds of philosophy that has remained until the present. The new theory of knowledge was to be based not on speculation but on the latest results of science.[12]

Hermann von Helmholtz argued that the future progress of philosophy depended upon "an exact knowledge of the processes of sensory perception."[13] In the latter half of the century, a new generation of philosophers emerged schooled in Johannes Müller's work on physiology, Wilhelm Wundt's experiments on just-noticeable differences, and Gustav Fechner's law of diminishing marginal response to psychophysical stimuli. Philosophers such as Avenarius and Ernst Mach began to argue that Helmholtz's *Physiological Acoustics* and *Physiological Optics* were more vital to philosophy than all previous speculations on the elements of Being or the spiritual necessities of the cosmos.

Efforts to establish philosophy on the solid ground of science were carried forth in new journals. In 1861 the first issue of the *Zeitschrift für exacte Philosophie* announced a new program for philosophy, one that was to draw from the "positive and empirical sciences" to achieve a new level of precision in the analysis of philosophical problems. The goal of the journal was to expose the errors of idealism and to establish in its place a "philosophical realism" that would combat the varieties of "loose and inexact" thinking in contemporary philosophical writings—including the "false speculations" of pantheism and materialism. Sixteen years later, in 1877, Richard Avenarius founded his *Vierteljahrsschrift für wissenschaftliche Philosophie* with even stronger claims to ground philosophy in the empirical sciences. The journal announced its intention to publish articles on "epistemology, philosophy of science and mathematics, psychophysics, psychology and anthropology, sociology and ethics, aesthetics, philosophy of language and of history." Conspicuously absent were religion and metaphysics. Philosophy, according to Avenarius, was to be "first and foremost a science—nothing more, and nothing less."[14]

In the course of the century, the increasing use of empirical methods by philosophers comes to be seen as a threat to more traditional academics. In the 1880s and 1890s, philosophers more comfortable with Plato and Aristotle come to fear the encroachment of experimental psychology into their discipline, as chair after chair in the philosophical faculties goes to students trained in the physiology of the eye or the anatomy of the ear.[15] (Typical of the age is that Wundt, Fechner, and their students would attempt to reduce human judgment to the ability to make fine distinctions in sounds, touch, and other perceptions, according to Ernst Weber and Gustav Fechner's principle of "just-noticeable differences." Lujo Brentano in 1908 would write a book claiming that subjective value theory could be grounded upon this principle, and that marginalist economic principles were derivable from this.) Empirical methods begin to penetrate other disciplines as well. Archival research comes to dominate the historical schools; biblical criticism applies the canons of exact historical scholarship to discover the "historical Jesus" (Strauss) or history *wie es eigentlich gewesen ist* (Ranke). With the growth of empirical methods, academic disciplines begin to draw in their boundaries, seeking refuge in progressively narrow, specialized research.

Max Weber reflected upon the specialization that pervaded so many fields of academia in the early years of the twentieth century. In his 1919 "Science as a Vocation," Germany's foremost sociologist wrote that science had entered upon a course of specialization from which it might never escape, and that whoever could not blind himself to fields outside his own might as well leave the work of science to others. It was only by strict specialization that the academic could achieve anything of lasting value. Specialization had been made necessary by the unceasing progress to which science is chained. The most one could hope for was to contribute in some small way to the growing stock of knowledge, which would then become obsolete in ten, twenty, or fifty years. In the field of science, Weber concluded, "only he who is devoted *solely* to the work at hand has 'personality.' "[16]

It was Ferdinand Tönnies, however, "father" of German sociology, who was to come to the conclusion that, by virtue of its ever-increasing specialization, science must be value-neutral. Tönnies claimed that it was natural for practical and theoretical spheres of science to have become separate from one another, for science, no less than any other evolving thing, cannot escape this process of continual division. Indeed, such differentiation "is the greatest law of evolution." Tönnies was not alone in this view, as "splitters" came to triumph over the "lumpers" in the spirit of Kant's dictum that "we do not enlarge, but disfigure sciences if we allow them to trespass upon one another's territory."[17] Even as new territories were being opened up, old ones were having their boundaries ever more tightly guarded.

Details of Weber's and Tönnies's arguments for value-neutrality we shall explore in a moment. Here, however, let me point out that the specialization that emerges in the second half of the nineteenth century has a political dimension—in art as well as science. Arnold Hauser in his *Social History of Art* has described the political aspects of specialization and of "art for its own sake," tying these to the philosophy of Kant in what could just as well be a description of "science for its own sake." The ideal of art "for its own sake" *(l'art pour l'art),* Hauser argues, sprang originally from Romanticism, where it represented one of the most powerful early weapons in the struggle for artistic freedom. But the political function of this ideal changes in the course of the century, as "art for its own sake" becomes an excuse for retreat from public life:

> For the romantics "l'art pour l'art" becomes the ivory tower in which they shut themselves off from all practical affairs. They buy the peace and security of a purely contemplative attitude at the price of an understanding with the prevailing order. Until 1830 the middle class hoped that art would pro-

mote its ideals, it therefore accepted art as a vehicle of political propaganda
... After 1830, however, the bourgeoisie becomes suspicious of the artist,
and prefers neutrality to the former alliance ... The middle class makes
"l'art pour l'art" its own; it stresses the ideal nature of art and the high,
superpolitical status of the artist. It locks him up in a golden cage. Cousin
goes back to the idea of autonomy in Kant's philosophy and revives the
theory of the "disinterestedness" of art, and here the tendency to specialization which becomes ascendant with capitalism proves useful. "L'art pour
l'art" is, in fact, partly the expression of the division of labour which
advances hand in hand with industrialization, partly the bulwark of art
against the danger of being swallowed up by industrialized and mechanized
life. It signifies, on the one hand, the rationalization, disenchantment and
contraction of art, but simultaneously the attempt to preserve its individual
quality and spontaneity, in spite of the universal mechanization of life.[18]

Much of the same could be said for science. Social and natural scientists alike were attracted to the neo-Kantian doctrine, both for its idealism and the autonomy it promised for sciences of the natural or social world. The Kantian separation of noumena and phenomena, as of theoretical and practical reason, provided the intellectual legitimacy for separate and autonomous spheres of science on the one hand and politics and religion on the other. Neo-Kantianism was a philosophy of compromise, allowing scientists to pursue empirical truth without disturbing religious belief, allowing faith in religious ideals without fear of contradiction by scientific evidence.

The age of specialized retreat is not, of course, without its critics. Auguste Comte in his *Cours de philosophie positive* warned against the rising tide of specialization, arguing that the time had come "to make the study of scientific generalities itself a great specialty." In 1835, the poet Charles-Augustin Sainte-Beuve accused his colleague Alfred Vigny of retreating to "the ivory tower" in what appears to be the first use of this expression in its modern sense. Friedrich Nietzsche even in his student days expressed his disgust for the narrowness of that "teeming breed of philologists, their mole-like activities, their full pouches and blind eyes, their pleasure in the worm they catch." In his eagerness to appear scientific, Nietzsche complained, the historian seeks to record history objectively and in its entirety. The "objective man" thus "waits until something comes along and then spreads himself out ever so tenderly, so that not even the lightest footfalls of ghostly beings slipping past shall be lost upon his surface."[19] Objectivity for Nietzsche was a euphemism for the inability of scholars to distinguish between the trivial and the significant; objectivity was merely "skepticism and paralysis of the will dressed up." Nietzsche (who fancied himself "the last unpolitical German") expressed his contempt for those who "eat of the dust of bibliographic trifles."

The Renaissance historian Jacob Burckhardt also criticized the narrowness of contemporary specialists, whom he called the *viri eruditissimi* and *capricorn beetles* and whom he compared with those who try to dig up a mountain: "They begin and dig a deep hole which, however, is nothing in comparison with the mountain. Death comes meanwhile, and what do they leave us? The rubbish lying beside the hole they dug." And again: "God, too, wants some fun at times, and then he creates those philologists and historians who think themselves superior to all the world when they have scientifically ascertained that the Emperor Conrad II went to the privy at Goslar on May 7 in the year 1030—and things of like world-wide interest."[20] Burckhardt's own *Cultural History of Greece* was greeted with suspicion by professional historians because, as he himself admitted, the book was a work of synthesis and interpretation and not the product of original research.

Specialization, associated with the new norms of narrow, empirical, and exhaustive research, served similar social functions as the ideology of neutrality. The scientist was not to bother himself with political affairs. His job was to follow the truth: it was not his place to challenge the politics of his patrons. But specialization was only one of several roots of the nineteenth and twentieth century ideology of neutrality, as becomes clear when we look at the debate over values that emerged in early-twentieth-century German political economy.

7

The *Werturteilsstreit,* or Controversy over Values

On January 3, 1909, a group of social theorists dissatisfied with the "political" character of Germany's Social Policy Association (Verein für Sozialpolitik) met under the guidance of Max Weber to found their own society, the German Society for Sociology (Deutsche Gesellschaft für Soziologie), in what is often recognized as the beginning of organized sociology in Germany. For nearly forty years (since the founding of the Reich in 1871), the Social Policy Association had provided the major forum for discussion of scholarly issues concerning social, historical, and political questions. The new Society for Sociology, breaking with the Association, was to be a strictly scientific society, and by this was meant, above all else, that it was to avoid all discussion of matters moral, political, and religious. Indeed, this position was set forth in the first statute of the new society:

> Under the name "Deutsche Gesellschaft für Soziologie" a society has been founded, which has its seat in Berlin. Its purpose is the promotion of sociological knowledge through the organization of purely scientific investigation and research, the publication and support of purely scientific works, and the organization of regular sociological meetings in Germany. The Society will provide all scientific directions and methods of sociology with equal space, and *rejects* the advocacy of all practical (ethical, religious, political, and aesthetic) goals.[1]

The rejection of concern for practical goals expressed in the German Society for Sociology's founding charter represented the culmination of a debate in the Social Policy Association, the so-called *Werturteilsstreit,* in which a group of young political economists, including Ferdinand Tönnies, Werner Sombart, and Max Weber, attacked the older generation of political economists for mixing facts and values, science and politics. The course of the debate provides some insight into why one group of sociologists found themselves so energetically opposed to value judgments in science. The debate is also significant, because it is here that expressions such as "value-free" and "value-laden" science come into popular usage.

The neutrality crusade of the German Society for Sociology was sparked by a growing dissatisfaction with the direction taken by conservative political economists in the Social Policy Association. The Association had been founded in 1872 by a body of reform-minded doctors, lawyers, professors, industrialists, and politicians concerned with constructing progressive social policy for the newly united German nation. Until the early years of the twentieth century, the Association gathered together Germany's most eminent social theorists and reformers, figuring as a forum for both discussion of political issues and development of social theory. It published an annual series of resolutions, which it presented to the German government and managed to exert considerable influence on matters of state policy.

The Social Policy Association remained the leading organ for social theory in all of Germany until the twentieth century. In the early years of the twentieth century, however, a younger generation of political economists led by Max Weber began to criticize the leadership of the Association for having failed to encourage independent scholarship—scholarship, that is, that was not subordinate to state policy. Weber, Sombart, and Tönnies all felt that the older generation of political economists, headed by Gustav Schmoller, had compromised its scientific integrity by acting as a servile advisory body to government in matters of state, industrial, and farming policy.

In 1909, at the Vienna meeting of the Association (dedicated to the problem of productivity in political science), a debate broke out over the proper role of value judgments in science, a debate which has continued even into the present and one which for many marks the transition from the "historical" or "ethical" school of political economy to value-free scientific sociology. Though the debate—subsequently labeled the *Werturteilsstreit,* or controversy over values—was conducted largely over the issue of the proper role of values in science, what was at stake was essentially a matter of the subordination of science to state policy. The central question that divided the young sociologists—Tönnies, Weber, and Sombart—from the older historical school of political economists—Schmoller, Philippovich, Herkner, and Gottl-Ottlilienfeld—was the question of what role values should play in economic science. What were the arguments?

In his paper delivered to the 1909 Vienna meeting, Professor Eugen von

Philippovich defended a conception of "productivity" in political science that would include moral as well as empirical judgments. It was impossible, he argued, to separate political or ethical concerns from questions of pure research. Indeed, "all economic research is ultimately subordinated to the goal of determining to what degree the well-being of the people is served by particular economic activities. It is an illusion to believe that the restriction of theoretical research to that which *is* can free one from conceptions of what *ought* to be."[2] Philippovich rejected Alfred Marshall's proposal to eliminate the word *productivity* from economic discourse simply because the word was associated with "ethical ideals." He also rejected the attempt on the part of classical British economists to measure economic productivity by the sheer quantity of goods produced, ignoring the equally important question of the *quality* of those goods and their contribution to the well-being of society. The very concept of productivity, Philippovich argued, implies ethical ideals. Is it not meaningful to ask whether goods produced in the service of immoral or unreasonable ends are "productive" in the same sense as morally useful goods? Should the fabrication of bombs for anarchistic attacks be considered "productive"?[3]

Professor Willie Herkner of Charlottenburg saluted Philippovich's call to explore the moral aspects of productivity, illustrating the importance of this dimension with a discussion of the working conditions in Germany's factories. The economic costs of labor, Herkner argued, are not simply those recorded by the entrepreneur; one must reckon as well with the depression and unhappiness that arises from restrictions on workers' freedom, the increased speed and intensity of the work pace, and the humiliation of being treated only as a number—these must be figured in the "costs" of production as well. Herkner maintained that those who (with William Petty) would consider under costs only that which can be "counted, measured and weighed" will of course exclude such vague and loose ideas from the "temple of science." This, however, would be a serious mistake; indeed, were we to exclude all questions of moral value and social well-being from our science, "we would very soon have to find some *new* science to deal with that which was lost from political economy."[4]

Central in the debate that divided the Social Policy Association was the question of the role of science vis-à-vis the state. The older historical school had long insisted on the importance of bringing science to bear on problems of public policy; Rudolf Hildebrand had gone so far as to argue that a particular discipline (such as philology) should be considered "not merely a science, but a handmaiden for the salvation of the nation." Political economy for these conservative paternalists presumed a harmony of state and social interests: capital and labor, males and females, Protestants and Catholics, each had its part to play in the prosperous growth of the nation. To abandon moral judgment would be to abandon the duty of the scholar to help maintain social order. This willingness to moralize on matters of social policy translated also into a special academic style. Joseph Schumpeter in the 1940s recalled how, as a student at the turn of the century, the *Kathedersozialisten* had lectured as if they were leading a political rally: "Lujo Brentano addressed his classes as he would have political meetings, and they responded with cheers and countercheers. Adolf Wagner shouted and stamped and shook his fists at imaginary opponents, at least before the lethargy of old age quieted him down."[5] Where would medicine be, Schumpeter asked, if physicians carried on in such a manner about the glories of healing?

The young sociologists wanted no part of such moralizing. And the issue was not a trifling one. Werner Sombart, in his contribution to the debate, declared that "today is a decisive day in the historical development of the Social Policy Association . . . It is the day that political economy decides whether it is to be a science." The ground over which the decisive battle was to be fought was "whether we political economists pose to ourselves the single problem of what *is,* or whether we at the same time pose the problem (or indeed pose as the only problem) that which *should* be." It was this split, Sombart declared, which had always been the decisive one in the Association; this was the question that would determine whether political economy was to be a science.

Sombart was willing to concede, with Schmoller, that "there is no nail driven into wood without ethical admixture." Sombart cited Weber's study of the influence of the Puritan ethic on the rise of capitalism as an example of the importance of ethics in economic life; he also made it clear that by excluding consideration of ethical questions from science he did not mean that ethical values play no role in economic life. "Rather, we mean by this that they should play no role in economic science. We mean that insofar as it is *science* that we are engaged in, we can neither be Puritans nor non-Puritans judging economic life; we must instead be objective observers or analysts. It is in this sense that we exclude value judgments from scientific observation." And why this Verbot? "For so long as value judgments play a role in scientific investigation, objective understanding of that which *is,* is not possible." Values are a source of bias, a hindrance to objective knowledge. For Sombart, as for Weber and Tönnies, scientific knowledge presumed a neutral ground on which to work; the intrusion of values could only hinder the pursuit of that truth.[6]

It was Max Weber, however, who was to provide the most comprehensive and articulate defense of value-neutrality. In his contribution to the debate, Weber suggested that there are only three legitimate ways a scientist can confront political interests. He can point out the contradictions between a person's values and their interests. He can also ask, empirically, what *means* must be used to achieve those interests. And finally, he may point out certain unintended consequences of pursuing and achieving those interests. These and only these can the scientist, *qua* scientist, undertake. The first is a question of logic; the second and third are questions of empirical fact. But whether these goals should be pursued in the first place, Weber says, is a question no science can answer. For "this is a question of conscience, of personal commitment, and in this realm science cannot tread."

For Weber, confusion of the separate realms of science and values borders on heresy—value-laden science he describes as "a thing of the Devil" and "original sin" *(Sündenfall)*. It is not that Weber underestimates the importance of practical or moral questions; indeed, his intent is precisely the opposite. Questions of value and practice, politics and world view, are for Weber of such "earthshaking importance" that he doesn't want to see them reduced to matters of a merely technical nature, questions of productivity and the like.[7]

Gustav Schmoller, dean of the older historical (or "ethical") school of economists, presided over the 1909 Vienna meeting that erupted into the *Werturteilsstreit*. But he himself did not enter the debate until shortly thereafter. In a 1911 article in the *Handwörterbuch der Staatswissenschaften*, Schmoller presented his conception of the place of ethics in economic science, claiming that values are indeed objective—in the sense that they are based on a universal "value-sense" *(Wertgefühl)* which has developed over millions of years of human evolution. Value judgments are objective, in the sense that they are a shared human product.[8] Schmoller admitted that supposedly objective values may in fact err; but he also expressed his confidence that human values "ripen" with time, becoming more and more humane through cumulative human experience. Schmoller conceded that religious, moral, or aesthetic beliefs may differ among peoples, but he also claimed that despite this diversity, there are always common values that link Catholic and Protestant, Christian and Jew, realist and idealist. The differences that separate men, he argued, are not so much differences in *ends* as in the *means* people use to attain those ends.

Schmoller warned that it would therefore be a serious mistake to attempt to exclude values from science, for the conclusions of ethical science are as vital as those of any other science, if not more so. Schmoller complained that those who advocated "pure theory" had had difficulty attracting students, and that those professors who did follow in the footsteps of Eugen von Böhm-Bawerk, Karl Menger, and others of the Austrian mathematical school had often been forced "to lecture to empty classrooms; their students memorizing equations from their mechanically copied notes, only to forget them within a couple of months after class is over, without having learned anything of real value." Schmoller claimed that it was impossible to eliminate teleology from science, no matter how hard one tried. Biology and geology even after Darwin had, in his view, preserved notions of directional development; Schmoller cited Wilhelm Wundt ("the premier scientist of the day") in support of the idea that *cause* and *goal* were complementary principles, each of which fulfilled the other.[9]

Others of the so-called *ethische Tendenz* or *Kathedersozialisten* ("socialists of the chair")[10] in German political economy expressed similar views. Oscar Engländer defended the search for "universal, objective, and ultimate values" as a legitimate task of economic science. Franz Oppenheimer described Kant's categorical imperative (to treat other persons as ends, never as means) as a universal standard on which one might found an ethical science of economy. F. von Gottl-Ottlilienfeld of Munich asked why Sombart singled out *value* as a term to attack—why not *capital* or *productivity* or any number of other words equally suspect? And what about that "fetish for facts" from which so many suffer?[11] Was that not at least as problematic as any of these other ideas?

For Weber and the young sociologists, however, the controversy over values had driven a permanent wedge between themselves and the older historical school. Earlier in the year (1909), Max Weber had already begun organizing a younger generation of political economists to establish a new forum for neutral, social scientific research, the German Society for Sociology. Throughout the winter of 1909–10, Weber corresponded with leading figures in Germany's new science of sociology—Ferdinand Tönnies, Rudolf Goldscheid, Werner Sombart, Georg Simmel, Ernst Troeltsch, Gerhart von Schulze-Gävernitz, and Robert Michels—to make arrangements for the new society.[12] For Weber, as for Sombart, the question of avoiding value judgments was the issue galvanizing the break with the Social Policy Association. In an unpublished manuscript circulated shortly after the Vienna meeting of the Association, Weber argued that the conservatives in the association were wrong to have imagined that scholarly expertise *(Fachkenntnis)* qualified one in any special way to participate in social policy discussions or to take a stand on practical problems.[13] Weber was well aware that the decision to exclude value judgments from scholarly discussions or classroom lectures constituted a political judgment. This, as we shall see, is the irony in Weber's position: neutrality represented (among other things) a critique of authoritarian scientism, of those who imagine that scientists are somehow specially equipped to deal with matters of moral or political controversy (see Chapter 10).

At the first meeting of the Society for Sociology in October 1910, in Frankfurt, the young sociologists hammered home the point that values should be excluded from science. Several sociologists took the floor, but it was Ferdinand Tönnies, author of the widely read *Gemeinschaft und*

Gesellschaft, who developed the point most forcefully in his paper on "The Paths and Goals of Sociology." The paper marks a turning point in the history of sociology, for it elevated the question of the place of values in science to the central methodological question of the new discipline.

For Tönnies, the separation of questions of science and politics, fact and value, was a natural and indeed inevitable process, closely linked with the growing specialization of academic inquiry. "Differentiation, separation, and the division of labor" were all part of "the law of evolution." The separation of theoretical and practical economy represented a natural extension of this process. Tönnies argued that a businessman should be a businessman, a scholar a scholar, but that the two should not be confused. One hardly expects a representative of an international trading company to discuss the merits of free trade in an objective manner. From the thinking scholar, however, one expects something more: "the thinker should stand *above* parties and interests." Eighteenth-century philosophers had presumed to discover objective ethical truths in the name of "natural law." But with growing skepticism on the existence of objective good or evil a pure science of sociology was born, separate from questions of a practical or ethical nature. And so,

> We leave all future programs, all social and political tasks, out of the game. Not because we look down on them, but rather because we recognize the insurmountable difficulties involved in establishing such ideas scientifically. Also because we expect from those of other persuasions—such as "scientific socialism"—that they will agree to keep the field of sociology *outside* such controversial issues so that we may restrict our inquiries to the much simpler problems involving knowledge of objective facts. And even if we admit that complete objectivity is unattainable, still we may strive for this goal with energy and determination, and through such efforts begin to approach this goal. This is our task. We want, in other words, as sociologists to explore only what *is,* and not what, according to whatever view, on whatever grounds, *should be*.[14]

For Tönnies, the problem of values was nothing less than the problem of objectivity. Social science was objective insofar as it avoided making judgments of value. The historian might rely on historical perspective to guarantee his objectivity; yet the sociologist, dealing with the present, could only rely on his restraint from value judgments. Only a concern with "pure factuality" *(reine Tatsächlichkeit)* could raise the study of social life above the conflict of parties, freeing it from the "crippling burden" of value judgments. Sociologists must strive for the precision of mathematics, the exactitude of astronomy. For "just as for the solar system there is but one sun, so for a system of science there can be but one sun: the Truth!"

Yet scientific sociology was neutral not in the abstract but rather in opposition to certain political ideas. Tönnies made this clear:

> As sociologists we are neither for nor against *socialism,* neither for nor against the *expansion of women's rights,* neither for nor against the *mixing of the races.* In each of these, however, we do find questions for empirical social research: in social policy, social pedagogy, and social hygiene. In this, sociology finds its proper boundaries without taking upon itself the task of furthering, or obstructing, particular ideas or movements.[15]

Neutrality, in other words, was proposed in opposition to socialism, the expansion of women's rights, and the mixing of the races. Each of these posed a major problem for social theorists in the early years of the century; to appreciate these problems, we must explore further the social and political background of European social theory.

At the end of his contribution to the debate in the 1909 meeting of the Association for Social Policy, the meeting that erupted into the *Werturteilsstreit,* Werner Sombart expressed one final justification for keeping science within the realm of politically neutral discourse. Science was to avoid all value judgments, for it was only on neutral, scientific grounds that a unity could be found amidst the diversity of social and class interests. Sombart contrasted the vision of a calm and unifying science with the specter of values in eternal opposition:

> In the face of growing personality and value-differentiation, we have the urgent task of finding a point around which we may once more unite, around which both believer and disbeliever, pantheist and atheist, social democrat and conservative, and whoever else, can unite their disparate views in the same way one unites over the solution of a particular historical or scientific problem. Such a unity, however, is excluded when we allow value judgments into scientific claims . . . [16]

The "value-differentiation" lamented in Sombart's remarkable passage was expressed in different spheres—in religion and in politics, in tensions between the sexes, in racial and national quarrels. The culmination of these conflicts in the first decade of the twentieth century led social scientists to search for some kind of order amidst what was widely hailed as the "fragmentation of the parties." Different solutions were proposed. Ernst Troeltsch championed the view that only an independent monarch could arbitrate Germany's difficulties in a disinterested fashion. Schmoller, Philippovich, and the *Kathedersozialisten* saw the resolution of Germany's troubles in economic growth, the benefits of which would filter down to all. For others, patriotism in the Great War would rally the classes of a divided Germany. But it was to *science* that Sombart looked for unity amidst the chaos of political strife. What was science that it could make such claims to arbitrate the great social conflicts of the day?

Sombart was not the first to find in science a source of political unity.

Thomas Sprat in his 1667 *History of the Royal Society* had spoken of the virtue of science in bringing men of different creeds into harmony with one another. Francis Bacon had proclaimed it the task of natural philosophy to return to that harmonious, prelapsarian state where humans had spoken a natural language "that did not incur the confusion of Babel." For Enlightenment optimists, to engage in science was to cultivate an attitude foreign to conquest and domination. Fontenelle thus claimed that "nothing can be more destructive to ambition, and the passion for conquest, than the true system of astronomy." Why covet the lands of the earth when, through science, one can bring to oneself the entirety of the cosmos? Adam Smith advised state-sponsored study of science as a cure for small-minded sectarianism: science would serve as "the great antidote to the poison of enthusiasm and superstition."[17]

The most powerful praise for science, though, is of nineteenth-century vintage. Thomas Macaulay recalled how, according to the early natural philosophers of Paris and London, perfect forms of government were to issue from the practice of science: "All classes were hurried along by the prevailing sentiment. Cavalier and Roundhead, Churchman and Puritan were for once allied. Divines, jurists, statesmen, nobles, princes, swelled the triumph of the Baconian philosophy." In an 1837 essay on "Francis Bacon," Macaulay celebrated the fruits of the Englishman's philosophy in one of the most spirited rhapsodies ever sung to science:

> Ask a follower of Bacon what the new philosophy, as it was called in the time of Charles the Second, has effected for mankind, and his answer is ready: "It has lengthened life; it has mitigated pain; it has extinguished diseases; it has increased the fertility of the soil; it has given new securities to the mariner; it has furnished new arms to the warrior; it has spanned great rivers and estuaries with bridges of form unknown to our fathers; it has guided the thunderbolt innocuously from heaven to earth; it has lighted up the night with the splendor of the day; it has extended the range of the human vision; it has multiplied the power of the human muscles; it has accelerated motion; it has annihilated distance; it has facilitated intercourse, correspondence, all friendly offices, all dispatch of business; it has enabled man to descend to the depths of the sea, to soar into the air, to penetrate securely into the noxious recesses of the earth, to traverse the land in cars which whirl along without horses, and the ocean in ships which run ten knots an hour against the wind. These are but a part of its fruits, and of its first-fruits. For it is a philosophy which never rests, which has never attained, which is never perfect. Its law is progress. A point which yesterday was invisible is its goal to-day, and will be its starting-post to-morrow."[18]

In Germany, the spirit of optimism rivaled that in England. Alexander von Humboldt praised Lorenz Oken's 1828 Assembly at Berlin (which later provided the model for the British Association for the Advancement of

Science) as "a noble manifestation of scientific union in Germany; it presents the spectacle of a nation divided in politics and in religion, revealing its nationality in the realm of intellectual progress."[19] Science, in other words, would succeed where religion and politics had failed.

The idea of science as a neutral arbiter can be divided into at least two elements. There is, first, the idea that justice is blind—along with the corollary notion that the essence of fairness lies in the perfect balance of opposing ideas. The idea of justice as blind is a comparatively new idea. In the seventeenth century, juries were composed exclusively of witnesses—those who were said to know most about the case. Those who knew nothing of the case were deliberately excluded. Today, the opposite is true: justice is blind, and those jurors are excluded who can be shown to have some knowledge or opinion on the case.

There is a second element in the ideal of science as a neutral arbiter, and that is the idea that it is to science that we should turn to solve social problems. Science brings prosperity, and prosperity brings social harmony. The prosperity argument was already popular by the middle of the nineteenth century, a time when progress in physics and chemistry appeared to guarantee a steady harvest of fruits. The London Exhibition of 1851 celebrated the role of science in producing economic abundance; in 1859, the Prince Consort expressed his hopes that the state would recognize in science "one of the elements of strength and prosperity." In 1873, the Duke of Devonshire's Commission on Education could report that the exclusion of science from the training of the upper and middle classes must be regarded as "little less than a national misfortune." Lionel Playfair expressed similar sentiments in his 1885 address to the British Association for the Advancement of Science, where he pointed out that "Few would ask now, as was constantly done a few years ago, 'What is the use of an abstract discovery in science?'" Scientific discovery was "the true foundation upon which the superstructure of modern civilization is built." Playfair left no doubts as to the practical advantages of "science for its own sake." France, he recalled, had suffered in 1870 at the hands of German armies for want of scientific skills. Steam and electricity, both offspring of pure science, had been of incalculable advantage for the economy of England. Science was "the source of wealth and power"; science should be studied "for its own sake and not for its applications," but science would bring both utility and strength.[20]

In Germany, the growing power of science was demonstrated first in the application of chemistry to agriculture. Early in the century, discoveries in the areas of chemical synthesis made possible the substitution of synthetic compounds for many traditional materials used in industry. In 1828, Friedrich Wöhler synthesized urea, proving that there is no "in principle" difference between the elements of organic and inorganic matter. In the 1840s,

Justus Liebig established his chemical laboratory at Giessen, launching what is usually regarded as the first systematic effort to apply chemistry to agriculture.

The most dramatic application of science, however, was in the production of synthetic chemical dyes from coal tars, a process whose significance lay largely in the fact that, for most of the century, the production of textiles was the largest manufacturing industry in Europe. Before about 1860, all colors used in the manufacture of textiles were derived from plant extracts. "Turkey red" was derived from the madder plant, vermilion from kermes, blue from woad, yellow from weld, and so forth. In the 1850s, chemists discovered that synthetic dyes could be extracted from coal tars. The consequences of this discovery, not just for the textile industry but also for the broader place of science in society, were dramatic. In 1868, Höchst and the Badische Anilin- und Soda-Fabrik (BASF) of Ludwigshafen began to hire university-trained chemists in an effort to develop a commercial process for the production of alizarin, an orange-red dye that had recently been synthesized from anthracene in Berlin. By 1871, within fifteen years of its beginnings, the German dye industry had gained an edge over its British and French competitors that would last through World War I. In 1881 Germany produced nearly half the world's dyestuffs; by 1900 its share had grown to nearly 90 percent.[21] Over the same period, Germany gained the lead in producing sulfuric acid, alkalies, soda ash, and bleaching powders.

The success of science in producing synthetic materials can also be seen in the number of chemists employed by industrial firms in the course of the century. The Bayer plant in Elberfeld hired its first university-trained chemist in 1874. Within seven years it had hired fourteen more, and by 1896 the company employed 104 full-time chemists engaged in research and production. In 1890, directors of the Bayer plant voted to provide 1.5 million marks to construct a new laboratory to house its growing research division; the company also acquired the personal libraries of August Kekulé, Viktor Meyer, and Henry Roscoe, in order to establish its own in-house library. By the end of the century, many other chemical manufacturers had established their own research laboratories with specialized libraries, training facilities, and seminars, where chemists would search for new dyes.

Dyes were not the only product of this research. It had been observed early on that unstable dyes could under certain circumstances explode violently; innovators consequently developed the field of synthetic explosives. Höchst, Germany's second largest dyestuffs manufacturer after Badische, was the first to prepare pharmaceuticals. The company produced analgesics as early as 1888 and soon thereafter began producing diphtheria and anti-tetanus sera. Höchst also pioneered the production of anesthetics such as

Novocain; in 1912 the firm employed more than 300 professional chemists producing not just dyes but explosives, photographic film, drugs, and other synthetic products.[22] By this time, Germany had become the world's undisputed leader in science and engineering: nearly half of all the world's scientific articles were being published in Germany—more in fact than in the U.S., France, Britain, and Russia combined.[23]

Scientists observing these changes wrote glowing accounts of the progress of their science. When A. W. von Hofmann opened the 1890 plenary session of the Gesellschaft Deutscher Naturforscher und Ärzte, he saluted the progress achieved in industry, trade, and agriculture through the growth of science: "How could business operate today without chemistry? How could shipping or industry or trade do without the physical sciences? The greatest revolution in all areas of life—from world-class businesses to the smallest companies—is that which has issued from the natural sciences!"[24] The German government had also begun to recognize the value of science for industry. Between 1890 and 1900 it provided the equivalent of about five million dollars for a crash program to develop synthetic indigo dyes of commercial value. Between 1866 and 1914 Prussian government support for German universities grew from two million to nearly twenty-seven million Reichsmarks, plus another nine million for the Technische Hochschulen. Most of this growth was directed toward the natural sciences: by the end of the century state grants, largely for chemical research, made up the bulk of German universities' incomes. Humboldt had argued that the man of science should live *for* science; support for science now meant that one could live *off* science as well.[25]

Such was the situation when Weber and his colleagues debated the place of values in science in the years leading up to World War I. Many believed that the progress of science would ensure prosperity for all time to come. The apparent paradox of science as both value-free and the source of human goods and virtues was resolved in the Baconian ideal of the "usefulness of useless knowledge": science, pursued for its own sake, would enable men to transcend their petty differences; science would triumph over war and social conflict as it had triumphed over ignorance and disease. Science was a harmonizing force, a unifying force. When Henri Potoniés, editor of the popular *Naturwissenschaftliche Wochenschrift*, proposed that all science be written in the "acceptable international languages of science" (German, French, English, and Italian), his goal was to ensure that national vanities be set aside in favor of "the neutral ground of science."[26]

Both sides of the *Werturteilsstreit* accepted this belief in the harmonizing force of science, even if they drew from it rather different conclusions. For Schmoller and his colleagues, social conflict would be resolved through economic growth: harmony was the product of abundance, just as strife was the product of poverty. Science would help to reduce social conflict through the discovery of objective ethical truths. For neutralists such as Weber and Sombart, the social effect of science was not so different. (Science in Weber's view would eventually become a great bureaucracy, where politics was filtered out through neutral administrative procedures.) Science united, where politics divided. Alfred Weber, younger brother of Max and a respected sociologist in his own right, argued in a 1909 essay that economists must remain "strictly neutral" for only then could they restore the "authority of true science" needed to demonstrate, by educating the populace, that there is "no real conflict between capital and labor" and that demands for better working conditions must await increased productivity.[27]

Scientists were to have a privileged position in the arbitration of social conflict. Why? Because the scientific attitude led them to value objectivity over error, balance over bias. According to J. T. Merz, this was what distinguished philosophy from science, for while "philosophy is interested, science is disinterested." In his comprehensive *History of European Thought in the Nineteenth Century,* Merz claimed that it was the triumph of modern scientific method to have achieved an "even and passionless tenor" in its judgments; the goal of the scientific mind was to acquire "an attitude as dispassionate and as evenly balanced as that of a judge to whose care the most momentous issues concerning life, happiness or misery are entrusted." It was to the nineteenth century that we owed the successful practice of pure science—science conducted "without fear or fervor"—science of the kind that had achieved such success in Germany in recent years.[28]

In Merz, two of our themes come together. On the one hand, there is the perception of the danger posed to science by its subordination to practical industry. This, Merz said, was a threat as great in his time as the threat of theology had been in previous centuries. But there is a second element. In providing an attitude "as dispassionate and as evenly balanced as that of a judge," science was supposed to provide a common ground from which to face the great and troubling questions of the day. In Sombart's words, science might provide "a point around which we may once more unite." Both of these themes—the threat of practical knowledge and the promise of a great and neutral arbiter—were put forth as reasons for why science should be value-free.

In one sense, these ideals might seem to contradict one another. Practical concerns were a "taint" to science, and yet scientific knowledge was to be of practical value in the solution of social problems. The contradiction disappears when we realize that both were consistent with a more general, liberal, principle of the relation of science and morals. The virtue of science, in this view, lies in its freedom from moral prejudice, its impartiality, its distance from the affairs of morals and politics. Science provides a forum for resolving moral conflict by transcending the prejudices that give rise to that conflict. Clarity is achieved at a distance; and from this distance, the free pursuit of science contributes to the social good as if by an invisible hand, one that works in science as in society as a whole.

The notion of science as a neutral arbiter flourished at a time when many believed that the progress of science was for the good of all and that science "does not take sides." This image would be shattered in the First World War, when thousands of soldiers died a miserable death from chemical weapons and people discovered what it was like to suffer the aerial bombardment of their cities. It was difficult to claim science as a neutral blessing after Fritz Haber and Karl Bosch synthesized nitrates (allowing the production of synthetic explosives), after Columbia University declared itself to be training "thinking bayonets" to fight against the Hun. Socialists, feminists, and racial theorists all challenged the ideal of value-neutrality, and it is to these challenges that we now turn.

8

The Social Context of German Social Science

The boundaries that isolate social science disciplines from one another are of recent origin. Prior to the end of the nineteenth century, it is meaningless to talk of "sociology" apart from "political economy," in the same way that one century before it made little sense to talk of the "scientist" apart from the "philosopher" (the English word *scientist* was not even coined until 1834).[1] As late as the first decade of the twentieth century, German social theory was done almost entirely under the rubric of historical or political economy *(Nationalökonomie)*; the idea of a separate science of sociology had little conceptual or institutional meaning.

If you look at the backgrounds of German sociologists, most taught political economy long before they came to the science of sociology. Lujo Brentano succeeded Wilhelm Roscher as professor of economics at the University of Leipzig; Max Weber succeeded Karl Knies at the University of Heidelberg. Werner Sombart, Ferdinand Tönnies, and Alfred Weber all began their careers teaching political economy; Georg Simmel lectured in philosophy. This contrasts with French and to some degree British and Italian traditions, where many sociologists came from engineering or medical backgrounds. Mandeville and Quetelet, for example, were trained as physicians; Auguste Comte studied engineering at the Ecole Polytechnique in Paris. Emile Durkheim (1858–1917) studied chemistry and became a founding member of *Isis,* the international review devoted to the history of science. Vilfredo Pareto (1848–1923) studied mathematics and classics before receiving a degree in engineering from the Turin Instituto Politecnico on index functions of equilibrium in solids.

By the mid-nineteenth century, however, this has begun to change. Sociology separates itself from political economy as a science of social relations, independent of economic foundations; political economy becomes "economic analysis," and a whole series of political and social issues are eliminated from the province of economics. By the late nineteenth century, social philosophy in Europe has divided in two: into economic theory devoid of social analysis, on the one hand, and social theory devoid of economic analysis, on the other.

The division of political economy into separate sciences of economics and sociology in the early years of the twentieth century is associated with the entrenchment of the ideal of scientific neutrality. Social theory in the eyes of the young sociologists might strive to become scientific, but to do so it must abandon its craft or practical-political origins and restrict itself to problems of pure theory. In the second half of the nineteenth century, one begins to see something new in the social sciences. Social science treatises begin to boast of being *pure* in the way, two centuries earlier, natural philosophers had praised that which was *new*. The year 1863 sees William Stanley Jevons's *Pure Logic;* in 1874 there is Léon Walras's *Theorie d'économie politique pure* and in 1879, Alfred Marshall's *Pure Theory of Foreign Trade* and his *Pure Theory of Domestic Values*. William Cunningham publishes a "Plea for Pure Theory" in 1892; F. Y. Edgeworth's "Pure Theory of Taxation" appears in 1897. In America, we have Lester Frank Ward's *Pure Sociology* in 1903. Franz Oppenheimer's *Theorie der reinen und politischen ökonomie* appears in 1910, and in 1920 a new edition of Tönnies's *Gemeinschaft und Gesellschaft* is given the subtitle: "Foundations of Pure Sociology."[2] With the premium given to purity, utility becomes as much a vice as a virtue: C. P. Snow, reflecting upon research in early years of the twentieth century, recalls that "We prided ourselves that the science we were doing could not, in any conceivable circumstances, have any practical use."[3]

It is important to recognize the special position of Germany in this regard. In the first half of the nineteenth century, positivist sociology came to dominate social theory in France, as did laissez-faire Manchester economic theory in England. Comte's "social physics" claimed to discover laws of motion of science and society in the transition from theological and metaphysical to positive modes of thought; Ricardo's *Principles of Political Economy* justified the imperial expansion of Britain in terms of supposedly natural and comparative advantages. In Germany, by contrast, social science in the first half of the century remained largely indifferent to these naturalist or positivist trends. Positivism was never as popular in Germany as it was France or in England; this lack of enthusiasm translated into a certain caution toward the new sociology as well. Wilhelm Dilthey in his 1883 *Einleitung in die Geisteswissenschaften* launched a savage attack on

Comte's sociology; neo-Kantian social philosophers such as Wilhelm Windelband and Heinrich Rickert rejected the naturalistic and universalistic aspects of Comte's philosophy. Tönnies represents an exception that, as it were, proves the rule. More than most of his neo-Kantian colleagues, Tönnies appreciated the positivism and naturalism of French and English social theory, though he was also critical of these movements. Tönnies was the author of a book on Hobbes and was well-versed in the writings of Comte, Herbert Spencer, and Lewis Henry Morgan; he also reported regularly to readers of *Schmollers Jahrbuch* on developments in the British eugenics movement. Even Tönnies, however, often cited as the "father" of German sociology, claimed never to have considered Comte's work—early or late—as genuine sociology "in a strict and scientific sense."[4] This is especially curious, given that it was Comte who coined the term *sociologie* in the first place.

Georg Lukács has suggested that with the decline of German idealism in the first half of the century, much of what remained of social philosophy (outside of Marxism) was largely a form of administrative and legal counsel to government; as late as 1859, Treitschke's *Gesellschaftslehre* could still contend that all social problems could be solved by purely administrative or legal means. ("Die Arbeiterfrage" was for Treitschke "eine reine Polizeifrage.")[5] The paternalism of the historical school fit well with this belief: the goal of social science was to maintain social order.

In the course of the nineteenth century, however, Germany undergoes a series of social and economic transformations that alter dramatically the shape and course of social theory. Foremost among these is that Germany becomes a major industrial power and begins to compete with England and France for colonial markets and raw materials. In 1870, Germany produced 1.7 million tons of raw iron, only one-quarter that of Britain. By 1913, Germany was producing 20 million tons, nearly twice that of Britain.[6] Industrial reform laws provided incentives for the formation of joint-stock companies; the rapid repayment of France's war debt of five billion francs further boosted economic optimism. The consequence of industrialism for the German academic was twofold: the retrenchment of German nationalism following Germany's colonial expansion, and the rise of a fear of socialism in consequence of new powers wielded by a growing industrial proletariat.

One might imagine that the rise of German industrial capitalism would have made Enlightenment ideals of *liberté, egalité,* and *fraternité* attractive to German intellectuals, and to a limited extent it did. The National Liberal party enjoyed support in Germany in the early years of Bismarck's rule (1878–1891); Max Weber actively campaigned for the National Liberals into the twentieth century. But many German intellectuals regarded Enlightenment ideals with a suspicion dating back at least to the Napoleonic wars. The growing economic competition with France and England made German intellectuals reluctant to embrace free-market ideals of government. German social philosophers rejected the individualistic, laissez-faire doctrines of British *Manchestertum* as part of the *apologia* Britain had developed to justify its economic and cultural dominance throughout the world.

The German rejection of French and British liberalism in favor of more particularistic, historical-relativist forms of social science should be understood in this context. German political economists found doctrines praising the virtues of free trade and free markets unattractive, in large part because Germans were not the ones who stood to gain from such freedom. Germany was a latecomer in the search for international markets. It was not until the second half of the nineteenth century that German overseas colonization began in a major way and by this time the colonial world was already ruled by England and France (and to a lesser extent Holland, Italy, and Spain). The German bourgeoisie found little of value in Ricardo's doctrine of "comparative advantage," according to which it was natural that "wine shall be made in France and Portugal, that corn shall be grown in America and Poland, and that hardware and other goods shall be manufactured in England."[7] German academics were also wary of doctrines preaching the inevitability of growth or progress, in view of the manifest dependency of that growth on historical contingencies. Germans found little to admire in the doctrines of Smith and Ricardo, Mill and Stewart—doctrines which appeared to justify the global economic dominance of France and England.

A second and more important consequence of Germany's industrialization was the creation of a large, urban industrial proletariat that was to exercise increasing political power in the course of the century. Germany's rapid industrialization drew large numbers of farmers into the cities to work in factories: in 1870 less than 5 percent of the German population lived in towns; by 1910 this had grown to more than 21 percent. Two-thirds of Germany's population lived on the land in 1871; on the eve of World War I the urban population was twice the rural population. In the thirty years between 1870 and 1901, Germany's working class grew from one-fifth of the population to one-third.[8] The relatively high literacy of German workers, combined with the extremes of class hierarchies and the rapidity of German industrialization, nourished the largest proletarian political movement in the world. The German government was understandably alarmed: in 1878 Bismarck instituted the Anti-Socialist Laws banning all socialist parties in Germany, especially the newly formed Sozialistische Arbeiter Partei, recently formed from the unification of August Bebel and Karl Liebknecht's Sozialdemokratische Arbeiter Partei and Ferdinand Lassalle's Allgemeine Deutsche Arbeiters Verein. Under the laws, socialist organizing and propaganda were declared illegal; in the twelve years the Anti-Socialist Laws were in effect (1878-1890), dozens of socialist newspapers were suppressed and thousands of socialist leaders were arrested or forced into exile.

In 1890 Bismarck's repressive laws were relaxed, and the mandate given the Social Democratic party (SPD) increased dramatically: in 1903 the Social Democrats won 83 seats in the Reichstag and 32 percent of the popular vote with an explicitly Marxist platform of the imperatives of class struggle and proletarian revolution. In the last decade of the nineteenth century, 425,000 German workers participated in strikes or walkouts. In the single year of 1905, more than a half-million workers laid down their tools and walked out, representing a total of 7,363,000 man-days lost to industry in one year alone.[9] Friedrich Engels in his introduction to Marx's *Class Struggles in France, 1848-50* chronicled the electoral success of the Social Democratic party, noting that since the introduction of general suffrage in 1866, the socialist vote had grown from 120,000 in 1871 to 1,787,000 in 1893—without coverage of the press, without even rights to assemble. With the help of the ballot box, the German proletariat was making a revolution: "The state was at the end of its rope; the workers, only at the beginning of theirs."[10]

By 1911, the German Social Democratic party had become not only the strongest party in Germany, but the strongest socialist party in the world (see Tables 2 and 3). Socialism posed a powerful threat to the German state in the first decade of the twentieth century; and it was socialism, more than anything else, that the sociologists had in mind when they criticized the

Table 2 German socialist vote, 1871-1911

Year	Vote	Seats
1871	124,655	2
1874	351,952	10
1877	493,288	13
1879	437,158	9
1881	311,961	13
1884	549,990	24
1887	763,128	11
1890	1,427,298	35
1893	1,786,738	44
1898	2,107,076	56
1903	3,010,771	81
1907	3,258,968	43
1911	4,238,919	110

Source: *The National Socialist Handbook*, no. 1 (Washington, D.C., 1912), pp. 23-25.

Table 3 World socialist vote, 1907–1912

Country	Year	Votes	Seats in parliament		
			Total	Socialist	Percent
Finland	1911	321,000	200	87	43.5
Sweden	1911	170,000	165	64	38.7
Germany	1912	4,250,000	397	110	27.7
Belgium	1910	483,000	166	35	21.0
Denmark	1910	99,000	114	24	21.1
Austria	1911	1,060,000	516	82	15.3
France	1911	1,106,000	584	76	13.0
Norway	1907	90,000	123	11	8.9
Switzerland	1911	105,000	170	15	8.8
Italy	1909	339,000	508	43	8.5
Holland	1909	82,500	100	7	7.0
Great Britain	1910	374,000	670	42	6.3
Russia	1906	(?)	442	17	3.8
United States	1910	642,000	394	1	0.25
Spain	1910	40,000	404	1	0.25

Source: Tabulated from *The National Socialist Handbook*, no. 1 (Washington, D.C., 1912), p. 27.

intrusion of value judgments into the sciences. Value-neutrality served two different, but related, functions. On the one hand, sociologists used value-neutrality as a tool to refute attempts on the part of Marxists to politicize social theory. "Scientific socialism" becomes one of the most common targets of the charge of "non-neutral" or "biased" social science. Sociology was also declared neutral in order to avoid the charge that *sociology* was simply another word for *socialism*. The confusion was a common one in the early years of the twentieth century. Intellectuals in the SPD argued that Marxism was the only true sociology; sociology was commonly looked to as the science of socialist movement. The sociologists of the Deutsche Gesellschaft für Soziologie wanted to separate clearly the science of society from the politics of social movements; neutrality served the young sociologists as both an offensive weapon and a defensive shield.

In order to appreciate the impact of the socialist movement on the German academic, it is important to recall the social standing of students and faculty at this time. In the years 1887–1890, fewer than one in a thousand students at Prussian universities were sons of workers (daughters were rarely admitted from any class). By 1900 this had increased only to one in a hundred; the overwhelming majority of students came from families in government, the military, and the business and landed classes.[11] The same was true for professors. German professors were generally drawn from the upper middle class, a class whose political fortunes changed dramatically over the course of the nineteenth century. The political engagement of Germany's liberal middle class culminated in the years leading up to the revolution of 1848—a period when liberals favored national unification to counter the arbitrary rule of local princes.

When the Frankfurt parliament established in the revolution of 1848 collapsed and the "professors' revolution" came to an end, bourgeois intellectuals abandoned politics during a decade of conservative counter-reaction. (Hermann Baumgarten, Max Weber's uncle, was one of the chief apologists for the retreat from politics.) The radicalism that had inspired the professors' parliament was lost, and there arose a new and more restricted conception of intellectual life in German society. In the truce with the state conceded by the liberal idealist, culture was to be divorced from politics: the intellectual *Bürger* was to work (or think) and the statesman to rule; politics was to be left to Bismarck and the politicians. Hence the origins of those apolitical professors whom Fritz Ringer dubbed the "German Mandarins"—loyal civil servants that formed the "intellectual bodyguard of the Hohenzollerns."[12]

The oft-cited maxim that the German academic was "apolitical" is true only in a restricted sense of that term. If there was one single issue on which German academics could agree, it was on the need to oppose the socialist movement sweeping through the working classes. Appreciating this is important for how one interprets the *Werturteilsstreit*.

The most common reading of the origins of the principle of value-neutrality in the social sciences points to the dissatisfaction on the part of the sociologists with the older historical school's moralistic subservience to government policy.[13] There is certainly some truth in this, and this was how the younger sociologists themselves often saw the problem. But it would be easy to overestimate the differences between the political attitudes of the two generations of political economists in this regard. Gustav Schmoller himself, in his opening remarks to the 1909 meeting of the Social Policy Association (the meeting that sparked the *Werturteilsstreit*), recalled that the *Kathedersozialisten* and the sociologists, whatever their other differences, "were united in protest against the imported Manchester theories, extreme radicalism, and the Social Democratic party."[14] While many of the young sociologists struggled against free-market liberalism, it was Social Democracy that became the primary target of their attacks. Weber's *Protestant Ethic and the Spirit of Capitalism* rejected the materialist conception of history; his 1904 defense of the ideal of objectivity in the social and economic sciences ridiculed historical materialism as a doctrine "found today only in the heads of amateurs and dilettantes."[15] Werner Sombart, in the years after the first (1905) Russian revolution, lectured union leaders to abandon political demands in favor of more cautious, short-term economic demands.

One should recall that German academics were employees of the state; this made it difficult for scholars to support anything that might be conceived as subversive. "Sociology," "social hygiene," and indeed "social *anything*" were all suspect. As early as 1887 Tomás Masaryk (Czechoslovakia's first president, after World War I) had noted that "many feel uncomfortable with the word *Sociologie*"; the word itself, he noted, "is an unfortunate one, for it reminds us of socialism."[16] Ludwig Gumplowicz, the first German to use the term *sociology* in the title of one of his books, protested often and loudly against the confusion of *Soziologie* and *Sozialismus*. Gumplowicz rejected Enrico Ferri's claim at the International Sociological Congress in Paris that "sociology must be socialist, or nothing at all." Gumplowicz asserted, by contrast, that "sociology is not a party program but rather a theory of society, a science of human societies. A political party platform is not a science; otherwise there would have to be a 'liberalogy' [*Liberalogie*] and a 'clericalogy' [*Klerikalogie*]."[17] Sociology for Gumplowicz had nothing to do with politics: "Sociologists can be either liberal or clerical, just as astronomers may belong to any of these parties."

The sociologists' fears of being associated with socialism were not unfounded. In 1894 Adolf Wagner, Gerhart von Schulze-Gävernitz, and Lujo Brentano (hardly radicals by contemporary standards) were all accused of spreading socialism and listed as prime targets of industrialist Karl von Stumm-Halberg's "Subversion Bill" (eventually defeated in the Reichstag). Fritz Ringer in his *Decline of the German Mandarins* describes the case of Leo Arons, a young physics instructor at Berlin, fired from his post for openly supporting the Social Democrats. Arons was never criticized for his scholarly work, which was up to par, but simply for his extracurricular activities. Prussian government authorities pressured the University of Berlin to have his *venia legendi* withdrawn, and despite support from the Berlin faculty (including, most prominently, Gustav Schmoller), Arons was removed after the passage in 1898 of a new law, a *lex Arons,* granting the state enlarged powers to remove teachers deemed politically unacceptable.[18]

The climate of fear and oppression was such that, prior to 1918, not a single German full professor was a member of the Social Democratic party. Those who sympathized with less radical causes also had difficulty obtaining appointments. In 1893, for example, Ferdinand Tönnies refused education minister Friedrich Althoff's demands that he quit the Ethical Culture Society and was thereby denied promotion to associate professor until 1908, after Althoff had died. In 1899, after the historian Hans Delbrück of the University of Berlin published objections to the government's policy of Germanization in Schleswig-Holstein, he was summoned before a disciplinary court, where he avoided being dismissed only after colleagues rallied in his support and he demonstrated his loyalty by presenting a speech in favor of government naval policy. When Max Weber retired from Heidelberg in 1899 he recommended Werner Sombart as his successor; Sombart's appointment was denied, ostensibly for his having been involved in publishing theoretical sociology of a politically suspect (that is, Marxist) nature. Fritz Ringer points out that Weber was himself involved in a protest against the removal of Robert Michels, a former student, accused of association with the Social Democrats.[19]

The kinds of fears German sociologists faced vis-à-vis socialism can also be seen in an exchange of opinions in the years 1919–1921 concerning the establishment of professorships of sociology in the new Germany. The first chair of sociology was not established in Germany until after World War I—in 1918—a time of highly politicized debates concerning the value and possible dangers of the new science.[20] In 1919, the soon-to-be Prussian Minister of Culture C. H. Becker wrote an essay advising the establishment of professorships of sociology to remedy the malaise he saw growing from the specialization of modern intellectual life. Becker argued that German science was suffering from bureaucratization and fragmentation; German science had lost its sense of purpose, its synthetic vision of the whole. Becker blamed both materialism and the growing interest in what was derogatorily labeled *Brotstudium*: narrowly vocational studies designed (in his terms) to satisfy more the stomach than the spirit. He also criticized the separation of the Technische Hochschulen from the universities, arguing that this had bred a generation of technical specialists ignorant of broader social life. Sociology was the answer to Germany's problems; sociology would help restore the "spiritual habits" Germany needed.[21]

For Georg von Below, the conservative professor of history at the University of Freiburg, sociology was less a remedy for Germany's problems than a cause of them. "The establishment of chairs of sociology at German universities, if they were occupied by socialists, would certainly invite a negative reaction." Von Below associated sociology with socialism—with that Enlightenment vision of social progress that tried to make religion superfluous. Spencer he associated with the evils of *Manchestertum,* Comte with the specter of Marx. Against Becker, Below warned that the establishment of professorships of sociology would open the door to dangerous currents of naturalism, positivism, and socialism.[22]

Fear of socialism (and of being labeled a socialist) was common among many social theorists at this time. Writers in the short-lived but ambitious *Monatsschrift für Soziologie* (1909–10) made this very clear. In a series of articles dedicated to the "Object and Aims of Sociology," the central question once again was the place of values in science. Abroteles Eleutheropulos, editor of the journal, asked whether the task of the journal was to be "a positive or an ethical one." Eleutheropulos described the views of various writers on this question, but those he found most repellent were those for whom the goal of sociology was the establishment of a moral order—especially those for whom "sociology that is not socialism is nothing!"[23]

By this time, sociological hostility to socialism had a long pedigree. In England, Herbert Spencer in both his *Principles of Sociology* and his *Introduction to the Study of Sociology* had criticized socialists for failing to ask what the consequences might be if "the material well being of the inferior is raised at the cost of lowering that of the superior." Spencer warned that "a gradual deterioration of the race" must follow in the event that "the superior, persistently burdened by the inferior, are hindered in rearing their own better offspring." He also claimed that one of the primary tasks of sociology was to prove that radical efforts to improve the human condition must meet with failure: "the study of Sociology . . . will dissipate the current illusion that social evils admit of radical cures." Spencer defended this view with the argument that society was "a growth" and not "a manufacture"; attempts to move society away from its natural course could therefore only result in failure.[24]

In France, René Worms, founder of the *Revue internationale de sociologie* in 1892 and the Société de Sociologie de Paris in 1895, proposed a sociology that was similarly designed to oppose the politicization of social life. In his methodological treatise on *La science et l'art en économie politique,* Worms lamented the fact that the public was so easily impressed by schemes for social reform, schemes which assume that society is sick and that the purpose of theory is to provide therapy. "Therapists" of this sort included the *collectivistes* (Karl Marx), the *nationalistes* (Henry George), the *blanquistes* (comité révolutionnaire central), the *guesdistes* (parti ouvrier français), the *broussistes* (fédération des travailleurs socialistes de France), and the *allemanistes* (parti ouvrier socialiste révolutionnaire), not to mention the various independent socialist groups. These, Worms said, are the reformers of the Left. And on the Right, there are the Christian socialists (Protestant and Catholic), the school of Le Play, and others wishing to return to industrial or familial forms of an earlier age. The error of all such systems, he argued, was that they are subjective. Would it not be better first to study the world as it is, before one decides how it ought to be? Worms considered classical economists to have tried to maintain their "equilibrium" and neutrality with regard to doctrines of both Left and Right, remaining true to the liberal ideals of the French Revolution. He also considered his own organismic sociology, following in the tradition of Spencer, Paul von Lilienfeld, and Alfred Fouillée, to be independent of all such reformist tendencies.[25]

Socialism, though, was not the only strain of politicized social theory that posed a challenge to social theory at the end of the nineteenth century. In the last decade of the century a culturally pessimistic, social Darwinian "racial hygiene" emerges in Germany, cultivated largely among doctors and lawyers and supported by the Pan-German League with funds from a variety of landed and industrialist groups. Racial hygiene was to resemble the socialist challenge in several ways. Like socialism, racial hygiene arose at least partly in response to increasing strength and agitation among workers; like socialism, racial hygiene purported to base its doctrines on the sure ground of science. Like socialism, racial hygiene was organized largely outside the universities. Finally, like socialism (though to a lesser extent), racial hygiene was put forward as both a conceptual framework and a practical program for the science of sociology.

It is difficult to imagine today how popular was the appeal of late-nineteenth-century racial hygiene—not just among social theorists but a broad class of professionals. Building on the increasingly popular ideas of Arthur Gobineau, Francis Galton, Georges de Lapouge, H. S. Chamberlain, and a pessimistic version of social Darwinism, the racial hygiene movement begins to gain respectability with the Krupp Prize Essay contest in 1900, the founding of the German Society for Racial Hygiene in 1905, and the appearance of the first issue of the *Politisch-anthropologische Revue* in 1902. Supported by landed Junker and industrialist groups and united against the common threat of democracy and socialism, racial hygienists advocated a program of state-supported "selection" to counter a perceived threat of the criminal, the insane, and the working poor to reproduce at a faster rate than the more "gifted" military, industrial, and intellectual elites. If the growing proletariat could not be stopped politically, then they might be stopped biologically: Alexander Tille, in his 1894 *Biologie eines Aristokraten,* advocated the sterilization of the unemployed poor to avoid political unrest; Hermann Siemens warned against the "Proletarization of our Future Generations" caused by over-breeding of the working poor.[26]

It is not surprising that Darwin was looked to for inspiration in social affairs. Darwin had demonstrated a profound historical link between humans and animals, ushering in a revolution that Thomas Huxley characterized as "a gigantic movement, greater than that which preceded and produced the [Protestant] Reformation." Soon after the publication of Darwin's 1859 *Origin of Species,* philosophers in Germany had begun to speculate on the implications of the theory for philosophy and social theory. Richard Avenarius, Ernst Mach, and Ludwig Boltzmann all explored the implications of Darwin's theory for the new science of epistemology. In 1891 Henri Potoniés announced to readers of his popular *Naturwissenschaftliche Wochenschrift* that the principles of Darwinism were valid "for spiritual, as well as for physical development of organisms." Forms of thought, as well as forms of the body, had arisen through the struggle for existence.[27]

Darwinism was not, of course, unopposed. Many conservatives equated

Darwinism with atheistic materialism, Haeckelian monism, or Bebel's evolutionary socialism. All of biology was rendered suspect: in 1879 Hermann Müller in Lippstadt was barred from teaching Darwin's doctrine of descent in Prussian schools; within a few years, the teaching of biology was barred entirely from Prussian secondary schools for fear of its association with materialism.[28]

By the end of the nineteenth century, social Darwinism and racial hygiene constituted major research agendas for the emerging science of sociology. Gumplowicz, Lilienfeld, and Albert Schäffle each in their separate ways sought to construct a sociology based on a synthesis of the organic metaphors of Comte and Spencer and the natural selection of Darwin. Alfred Ploetz and Wilhelm Schallmayer both considered themselves sociologists of sorts; H. E. Ziegler went so far as to argue that sociology must become a form of comparative anatomical ethnology.[29] German sociologists considered English eugenics of considerable interest for their own science: when Francis Galton founded the London Sociological Society in 1903, Ferdinand Tönnies devoted considerable space in *Schmollers Jahrbuch* to an analysis of the new movement. Austrian sociologists were equally fascinated by racial hygiene: Rudolf Goldscheid included a section for "social biology and eugenics" when he founded the Sociological Society in Vienna in 1907 (Paul Kammerer was elected secretary); Goldscheid had earlier discussed with Ploetz the possibility of establishing a sociological society in Munich, where questions of racial health were to be a central focus.[30] In 1910, Galton's *Hereditary Genius* appeared in German translation; the translators were the young sociologist Otto Neurath and his wife, Anna Schapire-Neurath.

Sociologists were sharply divided, however, over how and to what extent natural selection, the doctrine of descent, and racial or biological analysis more generally were relevant to the social sciences. Debate along these lines occupied much of the attention of sociological theorists in the two decades prior to the First World War. At the first meeting of the German Society for Sociology fully one-third of the entire session was devoted to a discussion of Alfred Ploetz's "racial hygiene," which Ploetz wanted to establish as the centerpiece of the new science of sociology. Tönnies and Leopold von Wiese supported the new Darwinian sociology. Weber, however, rejected Ploetz's program as an illegitimate intrusion of the natural sciences into the social sciences and, at least for a time, Weber's position held sway. At the second meeting of the society in 1912, where racial hygiene was again a major topic of discussion, Weber made clear his belief that Ploetz's "confused racial mysticism" could not be taken seriously.[31] Germany's foremost sociologist rejected as "totally unproved" Ploetz's claim that social development depended on racial health; he rejected Ploetz's idea of a *Vitalrasse* as the product of "unbounded subjective valuations" and labeled his newly founded journal (the *Archiv für Rassen- und Gesellschaftsbiologie*) "an arsenal of boundless hypotheses" containing "not a single fact that might be relevant to sociology." Weber maintained that sociology must consider social phenomena *sui generis,* and not as passive reflections of instincts or biological drives.

Despite views such as Weber's, racial arguments continued to play an important—though disputed—role in German social science, especially in anthropology.[32] In the 1920s and 1930s, with the growth of right-wing political movements, the racialist sociology of Ludwig Woltmann, Othmar Spann, and the racial hygienists was revived, sweeping neo-Kantian sociologies and the ideal of value-neutrality to the side (see Chapter 12). Sociologists looking to eugenics to unify theory and practice had little patience with value-neutrality: Goldscheid described Wert*freiheit* (value-freedom) as Wert*blindheit* (value-blindness); Alfred Grotjahn quit the Society for Sociology after Weber's critique. Fritz Lenz, Germany's foremost racial hygienist, wrote an essay in 1917 (reprinted in 1933) purporting to solve "the value-problem" by treating race as "the ultimate principle of value."[33] War, revolution, and economic crisis made the ivory tower retreat look like an out-of-date middle ground that would not hold.

Socialism and social Darwinism were two of the most powerful movements challenging academic social theory at the turn of the century. A third movement, loosely associated with the socialist movement, was the increasingly active movement for the emancipation of women. By the First World War, German feminists could count more than a quarter of a million women among their ranks, organized into two basic groups. One group supported equal rights as part of a general program of socialism (Clara Zetkin edited *Die Gleichheit,* the Social Democratic mouthpiece for the 175,000 women in the SPD), and the other constituted a more autonomous and liberal feminist movement that advocated reforms independent of socialist politics (the Verband fortschrittlicher Frauenvereine, the Bund Deutscher Frauenvereine). At the turn of the century 850 associations with a membership of nearly one million women were organizing for women's rights in Germany. Though Germany was slower than some European countries to achieve substantial political reforms (local suffrage, admittance to the professions, elimination of state-regulated prostitution), feminist agitation aroused sufficient fear within the Wilhelmine government that all political activity on the part of women was banned in most of Germany until 1918, twenty-eight years after even socialist parties had been legalized.[34]

As in the case of socialism, to appreciate the reaction of German academics to the women's movement one should understand the social composition of the German university. If there were few sons of workers among the German student population of the nineteenth century, there were fewer still daughters of any class. Prior to the twentieth century, women were admitted to German universities only on an exceptional basis and never with full standing. It is true that, as early as the eighteenth century, a small number of women had studied at German universities under special arrangements: in 1754 Dorothea Erxleben became the first German woman ever to be awarded a medical degree; in 1787 Dorothea Schlözer became the first woman to receive a Ph.D. For more than a century, though, few would be able to follow in their footsteps.[35] Regulations established at the end of the 1880s required that women wanting to attend a German university obtain special permission from the Education Ministry to enroll in a special (non-degree granting) course of study. Women were not allowed to matriculate as regular students until 1907, when the universities of Heidelberg and Freiburg admitted women. A decade later all German universities followed suit.

It is also important to understand the resistance most scholars had shown toward cultivating matters of the intellect. In the Enlightenment, European scholars laid the foundations for the exclusion of women from science in the face of attempts by women to gain access to studies. Jean-Jacques Rousseau, for example, in the notorious Book V of his *Emile*, articulated an elaborate and disparaging view of women's ability to participate in learning. Women's studies were to be confined to the practical; women might play a role in applying the principles discovered by men, or in making observations which lead men to discover those principles, but abstract and speculative truths themselves were "beyond a woman's grasp."[36] G. W. F. Hegel in his *Philosophy of Right* presented his conviction that though women are capable of limited education "they are not made for activities which demand a universal faculty such as the more advanced sciences, philosophy, and certain forms of artistic production." Women might have taste and elegance, but they could never attain the Ideal. Hegel proposed that the difference between men and women was like that "between animals and plants": women were like plants "because their development is more placid"; therefore "when women hold the reigns of government the State is at once in jeopardy, because women regulate their actions not by the demands of universality, but by arbitrary inclinations and opinions."[37] Such views become increasingly common in the nineteenth century, especially as access to university education becomes a central demand of European feminists.

Controversy over the admission of women to universities culminates at the end of the nineteenth century. In November 1895, Berlin newspapers announced that Heinrich Treitschke and Eric Schmidt had expelled several women from their classes, stimulating a renewed debate on the ability of women to pursue academic studies. In the following year, the theologian Arthur Kirchhoff invited 103 leading academics and intellectuals (all males!)

to comment on the "capacity and rights of the female sex" to pursue academic studies. The essays were published in 1897, in the most comprehensive survey of German academics' attitudes toward the admission of women to university studies at this time.[38]

The essays in Kirchhoff's *Akademische Frau* reveal a spectrum (though not a very broad one) of views on the educability of women. The physicist Max Planck wrote that in the rare event that a female student showing aptitude and drive asked to audit his lectures he was happy to comply, so far as this was "in accord with academic regulations." Planck warned, however, that such cases must always be regarded as exceptional; indeed, it would be a grave mistake to allow women into academic study on a par with men, given that nature has prescribed for woman the role of mother and housewife, and given that "natural laws cannot be ignored without severe damage which, in this case, will be expressed in future generations." (Like many of his colleagues, Planck was worried that study could lead to sterility.) "Amazons," he concluded, "in the intellectual realm, too, are contrary to nature."[39]

Wilhelm Ostwald in his contribution to the volume was broad-minded enough to realize that all humans have "a right to knowledge" and that for most disciplines, participation of women presented no specific problems. For his own field of expertise, however (experimental physics and chemistry), Ostwald was not so generous. These fields required extensive laboratory experience, and for certain "obvious" (but unspecified) reasons, women could not be allowed into laboratories. Individual professors, in his view, should decide whether women should be allowed into their laboratories. Ostwald himself, however, did not believe that there would be many women with sufficient willpower *(Willensdrang)* to pursue scientific studies. (Nor did he believe that university education was terribly important in science: he himself boasted of being self-taught.) Ostwald concluded by asserting that in the realm of "pure science" women must be regarded as "an exception" and will always be seen as such. This was not a bad thing, given that the "inner satisfaction" of being a mother or a nurse is "purer and stronger than that achieved through any scientific discovery." Ostwald felt obliged to object to anything that might prevent a woman from realizing the "exceptional and irreplaceable qualities" associated with women's traditional duties.[40]

Not all participants in the debate rejected the claim of women to university studies. Wilhelm Wundt, director of the Institute for Experimental Psychology at Leipzig, rejected the argument that there were already too many male applicants for academic positions and that women should be barred from competing with men on this ground alone. Wundt rejected this argument as "a brutal sexual chauvinism," which, like "class chauvinism," must be rejected on ethical grounds.[41] Felix Klein of Göttingen pointed out that six women were currently attending seminars in higher mathematics at his university and were performing equally with the best of the men in every respect. Klein qualified this only with the observation that they were all foreigners (two Americans, one British, and three Russians)—but surely German women, too, could do as well if only given the chance. Wilhelm Förster, director of the astronomical observatory at the University of Berlin, argued in this same vein that he had never entertained "even the slightest doubt that the intellect of women was equal to that of men." Women have "the patience and conscientiousness" needed to make fine and detailed observations, especially valuable for the science of astronomy. Förster explained the intellectual inequalities between men and women as a product of the injustices women suffer and their lack of education. The growing tension between the sexes he attributed to these same inequalities; he then concluded that the freshness of thought women bring to intellectual life could prove to be both a stimulant and a challenge for the world of men.[42]

Most of those writing for the Kirchhoff volume exhibited much more wooden thinking. Eduard von Hartmann argued that bringing men and women together in the flower of youth was "inadvisable," given what might transpire between them. Otto von Gierke argued that access to academic study would erode the health of the family and hence the strength of the nation. Wilhelm von Bezold (the art historian) claimed that women have problems with "color composition" and cannot master "the principles of perspective." August Dorner (the theologian) maintained that studies could damage a woman's ovaries, thereby reducing Germany's already declining birth rates. Wilhelm Lexis (the historian) argued that university training for women would only sharpen the already stiff competition for jobs, displacing men. Carl Stumpf, director of the psychological seminar at the University of Berlin, argued that higher learning would make women unfit for marriage: after all, he mused, "pale cheeks, bespeckled eyes, and frazzled nerves do not exactly inflame the passions of the male of the species!"[43]

Still others claimed that academic study for women would violate women's nature, God's will, or Darwinian selection. For F. A. Weber, women simply lacked the requisite "intellectual energy." Georg Runze argued that women "think, work, perceive, and live" in ways that bar them from the world of the spirit. Johannes Conrad argued that women excel in understanding and memory but fail in the realm of synthetic ability and concentration. Men specialize, women generalize. University education might be appropriate for those women who fail "to find shelter in the harbor of marriage," yet these would always be exceptions. Otto von Gierke argued that women are intuitive and subjective, men analytical and objective; man's sphere is public, woman's is private. For the anatomist Karl von Bardeleben of Jena, pregnancy, menstruation, childcare, menopause, nervousness, and hysteria combined with inferior skeletal, muscular, vascular, and coronary strength to prevent women from practicing medicine satisfactorily. Given that even primary school had "disastrous" effects on the female backbone, chest and pelvis, how much worse might the trauma of university studies be?[44]

Such were the arguments used by German scientists concerning whether women should be admitted to the universities. (No one in the Kirchhoff survey seems to have shared the extraordinary views of Max Funke, who asserted in a popular book that differences between the sexes were due to the fact that men and women were separate species, with women constituting the "missing link" between man and ape!)[45] Interestingly, sociologists did not differ fundamentally from their forebears of the historical school in this regard. In fact, the supposed inability of women to exhibit a detached attitude toward objects of investigation was commonly used by sociologists to characterize women as inherently unable to do science.[46] If we turn now to sociologists involved in the *Werturteilsstreit*, we can see that the *Frauenfrage* and the question of the role of values in science were curiously entangled with one another. Consider the views of three of the greatest sociologists of the age: Tönnies, Simmel, and Weber.

Ferdinand Tönnies in his 1887 *Gemeinschaft und Gesellschaft* claimed that "women are usually led by feelings, men more by intellect." Men alone are capable of calm and abstract thinking; men alone are able to reason and deliberate in a logical manner. Women are "naive and frank, soft and sensitive, given to outbursts of passion and emotion"; men, by contrast, are objective, hard, and courageous. Tönnies's famous distinction between *Gemeinschaft* and *Gesellschaft* is (among other things) a gendered distinction: *Gemeinschaft* is the sphere of woman, as *Gesellschaft* is the sphere of man; man is to culture as woman is to nature.[47]

Tönnies did not believe that these differences in men's and women's intellectual abilities were inborn. Rather, they derived from physical differences which have channeled men and women into different processes of historical development. Differences in physical strength determined the original division of labor between the sexes. Naturally weaker and less mobile, women remained confined to the inner circle of home life as curators of family possessions and preparers of food. Men, being stronger and nimbler, naturally "directed their energy to the outside." While women stayed at home, men worked the fields and protected the home from enemies.

Having been channeled into different types of work by differences in physique, men and women then developed the cognitive abilities required to carry out their specific tasks. According to Tönnies's history, as the male became an active hunter his mind sharpened. In dividing the bounty of his hunt, the male learned to make comparisons of size, weight, and value of captured animals or other human beings, land, or tools. He learned to measure and to weigh, to balance debts and credits, and from these skills gradually developed the ability to think abstractly and to exercise the "rational will" on which all science is based. Although women also inherit the ability to manipulate symbols, they generally remain anchored in the instinctual thought processes suitable to their labor as mothers. Women retain a "natural will" essential to their home life. Theirs is the realm of direct and immediate experience, deriving from their more passive, "constant and limited activity." Women are receptive and sensitive. They enjoy the present rather than plan for the future. They develop not a scientific but a moral judgment, an ability to distinguish good and evil, beauty and ugliness, which, as Tönnies puts it, "in no way corresponds with the recognition of objects and processes (objective knowledge)." The spirit of woman is one of "community" *(Gemeinschaft)*, the spirit of man one of "civilization" *(Gesellschaft)*.[48]

Georg Simmel, one of the founding fathers of German sociology along with Tönnies, Weber, and Sombart, provided a more subtle analysis of the alienation of science from women and vice versa. In his 1911 *Philosophische Kultur,* Simmel argued that science itself is masculine, that objectivity and neutrality are attitudes of the male spirit, and that women tend to identify too much with their surroundings to allow them to develop an objective attitude toward the natural world. Masculinity and femininity represent opposite poles in the life of the human species—dimensions we measure by values that are masculine in their very essence:

> The values of artistic achievement, of patriotism, morality, and the specifically social ideals—these, together with the sense of justice we associate with ethical judgment, the objectivity of theoretical knowledge, and the power and deepening of life—these are all human *in general* (so it is claimed, and so in fact is the case) and yet in their actual historical formation are completely masculine. If we consider the simple idea of the "objective" for what it is, we must conclude that in the historical life of our species, objective = masculine.[49]

Simmel argued further that the male of the species in modern life is not just *superior* to the female, he is Man *in general.* As a result, the relation between man and woman is not simply one of dominance and subordination (master/slave); there is a further element. For it is the privilege of the master that he need never remind himself that he is master. And it is the burden of the slave that he (but here, of course, she!) can never forget his (her) servitude. It is thus a rarer thing for a woman to lose her sense of herself as a woman than for a man to lose his sense of himself as a man. In consequence, there are fundamental differences in the psychologies of the sexes. A man will often think in purely factual terms, without his masculinity entering into his thought at all. A woman, however, will never leave her feeling as a woman. In other words, it is easier for a man to shed his masculinity than for a woman to shed her femininity. (Simmel defends this notion with the argument that less is at stake for a man in his relation to a woman than for a woman in her relation to a man). It is therefore easier for a man to raise himself into the sphere of "neutral objectivity" *(neutralen Sachlichkeit und Gültigkeit),* where the specifically masculine colorings of his thought are repressed or subordinated. Women, by contrast and in consequence, will often perceive that certain institutions, efforts, judgments, and interests are typically male—often the very same institutions or judgments that men naively find benignly neutral or matter-of-fact *(einfach sachlich).*

There is a further aspect to this arrangement. As Simmel argues, systems of domination have always sought to clothe themselves in the guise of neutral and objective legality—so that power appears as justice, force as law.[50] The history of political formations, religious and economic doctrines, and family law is full of examples of this. The patriarch provides himself with assurances that his power is based not on the arbitrary rule of force but rather on the common interests of the family, or on some other objective and impersonal legality. The psychological superiority of men over women is thus transformed into a logical superiority: as power relations are disguised, the very opacity—apparent neutrality—of these relations is converted into masculinity, so that objectivity itself—or legality, or justice—becomes a masculine attribute, deficient in the female, fulfilled in the male.

Simmel's treatise is one of the more insightful socio-psychological speculations on the *Frauenfrage* in the years surrounding World War I, prior to the time when the women's movement would subside in the face of economic depression and fascism. Paradoxically, there remains some ambiguity whether Simmel in his *Philosophische Kultur* was defending, attacking, or simply describing and explaining the inferior objectivity of women. Significant for our purposes, though, is that both Tönnies and Simmel expressed the science-nonscience distinction (and more specifically the science-values distinction) as a male-female distinction. The ideal of neutrality (or objectivity) was defined in such a way as to exclude women from science. Central for both Tönnies and Simmel was that whereas men are objective, women are subjective. Women cannot achieve the distance or detachment from objects necessary to pursue objective science.[51]

The views of Tönnies and Simmel reflect those of many sociologists at this time, but not all.[52] Among the founders of German sociology, Max Weber was probably the most sympathetic to the feminist cause. His wife, Marianne, was active in the Heidelberg feminist movement and claimed to have been the first woman to have attended philosophical lectures at the

University of Heidelberg (in 1896). In 1907 Marianne published a history of the legal status of women,[53] and in 1910 she organized a meeting of the Bund Deutscher Frauenvereine in Heidelberg. Feminists Gertrude Bäumler (editor of the five-volume *Handbuch der Frauenbewegung*) and Marie Baum were visitors to the Weber home in Heidelberg. As early as 1896, at a Heidelberg meeting on the question of whether women should be able to study at universities, Max defended women's rights with such energy and apparent success that Marianne declared him to be "more of a feminist" than she was. In subsequent years, Max criticized the Mutterschutz movement (advocating free love and the rights of women to bear children out of wedlock) for being based on a "crass hedonism" and "an ethic that would benefit only men." Weber also defended his wife against anti-feminist attacks, as when, following the 1910 meeting of the Bund Deutscher Frauenvereine, he challenged one particularly nasty critic of the meeting to a duel for having insulted those in attendance as "a bunch of Jewesses and spinsters." (Weber, a dueling enthusiast, sported a prominent facial scar from his student days.)[54]

Weber nevertheless also showed an insensitivity to women's problems not untypical for men of his era. There is virtually no mention of the position of women in any of his massive writings on history, sociology, economics, or law. He was also not entirely comfortable with his own wife's intellectual pursuits, as can be seen from a long and rather paternalistic letter he wrote to Marianne in 1893, shortly before their marriage:

> Shall I send you the Bebel? If you want it, I shall do so right away, for I do now consider myself your guardian. Or shall we read him together at some future time? . . . Above all, take care of your body now. You must get stronger and look outward rather than inward, into your interior as well as mine, and you must not think with such contempt of those who are "only housewives." I mean this for your own good . . .
>
> I almost fear you misunderstand me to say that I am taking back or watering down my earlier request that you should make the greatest possible demands upon me regarding discussion and sharing of intellectual interests.
>
> On the contrary, my child, this is the situation: to prevent such a companionship in the "intellectual" sphere from jeopardizing your position, I must never get the—unconscious—feeling that . . . you are entirely dependent on me. And this very situation, so it seems to me, could easily arise unless I get the feeling that in your practical affairs you have an independent sphere of activity that is controlled by you and fulfills your *practical* interests as much as my teaching profession or some other vocation that might be in store for me fulfills mine.

Marianne indicates in her biography of her husband that she did not take this particular piece of advice very seriously but rather "did what her conscience prompted her do," and continued a creative and independent intellectual life of her own.⁵⁵

Max Weber, again, was more broad-minded than many in his views on "the woman question." For many sociologists at this time, the principle of value-neutrality represented both an attempt to avoid taking a stand on women's rights and a disparagement of women's capacities to produce science. Impersonal detachment was both the essence of the scientific attitude and a masculine virtue. Women had failed in their bid to enter science, not from particularities of their social place, but from failures of their psyche. The exclusion of values from science worked to reaffirm the exclusion of both women and "the feminine" from science.

Such was the political climate faced by the sociologists gathering for the first time in 1909 under the banner of neutrality. After two decades of banishment, the Social Democratic party had become the largest political party in Germany. The largest feminist movement in the world had begun to raise increasingly militant demands to the Wilhelmine government. The eugenics movement had begun to lay claims to the entire field of sociology.

In the face of such diverse pressures, it was not clear in the closing years of imperial Germany which direction social theory in general and sociology in particular would go. There were a number of options. Religious, racialist, formalist, and social Darwinian sociologies filled scholarly journals, each putting forth its program for the social sciences. For Ludwig Gumplowicz, sociology rooted itself in the positive analysis of racial and national struggles; for Marxists, sociology was part of a larger philosophy of history. Georg Simmel spoke of sociology in terms of the "geometry of the social world"; Max Weber referred to the science whose method was *Verstehen*. Racial hygienists envisioned a sociology based on efforts to control the breeding of inferiors; conservatives of the historical school hoped that sociology could continue to serve as a guiding light for government in matters of social policy. In 1923, Kurt Singer could describe nineteenth-century sociologies founded on biology, mathematics, ethnography, epistemology, psychology, and politics.⁵⁶

Each of the movements I've described—feminism, racial hygiene, and social democracy—posed a challenge to fundamental aspects of German society. In the imagination of the academic, these various challenges were often linked with one another. Otto von Gierke rejected feminist demands for bringing women into the university on the grounds that this would create a "new and revolutionary student proletariat." Max Weber rejected social Darwinism and Ostwald's "cultural energetics" as forms of "intellectual imperialism" comparable to Marxism. Materialism according to Arthur Kirchhoff took root in the festering sores of poverty and ended up destroying the family.

Several of the alternatives to neutralist social theory did in fact look to one another for support. By the turn of the century, the strongest feminist movement in Germany was that associated with the Social Democrats. August Bebel drew from social Darwinism in his justification of socialism and helped to link socialism with feminism with his 1883 *Die Frau und der Sozialismus*. Certain feminist groups shared with the progressive wing of the Society for Racial Hygiene common concerns for reform in child care and provisions for unwed mothers. Several early members of the racial hygiene movement (Ploetz and Schallmayer, for example) and a number of the nordic supremacist "social anthropologists" (Woltmann, for example) expressed support for Social Democracy early in their careers.

Academics occasionally expressed sympathies with one or another of these movements. As employees of a cautious and repressive state, however, German academics had little room to maneuver. Like all other government employees, academics swore an oath of allegiance to the king of Prussia promising to uphold the laws and ideals of the Reich. Academics could be fired for advocating political reforms. The rise of the ideal of "neutral science" should be understood in this context. Value-neutrality provided a means of escape from the explosive questions of the day. In the words of Fritz Stern, the German academic, surrounded by political turmoil, built moats instead of bridges.[57]

Neutrality served as a means of ensuring the autonomy of the new science of society. It was in this sense that Tönnies declared that sociologists would consider themselves "neither for nor against socialism, neither for nor against the expansion of women's rights, neither for nor against the mixing of the races." Neutrality served to defend against the charge that *sociology* was just another word for *socialism*. But institutional autonomy was not the only social function served by the principle of neutrality. Neutrality was not just a shield, but a weapon: neutrality was used to thwart attempts on the part of feminists, social Darwinians, and (especially) socialists to politicize social theory. It was largely in reacting against these movements that sociologists formulated the ideal of value-neutrality.

9

Neutral Marxism

The association between sociology and socialism that conservatives feared in the first decades of the twentieth century was not a spurious one. Many of the young sociologists were enamored of the new socialism, especially in its more liberal forms. Of even greater consequence, though, was the fact that several prominent members of the SPD (August Bebel, Karl Kautsky, and Eduard Bernstein, among others) embraced the new science of sociology, attempting to bring it into accord with the materialist interpretation of history in particular and Marxist principles more generally. Eduard Bernstein held up sociology as the "pure science" whose application was "socialism"; Max Adler, Nikolai Bukharin, and Karl Kautsky all wrote major treatises on sociology. Werner Sombart, whom Engels considered "the first German professor to have appreciated the theoretical Marx," was one of the founders of the German Society for Sociology; indeed, several other sociologists (Oppenheimer and, with less enthusiasm, Weber) declared at least partial sympathy with social democracy in one form or another when it had become respectable to do so. In Austria, too, Marxism was closely identified with sociology. Max Adler was both a leading theoretician of Austria's Social Democratic party and a co-founder of the Austrian Society for Sociology; he was also the first professor of sociology at the University of Vienna. The close association between sociology and socialism in Austria at this time may help to explain the perception of a later generation of Austrian philosophers (notably Karl Popper, Ludwig von Mises, and F. A. Hayek) that Marxism and positivism were closely linked.[1]

It is important to recognize, however, that the socialism with which many young sociologists were to identify changed in the early years of the twentieth century. In the first decade of the century, one sees the emergence of an "academic Marxism" that conceived the essence of the Marxian doctrine as a set of neutral, sociological tools (social history, economic determinism, and so forth) one might apply to the analysis of history or society—quite independent of one's political perspective. Marxism in this view represented a "natural science of society," a set of universal and deterministic laws of the evolution of society, guaranteeing the emergence of specific social forms. These two trends—the reduction of Marxism to a set of historiographic or analytic tools and the interpretation of Marxism as a natural science of society—are perhaps best exemplified in the works of the socialist-turned-sociologist Werner Sombart, the "revisionist" SPD leader Eduard Bernstein, and the Austro-Marxian Max Adler. Each of these men would become strong advocates of the principle of neutral science; each would become instrumental in defusing more radical elements of the social democratic movement. We look at each in turn.

Werner Sombart was born in Ermsleben, south and west of Magdeburg, in 1863. His father, like Weber's, was a representative of the National Liberal party in the Prussian parliament and German Reichstag; the elder Sombart was also a co-founder of the Social Policy Association and a Junker industrialist—a combination that was to prove fateful in German political formations of the late nineteenth and early twentieth century. After studying with Gustav Schmoller and Adolf Wagner at the University of Berlin, the young Sombart graduated in 1883 with a thesis on the Roman Campagna. At the age of twenty-seven, Sombart was appointed associate professor of political economy at the University of Breslau, where he was soon branded the "red professor," not just for his exposition of the Marxian theoretical corpus but also for his unorthodox teaching methods, including excursions with students into mines, factories, and farms.[2]

Sombart's 1888 doctoral dissertation represents one of the first German theses to draw upon the methods of Marx. The young economist analyzed the landed aristocracy in Italy as a purely parasitic class, having lost all interest in the arts and sciences, scarcely familiar even with their own land, and feeling no obligation to participate in either government or the military. The landed aristocracy Sombart makes responsible for the poverty of the Italian peasantry; indeed this, he points out, was the achievement of Marx: to have recognized that poverty is often the *product* of wealth and that conditions supporting the accumulation of wealth can also be conditions for the creation of poverty.[3]

By the 1890s, Sombart had gained a reputation as something of a maverick in the sphere of German academe. Friedrich Engels, in his afterword to volume 3 of Marx's *Kapital* (published in 1894), praised one of Sombart's articles in *Braun's Archiv* as "admirable for the fullness of its exposition of the Marxian system. It is the first time that a German university professor has managed to see in the whole of Marx's writings that which Marx himself really intended, recognizing also that criticism of the Marxian system cannot consist in a rejection of the system—from whatever point of view—but can only consist in its further development."[4] Sombart, according to Engels, appreciated unlike any other German professor of his time the "theoretical Marx," the Marx of historical materialism and the doctrine of surplus value, the Marx of the declining organic composition of capital and the falling rate of profit. But it was also this very reliance upon the purely theoretical Marx that was to earn him the reputation of being an "overly academic" and ultimately pseudo-Marxist among more activist elements of the socialist movement.

In the fall of 1895, Sombart delivered a series of lectures in Zurich on "Socialism and Social Movement," published the following year as a small book. The purpose of his lectures was to argue that the socialist movement was not just a manifestation of anarchy or civil war, something to be associated with "gasoline and dynamite," but rather a social phenomenon with deep historical and economic roots. Improvements in transportation and communication—telephone and telegraph, newspaper and railway—these were the root cause of modern social disaffections. Innovations such as these had brought men closer together, introducing a continuous state of flux into science, art, morals, religion, and the economy. The resulting disorientation had produced much of the uncertainty and social protest of the times: "Edison and Siemens," Sombart wrote, "are the spiritual fathers of Bellamy and Bebel."[5]

Sombart's book was to become immensely popular both in Germany and abroad. By 1924, the tenth edition of what was now a two-volume work had been translated into twenty-four languages; Sombart himself had become a popular sensation, delighting audiences with lectures on the history and causes of the current turmoil in Germany. According to Ludwig von Mises it was Sombart who, more than anyone else, "introduced Marx into German science, and made Marx familiar to German thinking." And it was Sombart who, according to his biographer, was to become the only German professor of his time ever to fill the Philharmonic Hall of Berlin to capacity for a series of lectures.[6]

Already by the turn of the century, however, Sombart had begun to grow dissatisfied with the organized socialist movement. As state power became a serious possibility for the Social Democrats, Sombart sided with the parliamentary, reformist wing of the party against calls for revolution, a move which was to bring him criticism from radical elements of the party. Lenin, Rosa Luxemburg, and Karl Liebknecht all saddled Sombart with the charge that his was a purely theoretical Marxism, purged of all revolutionary practical impulse. As early as 1892, Sombart had argued that Marx's work was of a "purely theoretical character," meaning by this that it made no reference to what is good or moral, but only to what must happen through the laws of history. Marxism was "value-free science": "The entire Marxian system" he declared, "contains not a single grain of ethics."[7] Even in his popular book of 1896, the purpose of his writing had been to present Marx as a purely analytic "tool" that might be used to understand the social movements threatening Germany. The motto for this book was *je ne propose rien, je n'impose rien, j'expose* (I don't propose anything, I don't impose anything, I just expose), and this Sombart held to characterize the Marxian theoretical method.

Marxism differed from other socialist systems, according to Sombart, in having an "anti-ethical bias." Marx never maintained that surplus value was not "owed" to the entrepreneur, nor did he claim that workers have a "right" to the full product of their labor. Socialism for Marx, in Sombart's view, derived not from ethical motives but from the workings of social and historical forces, from conflicts arising from the concentration of private property in the hands of a few and efforts to socialize the means of production. What distinguished scientific from utopian socialism, Sombart argued, was the realization that social movement depends ultimately not on ideals but on realities: socialism had become a movement, a reality, and not just an idea. This was what distinguished modern Social Democrats from the Insurrectionists of Lyon, or the Chartists, Putschists, Clubistes, Blanquistes, and other revolutionaries of former times. Marx's "realism" was supplanting the utopian visions of Robert Owen, Charles Fourier, and the early workers' movement; socialists had realized that "interests rule the world" and that one must arm oneself with weapons stronger than Owen's "eternal love."[8]

One of the primary effects of Sombart's work was to make Marx's teachings respectable in academic circles. Part of this legitimacy derived from the fact that Sombart linked "scientific socialism" with the need for gradual, parliamentary reform: evolutionary rather than revolutionary change. Scientific socialism recognized the possibility of slow and steady social progress, involving redistribution of the social product not by violence but by legislation.

This was only one of several revisions Sombart would make of Marxism. He also argued, for example, that socialists must realize that it is relative and not absolute deprivation that incites social movement.[9] Foremost in his appeal to bourgeois sensibilities, though, was his insistence that sociology was not a political program. As early as 1893, Sombart had advocated a strict separation of science from political issues: "Influencing the course of day-to-day politics can no longer count as the primary purpose of science. The central focus of science must be knowledge of the main trends of our scientific and social development: the purely theoretical questions of 'where from?' and 'where to?'"[10] This represented a change, he argued, from the task of theory in former times. Social scientists had once studied advanced cultures in order to gain insights into matters of social and economic policy. Today, however, the focus of scientific inquiry has shifted from advanced to primitive forms of social organization. Today it is these lower forms of culture that are of interest to the sociologist—not because they are useful, but for the same reasons a biologist studies the lower forms of life: to establish certain principles.

Such was the position Sombart and the other sociologists were to celebrate in 1910, at the founding of the German Society for Sociology. Sociology had reached a stage where it could be considered scientific. Sombart recognized the political value of such a view in helping make his "socialist" theories palatable.

Sombart's advocacy of Marxism originally caused him difficulty in finding academic appointment. Weber had recommended the young Marxist to succeed him at the University of Freiburg when he retired in 1899, but the government refused. Sombart failed at several attempts at promotion until 1906, when he finally obtained a position at the Berlin College of Commerce. But his advocacy of sociology as a purely neutral science, along with his vociferous distinction between economics as a practical art and as a theoretical science, helped immunize his science from criticism and helped him win a reputation as one of Germany's most "respectable" Marxists. Sombart was well aware of the political favor neutrality had brought him. In 1929, at a meeting of the German Society for Sociology, Sombart reflected on the early years of his career, when he first realized the value of separating science from politics:

> I was at that time a convinced Marxist, and was at the same time a professor at one of the monarch's Prussian universities. I tried to resolve the tension that resulted from this contradiction by the following perception that I had at this time: value judgments do not belong in a science. Hence I can carry on scientific work quite independently of the personal values I happen to hold.[11]

By the time of World War I, Sombart had become so disillusioned with socialist, liberal, conservative, and clerical parties that he began to cast about for alternatives to class-oriented politics. Sombart never completely abandoned his fondness for a certain form of socialism, but his socialism, like that of Weber, was one that he was to identify increasingly with Germany's national cause. Sombart's particular development in this regard was dramatic. In 1905, he was the most widely read socialist author in Russia after Marx. Within a decade, however, he had become active in defending Germany's position in the war, arguing, for example, that while the English were "merchants," the Germans were "heroes." Sombart's drift to the right continued into the 1920s. In 1924 he rejected entirely the ideas of that "Godless, inner-torn and liver-sick" Marx; the German proletariat now represented little more than "Bentham and cotton dust." Finally, in the early years of the Nazi era, Sombart published his *Deutscher Sozialismus*, arguing for a third, "German," socialism distinct from either SPD or Bolshevist variants.[12]

Even in his earlier work, however, it is not hard to find hints of his later dissatisfaction with socialism. In his 1896 *Sozialismus und soziale Bewegung*, Sombart had described Marx's failure to appreciate the fact that the history of the world is not just a history of class struggles, but also of the struggles of nations against one another for "a place around the feeding trough." There are thus two struggles: "there is the *social* struggle, which Marx described in his discussion of classes. Then there is the *national* struggle: the struggle of one community against another for power." Both were important. Sombart left no doubt as to what he meant by the "national struggle." Warning against the rising power of Russia, Japan, and China, he predicted that the European proletariat might one day have to unite against this threat, "should the coolies swarm over us like rats." Sombart warned that we should not forget that "the history of the world consists in the interplay of both these poles—the national and the social."[13] Such were the kinds of views that would make the aging Sombart a fellow traveler of the Nazi movement (see below and Chapter 11).

Sombart's revisionist view of the national question was not the only point where he broke with orthodox Marxism. He also played an important part in supporting the trade union movement, which was to become powerful in the first years of the century and which, at least in the eyes of radicals within the Social Democratic party, was ultimately responsible for the notorious schism that divided the party into revolutionary and reformist tendencies. Sombart sided with reformist elements in the labor movement advocating the abandonment of revolutionary ideals in favor of more cautious efforts to increase wages and improve working conditions through negotiation with management. In 1900 he advocated "political neutrality" for the trade unions; in 1905 he lectured to union groups to try and convince them of the utility of separating economic demands from the "utopian" political demands associated with socialist agitation.[14]

Sombart abandoned his support for the trade union movement after the war, but his interest in the problem of nationalities continued throughout his career. In his 1911 *Die Juden und die Wirtschaftsleben*, he tried to explain the origins of capitalism in terms of the rise of Jewish usury. Relating his theme to the work of his friend, Max Weber, Sombart declared that in its purest form "Puritanism *is* Judaism." In 1928, Sombart edited a collection of essays on *Volk und Raum* asking: "Can Germany maintain a growing population within its existing borders?"—he was doubtful.[15]

With the triumph of fascism in 1933, Sombart spearheaded efforts to reorganize the Social Policy Association. Early in the year he noted that, according to certain unidentified "voices," the Association had become "boring and uninteresting." Sombart ordered a meeting of the board of directors to discuss its reorganization, and on March 3, 1933, he asked that the lectures and debates of the Association henceforth be presented "to the broadest possible public" and that such debates be "once again be infused with soul ... in accordance with the National Socialist world view." (Private workshops, however, were still to be conducted in a manner as to exclude "problems of *Weltanschauung*.") In October Sombart sent a letter to all members of the Association informing them that, in accordance with the provisions of the *Gleichschaltungsgesetz* (requiring proof of "Aryan ancestry"), professors Emil Lederer and Ludwig von Mises had resigned their posts within the Association.[16]

Already an old man by the time of the Nazi seizure of power, Sombart did not play a major role in the sociology of the Nazi era. He never joined the Nazi party, yet he was not entirely unsympathetic with certain prejudices of the new regime. In his 1938 *Weltanschauung, Wissenschaft und Wirtschaft*, he contrasted the "keen Jewish intelligence of a Ricardo, Marx or Keynes" with the "deep German obscurity [Unklarheit]" of a Müller, Knies, or Schmoller. Science in many cases, he wrote, is bound to the "blood" or "spirit" of a people. But Sombart also questioned whether this must and should be the case. Scientific results, after all, are either true or false, and the *Weltanschauung* (world view) of a researcher should have no bearing on this. Sombart pointed out that one can investigate the history of the middle class under capitalism "independent of whether one desires the fastest possible demise of this class (as does the Marxist) or makes it the heart of the people's economy (as does the National Socialist)." Sombart concluded with a warning against a "knee-jerk" National Socialism, requiring that one blindly strive to make all things as "German" as possible. What kind of work would Dürer have produced, if he had asked himself with each stroke of the pen whether he was painting *Deutsch*? A good German, Sombart wrote, will produce good work, and we should not demand more than that.[17]

Sombart's conception of Marxism as a neutral science of society can be compared with Eduard Bernstein's revisionist Marxism, the official doctrine of the Social Democratic party after World War I. Like Sombart, Bernstein developed his ideas at a time of profound changes in the German socialist movement. Social democracy was attaining steadily greater influence, and not at a time of increasing immiseration, as Marx had predicted, but at a time of unprecedented prosperity. Buoyed by colonial expansion, capitalist economic growth appeared unlimited. Class inequalities appeared even to have softened. Bourgeois intellectuals were joining the movement, and the German government had been forced to agree to many socialist demands—notably the innovative medical insurance and social security initiatives of the 1880s and 1890s.

It was in this context that Eduard Bernstein (1850–1932), one of the most powerful figures in the socialist movement (Engels had named him executor of his estate and trustee—with Karl Kautsky—of his literary *Nachlass*), undertook his revision of Marxism. Bernstein argued that Marx and Engels had oversold the inevitability of class struggle and that many traditional socialist goals could be better realized through peaceful than through violent means. Socialism was not a natural inevitability but a choice that humans had to make—socialism was ultimately a matter of ethics and of education. One had to distinguish the separate tasks of science and of ethics in this struggle, and here is where we encounter the question of value-neutrality.

In his 1901 essay, "How Is Scientific Socialism Possible?" Bernstein presents what is probably the most subtle and comprehensive defense of the neutrality of science by a socialist theorist. In his paper, delivered originally to a meeting of the Social Science Student Association in Berlin, Bernstein suggests that there was a fundamental ambiguity in Engels's writings concerning whether socialism should be considered a science or an ethical ideal. On the one hand, in his preface to the second German edition of Marx's *Poverty of Philosophy,* Engels had argued that socialism is a doctrine designed to end exploitation. And yet, in his 1895 preface to Marx's *Class War in France,* Engels makes clear his view that "scientific socialism" has nothing to do with ethics. How, Bernstein asks, does one reconcile these two views? Is socialism a science? And if so, what does this say about the place of ethics in "scientific socialism"?

Bernstein claims that one cannot derive socialism from an analysis of surplus value, or indeed from any other empirical facts about the world (he notes that Engels recognized this fallacy as well). The discovery of surplus value in capitalist economies does not prove that socialism is good, any more than the recognition that slaves produce more than they consume is necessarily an indictment of slavery. It is possible, in other words, to admit that there is such a thing as surplus value and to deny that it implies any form of exploitation (this, Bernstein points out, was the position of the Anglo-Austrian marginalist school). Conversely, it is possible to denounce capitalist exploitation without appealing to any particular theory of surplus value (this was the position of Owen and the utopian socialists). In light of this apparent independence of description and evaluation, one must then ask what it means, or whether it is even possible, for socialism to be "scientific."[18]

Socialism, Bernstein suggests, is not just a "movement," but movement toward a goal. There is a moral interest involved in socialism, for without interest there can be no action. There is also a science appropriate to this struggle of interests, and that science is sociology. Sociology is the "pure science" form of that which, when "applied," becomes socialism. Socialism is simply applied sociology.[19] The problem is that philosophers have insufficiently distinguished socialism as a science and socialism as an ethical (or political) ideal. Bernstein appeals to Francis Bacon to distinguish science from "affairs of state." Whereas the former concerns itself only with relation and motion, the latter concerns itself with questions of authority and opinion. Science, Bernstein says, is simply "ordered knowledge," and thus can have nothing to do with politics. This is as true for the science of society as it is for the science of nature: "No one today would speak of a 'liberal physics' or a 'conservative chemistry.' But should this be any different for the science of human history and of human institutions? I, for one, say no, and cannot consider a liberal, conservative, or socialist social science anything but a contradiction in terms."[20] Bernstein argues that science must be apolitical, in the sense that it must remain agnostic about the ultimate human consequences of actions or events in the world. The goal of science is knowledge, not practice. Science is *tendenzlos,* it belongs to no party or class.

Where then does this leave the idea of "scientific socialism"? Socialism, Bernstein concludes, is not a science; indeed, "no 'ism' can be a science" *(kein Ismus ist eine Wissenschaft).* The very notion of "scientific socialism" (or any other confusion of science and politics) is dangerous, he argues, both for science and for politics. For science, the danger lies in the threat such a confusion poses to the autonomy of scientific knowledge. For politics, the danger is that we confuse a living, moral, struggle with preordained laws of nature. Scientific socialism, Bernstein concludes, somewhat quixotically, "is possible in the measure it is necessary—that is, in the extent that one can reasonably expect from a movement whose task is to create something new."[21]

Bernstein was the leading representative of what he himself christened the *revisionist* school of socialist thought. Rejecting what he called the *catastrophe* theory of socialism—the theory that the bourgeois economy would eventually collapse under its own weight—Bernstein argued instead for an evolutionary socialism, and supported this view by referring to Engels's 1895 contention that the primary task of the SPD was to work for an increase of votes. The *Communist Manifesto* was correct in its analysis of the tendencies of bourgeois society, but it needed revision in its estimation of the timetable of revolution and the means by which change would occur.

Bernstein also thought it crucial to separate the *facts* of socialist analysis from the *goals* of the socialist movement. This, too, required a revision of Marxist doctrine. Marx and Engels had wrongly based their hopes for a socialist society not on moral judgment, but on certain supposed "laws of history" guaranteeing the triumph of particular social forces. Bernstein argued instead that socialism would be a product not of knowledge but of "the will"; he defended this position with the argument that whereas science merely determines the facts, it is the will that creates goals and moves one toward those goals. Socialism must in this sense be critical *(kritisch)* but not scientific *(wissenschaftlich)*; science in itself is politically neutral, and can be neither socialist, conservative, nor liberal, but simply true or false.[22]

Bernstein separated the goal of socialism from the science of social forms in order to allow a certain flexibility in socialist analysis. By distinguishing facts from values we can preserve socialist goals, despite changes we may be forced to make in socialist analysis. After all, Bernstein pointed out, it was hard to deny that many of the predictions of the founders of socialism had failed. The number of propertied had grown, not declined. The privileges of the bourgeoisie were slowly being supplanted by democratic reforms. Legislative reforms, combined with the rise of the trade unions, were eliminating the need for violent action. Bernstein noted that Engels recognized this himself, when he argued in the 1895 preface to Marx's *Class War in France* that social democracy would flourish "far better by lawful than by unlawful means and violent revolution."[23]

Bernstein, like Sombart, agreed with Engels that one should not take refuge in "utopian" socialism. German social democracy, he claimed, is founded upon what Marx and Engels considered "scientific socialism." But as in all sciences, "a distinction can be drawn between pure and applied science." The first consists of a set of universal principles derived from a series of appropriate experiences. The principles of pure science will of course change as we acquire new knowledge and as social circumstances change. But we will never be able to derive ethical ideals from our science of society. Bernstein says it is high time to distinguish the "pure science" of socialism from its practical applications: indeed, "a systematic stripping of its applied parts from the pure science of Marxist socialism has not hitherto been attempted." Early hints of this, Bernstein writes, can be found in Marx's preface to his *Critique of Political Economy* and also in the third part of Engels's *Socialism, Scientific and Utopian*. This "pure science of socialism" consists of two parts: (1) the program of historical materialism (the theory of class warfare), and (2) the theory of surplus value (along with the historical movements that can be derived from this theory). A deterministic materialism is the centerpiece of this pure science of socialism: mechanical forces determine in the last analysis all occurrences, even those which appear to be caused by ideas. Application of this principle to history implies "a belief in the inevitability of all historical events"; the materialist is "a Calvinist without God." One therefore had to overlay on top of this materialism a separate sphere of ethics, an ethics that could not be derived from science.[24]

Marxism for Bernstein must consist in a (neutral) deterministic science and a set of (moral and political) ideals. These together constitute the system that is socialism, but to confuse the two or to imagine that one might be grounded on the other would be a great mistake. That is why science is and must remain neutral with regard to political ideals. The "pure science" in socialism must be distinguished from the "applied." Circumstances in society change, and the pure science of socialism—sociology—must adapt to those changes. So long as science is neutral, though, the fact that society and therefore the science of society *changes* does not mean we must abandon our ethical ideals. We may still preserve our ideals, even if our predictions turn out to have been wrong.

Consider briefly one final advocate of value-neutral Marxism. Max Adler, leader of the Austrian SPD, advocated a view of science and ethics similar to that of Bernstein and Sombart. A follower of Ernst Mach, Adler was eager to cleanse Marxist philosophy of all "metaphysical" elements in favor of Mach's "critical" (that is, Kantian) philosophy. With Engels, Adler defined materialism as that which we know owing to the latest results of science—a philosophy according to which, in Engels's words, "with each great scientific discovery, materialism must change its form." Adler agreed with the opinion of Masaryk, in 1907, that "the Hegelian dialectic, even in its materialist form, is nothing but hocus-pocus"; Adler seconded the view that Marx and Engels had "woven their materialist rug with metaphysical spiderwebs."[25]

As an alternative to this "metaphysical" Marxism Adler proposed a "positive" Marxism, one that would not simply repeat Marxian dogmas but would push forward Marxism as a science. The task was to see Marx not as always correct but as fallible; the positive philosophy of Marx was to be united with the critical philosophy of Kant. Marxism, in short, was to become a science. Adler praised Karl Kautsky's 1906 book, *Ethics and the Materialist Conception of History,* for its attempt to establish a socialist ethics on a "natural-scientific basis"—that is, on the "dialectical-historical" method developed by Marx and Engels. Adler reserved judgment on whether Marx, Engels, and Joseph Dietzgen were to be considered materialists. They were all "positivists" insofar as they declared themselves the enemy of all metaphysics—materialist or otherwise. If they were materialists, this meant nothing more than that they were dedicated to the relentless tracing of all phenomena back to causal laws. Each was devoted to the "radical and methodical exclusion of all religious and speculative belief in miracles"; each was devoted to eliminating knowledge that could not be derived from "the fair and fertile earth of experience." Adler claimed that this was also the principle to which the critical philosophy of Kant adhered "with all its heart."[26]

Karl Kautsky was more cautious in his endorsement of positive science than his Austrian friend Adler. In his 1906 treatise on ethics, Kautsky argued that even though materialism (in the sense of a naturalistic world view) had "always stood in opposition to oppression," one must nevertheless be cautious in interpreting the most recent trends in science. Natural science in its nondialectical form, he argued, the form of Ludwig Büchner and Ernst Haeckel, had made its peace with the bourgeoisie in recent years and now stood allied with this class against social democracy.[27] Both Kautsky and Adler still felt, however, that it was important to distinguish sharply between "normative" and "scientific" judgments. Adler's separation of science from values rested upon the neo-Kantian notion that neither of these realms may inform the other. Theoretical knowledge has always to do with existence or occurrence *(Sein oder Geschehen)*; ethics, by contrast, has to do with problems of good and evil. Ethical knowledge transcends the "how" and "why," because the good is not that which is but that which ought to be.[28] Theoretical knowledge has only to do with "what is," not with "what ought to be."

To summarize, neo-Kantian Marxists of the early twentieth century rejected the conception of Marxism as a "natural science of society"; they rejected the assumption that there are invariant historical laws, along with the optimistic faith that "what is and will be, must be and is good." Sombart, Bernstein, and Adler drew from Kant to argue that Marxism must be both an ethical and a scientific system, but they also cautioned against confusing these two spheres. Bernstein distinguished between Marxism "pure" and "applied"; Adler distinguished between normative judgments and scientific facts. Neo-Kantian Marxists separated the science of Marxism from its ethical content, in order to preserve socialist ideals even if particular predictions of socialist theory proved to be false, and to avoid the fatalistic quietism associated with the view that a given social order or sequence of social events is natural or inevitable.

I do not want to leave the impression that value-neutrality was the predominant interpretation of Marxist social theory in the early years of the twentieth century. The radical separation of the science of Marxism from its politics did not go unchallenged. The *Welt am Montag* warned that Bernstein's ideas would "rob workers of faith in the scientific nature of socialism";[29] Rosa Luxemburg criticized Sombart for making the goal of socialism a "purely subjective" one. Nikolai Bukharin distinguished between the fact that all social life is determined (there are no "accidents" in history) and the fatalism that says that regardless of how we act, history will run its preset course and all we must do is sit back and wait. Alexander Bogdanov defended a "proletarian science" that would bring political goals to bear on the structure of scientific inquiry;[30] Georg Lukács contrasted the positivist-empiricist reduction of Marxism with a more dialectical approach, stressing the role of human agency and consciousness in social movement.

What is often overlooked, then, is that the ideal of *Wertfreiheit* was as much a statement about *values* as it was about science. Indeed for some, neutrality was needed to prevent the abuse of science. To understand this caution, we must explore why Max Weber—the most subtle thinker among the early sociologists—argued that science is and must remain value-free.

10

Max Weber and *Wertfreie Wissenschaft*

The proponents of the new "scientific sociology" claimed that their intent, in Sombart's words, was not to propose or to impose but only to expose the truth. As we have seen, this value-neutrality was deployed, among other things, to defend against charges of socialism. The banner of neutrality was raised not in the abstract, but in order to define a political stance vis-à-vis concrete political movements. This can also be seen in the work of Max Weber, chief exponent of value-neutrality in the social sciences. It was Weber who popularized the notion that science is and should be value-free; Weber set the tone for the debate that carried across the Atlantic and into American social and economic thought. His position is complex; he was probably the most subtle of the early German sociologists. He has been called the "Marx of sociology"; as we shall see, however, he was a Marx without a *Manifesto*.

Max Weber was a polymath of first rank. His empirical studies show him to have mastered an astonishing breadth of historical materials, ranging from Chinese bureaucracy to the sociology of music, from the religion of India to the economy of ancient Egypt. In his *Sociology of Music*, Weber described how competition among the princes and prelates of Renaissance Italy, France, and Germany shaped musical traditions, why the hollow fifths of Puccini's early operas were banished from music by a college of musicians in Florence, how sonority and tonality derived from the particularities of instrument design, and how the tempered scale of Bach's *Well-Tempered Clavier* emerged as a product of sociological changes. Weber was as comfortable writing on the politics of Egypt of the New Kingdom or republican Rome as on medieval law or the ancient religions of India and China.[1]

But Weber also advised a cautious, empirical methodology, one that excluded value judgments from the practice of science. In 1904, as he along with Edgar Jaffé and Werner Sombart took over editorship of the *Archiv für Sozialwissenschaft und Sozialpolitik,* Weber articulated a new program for the journal and the social sciences in general, in an essay titled "Objectivity in Social Science and Social Policy." Let us look in some detail at the defense of neutrality Weber presents in this essay, often considered the key document in the *Werturteilsstreit.*

Every science, Weber argues, originates with practical problems—and social science is no exception. The original task of social theory was to evaluate policies proposed by the state, as was typically the case in the Social Policy Association. But after a certain point a "mature science" moves beyond the practical problems that first give it life. Scholars begin to ask questions of a more theoretical character. Weber complains that social theory has not yet made this transition from a craft tradition to a theoretical science. In particular, the rise of a new science of society has not been accompanied by a sufficiently strong separation between "what is" *(das Seiende)* and "what ought to be" *(das Seinsollende),* between the evaluations of practical problems on the one hand and science for its own sake on the other. Against this separation, Weber claims, there have worked two opinions: first, the belief that there exist regular and inevitable natural laws of social life; and second, the belief that there is some developmental principle that dominates economic processes. Such views postulate that "what ought to be" can be equated with either the unalterably real *(das Seiende)* or the inevitably becoming *(das Werdende).*[2]

It is against both opinions, representing scarcely disguised versions of Mill's and Comte's positivism on the one hand and Marx's and Hegel's developmental theodicy on the other, that Weber raises the banner of neutrality. Positivism erred for having assumed a unique and unilinear sequence of modes of thought; Marxism erred for having assumed that societies proceed through a series of invariable economic transformations. Both also wrongly assumed that there is some necessary relation between "what is" and "what ought to be." In fact, Weber argues, the moral order cannot be derived from the movement of history or any other facts of social life. The real is not the rational; the moral order is no necessary ally of either *status* or *fluxus quo.*

Morality, for Weber, represents the casting of ideals rather than the formation of allegiances. In this, Weber's critique of naturalistic social theory—a kind of secular theodicy—reproduces that launched by Hume a century and a half earlier (see Chapter 4). I use the term *theodicy* as it occurs in Leibniz's book by that title, generalized to include not merely the belief that God has taken all things into account in his design of the world (which is more properly a static "optimism"—that is, the belief that all things are for the Good). I mean as well the historicist version of theodicy—for example, that of Hegel's *Phenomenologie des Geistes*—according to which all things are either already for the best or are becoming so in the course of historical development. It is in this more general sense that much of modern social theory (not just the Marx of *Kapital,* but social Darwinism, neoclassical economics, Comte's and Turgot's historical stages, and so forth) has been subject to the charge of theodicy.[3]

In practice, Weber argues, pure science is an unattainable goal. Moral or political values constitute a kind of unavoidable contamination in science, a contamination which always takes its toll as one of the "most dangerous confusions of our time."[4] Weber asserts that a scholarly journal should nonetheless strive to minimize the moral or political judgments in what it publishes. The *Archiv* "has never been, and should not become, a place where polemics against particular political parties is carried on; nor should it provide a platform from which political ideas should be advertised. There are other organs for this." The *Archiv,* Weber asserts, has never been a socialist journal; nor will it in the future become a bourgeois one. This does not mean that one cannot be critical, he explains. But the true task of critique is not to ally oneself with this or that party, but rather to discover what means are required to reach certain ends, and with what other goals those ends are in conflict or agreement.[5]

In these terms we can understand one of the central elements in the Weberian conception of social science. Science can provide only facts; values are a matter of personal faith. Science, if it is to be objective, must be value-neutral. This does not mean the abandonment of ideals but simply their separation from science. The scientist, in other words, must wear two quite different hats: one as citizen with values, one as scientist with none. Both cannot be worn at the same time. Nor are there mechanical schemes for deriving values from facts: one cannot average or "balance" values to eliminate them; the golden mean does not eliminate bias. The task of science must be restricted to (1) the investigation of the effectiveness of alternative means for achieving given ends; (2) the exploration of unintended consequences of pursuing certain ends; and (3) the logical analysis of political goals to expose the value premises upon which they are based.[6] Beyond this the social scientist, *qua* scientist, must not go.

Weber's defense of neutrality must be understood against the backdrop of the social and political context of German social science at the end of the nineteenth century. Consider first his own early research on agricultural workers in northeastern Germany. Since the middle of the century, state authorities had realized that social science methods (especially statistics) could be used to find out valuable information concerning the life and attitudes of German citizens. In 1847, for example, after particularly severe crop failures in northeastern Germany, Prussian agricultural authorities commissioned a survey to find out how peasants were coping with the shortages. The commission investigated, among other things, family wages, spending patterns, and the ability of families to meet basic needs.

Government-sponsored surveys in the latter half of the century continued to measure the attitudes and values of workers, the primary purpose being, in the words of one Social Policy Association member, "to find out where the loyal citizen's shoes are pinching." In 1872–1873, after the formation of the Second Reich, legislation was passed to relieve some of the more oppressive conditions of German industry, establishing shorter work hours, restrictions on child labor, and minimal health security. In 1875 the Reichstag, together with the Congress of German Landowners, commissioned the first great survey of factory workers and apprentices to investigate worker response to legislative reforms. The survey included questions on things such as workers' income and savings, illegitimacy, the use of local libraries, emigration rates, and attitudes toward socialist propaganda.[7]

It was in this context that Weber began his first empirical research on the situation of agricultural workers in East Prussia. In his report to the Social Policy Association, Weber explained that the point of his survey was not to ask "do they live well or badly?" or "how can we help them?" but rather "will the state be able to rely on them for the solution of political problems it will face in the near future?" The problem, as described by Weber, was that the *Deutschtum* of the region was being threatened by the influx of Poles and Russians willing to work for low wages under poor conditions. German workers had begun to leave the land in order to free themselves from the Junker estates, where they were treated as serfs. Weber blamed this exodus on "agrarian capitalists" who, by employing cheap foreign labor, were displacing the German peasant. The solution he proposed was to halt the flow of Polish and Russian laborers into German territories east of the Elbe. Only then could Germany's cultural and strategic integrity be guaranteed.[8]

Weber returned to this theme in his 1894 inaugural address at Heidelberg, where he called for the German border to be sealed against Russian and Polish penetration.[9] Weber's nationalism so impressed Friedrich Naumann, head of the Evangelisch-Soziale Verein, that he wrote in the July 1895 issue of *Die Hilfe:* "Is he [Weber] not right? Of what use is the best social policy if the Cossacks are coming? Anyone who wants to concern himself with internal policy must first secure fatherland and frontiers; he must consolidate national power. Here is the weakest point of the Social Democratic party; we need a socialism which can be administered . . . Such a socialism does not exist; such a socialism must be national."[10] Weber's Russophobia continued throughout his life. After World War I he forecast

American world domination, hoping at the same time that that domination "would not be shared with Russia." The Russian danger had been banished only "for the present, not for ever. For the moment the hysterical and disgusting hatred of the French" was the main danger.

Weber's distrust of Poland and Russia was something he shared with other nationalists of his time. Until 1899 he was a member of the conservative Pan-German League; he broke with this group when they refused to take a stronger position advocating the expulsion of Polish labor from German territories. Weber never did accept what he saw as extreme elements of conservative nationalism, though. He criticized those who voted for the Anti-Semitic party "because they didn't know any better," and expressed his distaste for those idealists who, under the influence of Heinrich von Treitschke, "have become mystical, nationalist fanatics"—those whose political interest "is very recent . . . and who make the greatest noise."[11]

Weber was critical of social Darwinists who attempted to derive ethical or political propositions from conditions of the empirical world. Still, in outlining his version of militant nationalism, he drew upon Darwinian imagery common to the times. In his 1894 inaugural address he argued that it was wrong to imagine that peace and happiness were in store for the future; the "population problem" alone made sure of that. On the contrary, "only in a hard struggle between man and man can elbow room be won in our earthly existence."[12] Twenty years later he put this in even stronger terms:

> Only those peoples who've achieved political maturity are a "master people" *(Herrenvolk),* a people whose control lies in its own hands . . . Master peoples alone are qualified to intervene in the course of world history. If peoples who do not possess this quality attempt this, other nations will revolt against them and their attempt will lead them to internal collapse.[13]

Political power and military force were closely linked for Weber. The central feature of the modern state was its monopoly over the exercise of legitimate force in a given territory. The feudal state, by contrast, was characterized by a system of private ownership of the means of military violence (self-equipped armies). The right of a father to discipline his children derived from the former right of heads of households to life-and-death rule over children and slaves. (Weber argued that it was Luther who removed the responsibility for violence from the individual and placed it in the hands of the state.) A state not grounded in violence, according to Weber, was not a state at all but anarchy. Violence was in fact the key to political power more generally: "All political formations are formations of violence."[14]

Weber's conception of the strong national state held important consequences for both his theory of values and his practical political position. National allegiance he identified in 1894 as the ultimate value-base of economic science: "Economics as an explanatory and analytic science is *international;* but as soon as it expresses values it becomes bound up with our own personal and national identities . . . The economic policy of a German state, like the value-standard of a German economic theorist, can therefore only be German." Science in this sense is the servant of politics, the "political interests of the nation." Weber claimed that the ultimate value-standard for economic inquiry must be *Staatsraison*—the interests of the state.[15]

It might seem odd that Weber should defend science as both neutral and in the service of the state, but that was precisely the positivist conception, the conception articulated in the Vienna School and in Wittgenstein's conception of philosophy as a purely formal science, one that can do nothing but judge the logical consistency of propositions. It is the conception criticized in the Frankfurt School as "instrumental reason"—science which purports to serve no goal but which *de facto* serves the interests of those who manage to initiate research in the first place.

The political function of Weber's neutralism can be seen in the empirical research he undertook for the Social Policy Association. From 1909 to 1911, at the height of the debate over values, Weber was involved in a survey of industrial workers' attitudes at the suggestion of his brother, Alfred Weber. The intent of the survey was to combine questionnaire data with factory records in order to gain some sense of workers' attitudes toward their work. In fact, the survey never made much progress, as most workers refused to fill out the forms. Interesting for our purposes, however, is that Weber defended his project as a neutral piece of research: "The issue is *not* how social conditions in industry are to be 'assessed,' or whether the situation in which large-scale industry places workers today is satisfactory or not, or whether anyone, and if so who, should take the 'blame' for any unsatisfactory aspects, or what could or should be done to improve it and in what way." True research should concern itself with none of these things:

> No, it is exclusively a matter of the unbiased, objective statement of facts and the ascertainment of their causes . . . The whole problem at issue—it does not seem superfluous to stress this to my colleagues too—is, politically speaking, a totally *neutral* one by its very nature. From this it follows, for example, that when a researcher meets complaints from the workers about conditions in factories (wages, conduct of foremen, etc.), this would *not* within the terms of the present survey concern him as a practical "issue" on which he would have to pronounce judgment: rather it would be taken into consideration simply as a phenomenon attendant upon certain technical, economic, or psychological transformations whose progress it is his business objectively to *explain*.[16]

Science was to be neutral, but by not challenging the purposes to which this research was put, the scientist ended up serving the interests of those who had defined "the terms of the present survey."

Weber himself did of course hold clear values concerning the nature of German society. In accordance with his conception of national interest as the ultimate standard of value, Weber judged economic classes according to their fitness to rule. Neither the Junker aristocrats nor the working-class Social Democrats were, in his view, fit to rule. In 1894 he warned, "It is dangerous, and in the long run contrary to national interests, if a class in economic decline holds political power in its hands. It is even more dangerous, however, if those classes into whose hands economic power and therefore political authority is shifting [namely, the industrial proletariat— R.P.] are politically immature. Both are threatening Germany at this time and, in truth, they provide the keys to the present danger of our situation." Weber felt that none of Germany's classes was competent to rule. The SPD was "less politically mature" than the "clique of journalists who lead the party want to make it seem"; Germany's landed elite had fossilized into a conservative and unimaginative social force. The middle classes were hardly any better. The German *Bürger* had lost interest in politics after the national unification in 1871—a time when, at least for this group, "history appears to have come to an end." With neither the conservative aristocracy nor the urban proletariat fit to rule, and the middle classes uninterested in politics, Weber saw little hope for Germany's political future. "An immense task of political education lies before us, and this must remain the ultimate aim of our science."[17]

Weber greeted the outbreak of war in 1914 as one sign of hope for Germany. Consistent with his usual spirit of *Realpolitik,* Weber argued that force was the last argument of any policy; he trusted the leadership of the army and worried only that the intervention of the Kaiser might ruin the show. With the war in full swing, Weber shared with his friends his excitement. On October 15, 1914, he wrote to Tönnies: "This war, even with all its ugliness, is great and wonderful; it is a worthy thing to experience."[18] Six months later, at the age of fifty and now in charge of reserve military hospitals in Heidelberg, Germany's foremost sociologist had not yet changed his mind. Writing to his mother, Weber noted that though "this great and wonderful war" promised to pass him by, this did not matter. For "we have proven that we are a great and cultured people [*grosses Kulturvolk*]: human beings who live amidst a refined culture, who then stand up against the horrors of war (which is hardly an achievement for a Senegalese Negro!) and then return from war, basically decent, like the great majority of our men— this is genuinely human." Even with the death of his friend, Emil Lask, and his own younger brother, Karl, Max continued to believe the war was for a good cause. On September 4, 1915, he wrote to his mother, assuring her that her son's death on the front had been "a beautiful death, in the only place today where it is worthy of a man to die."[19] Weber subsequently argued for German military bases as far as Warsaw and even further north; for the Western front, he advised that the German army occupy Liège and Namur for twenty years.

Weber occupied a position somewhere on the center-right of the German political spectrum, typical for an academic of his time. He voted National Liberal following in the footsteps of his father, who had been active both as a councillor for the city of Berlin and as a member of the German Reichstag. As already noted, Max abandoned the nationalist Pan-German League only in 1899, protesting that his voice was not heard. He campaigned for the coal magnate Karl Baron von Stumm-Halberg in the Saar on a platform supporting legislation authorizing the arrest of trade union leaders in the event of strikes. Early in his career, Weber supported Bismarck's extension of the Anti-Socialist Laws on the grounds that the Social Democrats "by their manner of agitation were going to compromise fundamental institutions of public life." The only alternative Weber saw to banning the Social Democrats was to restrict certain freedoms of speech, assembly, and association in German society as a whole. The young Weber considered this latter a more egalitarian solution to the problem: it was preferable, as he put it, "to muzzle everybody than to put some in chains."[20]

Democracy for Weber was only one among several means to ensure the strength of a nation. Though he considered monarchy the most appropriate form of government for Germany, he asserted that he would not support the Kaiser's war if it were fought "for anything but national goals"—not, in other words, if it were fought over, say, whether it was better to be ruled by parliament or a monarch. Weber made this clear in a letter to his wife, Marianne, where he claimed that he didn't "give a damn what *form* of government a country has, so long as it is run by competent politicians and not dilettante incompetents such as Wilhelm II." (Effective government for Weber exhibited two elements: an impersonal and politically neutral bureaucracy, and a personal and charismatic leader.) Weber was to maintain this position. In a letter to Friedrich Crusius of December 26, 1918, the German sociologist wrote that despite the war and the fatherland's defeat, he still believed in the "indestructibility of Germany" and considered it "as never before, a gift from heaven to have been born a German."

After the war, despite his support for the new Democratic party, Weber's view of democracy remained a very qualified one. In an interview with Erich Wilhelm Ludendorff (where Weber tried to get the German general to take the blame for the war and resign), Weber described his conception of democracy as one where "the people choose a leader in whom they trust. Then the chosen leader says, 'now shut up and obey me.'

People and party are no longer free to interfere in his business. Later the people can sit in judgment. If the leader has made mistakes—to the gallows with him!" To which General Ludendorff (later an avid fan of Hitler) is said to have answered, "I could like such democracy!"[21]

Weber's distrust of social democracy was shared by many liberals of the Bismarck and Wilhelmine era, people for whom socialism was feared as the greater of alternative evils. Eugen Richter in 1878 advised that liberals, faced with the choice between social democrats and conservatives, should not hesitate to choose the latter. Weber himself feared that socialism would burden the economy with the same kind of bureaucracy that encumbered the political order, and this he did not want. A romantic and an individualist, he resented state appropriation of exclusive rights to bear arms or administer political structures. Fearing the encroachment of state bureaucracies into independent economic activity, Weber warned that Germany was witnessing the "dictatorship of the bureaucrat, and not of the worker," on the march.[22] (Weber in his contribution to the *Werturteilsstreit* had argued that the Social Democrats, unlike the parties of the middle, had no genuine *Weltanschauung*.)

Weber was one of the earliest to recognize the growing bureaucratic character of the Social Democratic party in the years prior to the First World War. But it was not just his distrust of bureaucracy that led him to oppose Marxist-inspired socialism. Against historical materialism he argued that history is more complex; his *Protestant Ethic and the Spirit of Capitalism* sought to show that religious ideals can be as consequential as material forces. Against the idea of class struggle he argued that competition among national and ethnic groups was ultimately more important in shaping history. The essence of capitalism lay not in the system of buying and selling labor for a wage, but rather in the extent to which social and economic life has been "rationalized"—that is, mediated through various forms of calculation. (Unlike Marx, Weber believed that rationalization—like bureaucracy and alienation—was an ineradicable part of "modernity," and not simply a feature of capitalist economic formations). In Lukács's terms, Weber attempted to de-economize and spiritualize the phenomenon of capitalism.[23]

Weber's sense of compromise and *Realpolitik* allowed him a certain ideological flexibility as political winds shifted in the early years of the twentieth century. As the Social Democrats emerged as Germany's most powerful political party, Weber took a more conciliatory line and hoped for a bourgeois cooptation of the party. In 1907 he pointed out that conflicts were brewing between the party leadership and more radical elements of the party. He also predicted that if the SPD hierarchy chose to support the war, then severe difficulties could be expected from the radical wing of the party. As it turned out, Weber's predictions were realized in

1914 when the party leadership voted to support mobilization, and again in 1918 when the SPD formed a coalition government with the Catholic Center party, spawning the "great schism" of the party into the revolutionary Spartacus (communist) party and the more reformist SPD.[24]

In October 1918, as workers' soviets were set up throughout southern Germany, Weber was invited to join the Heidelberg Workers' and Soldiers' Soviet. But Germany's leading sociologist refused to support the revolution, calling it "a kind of narcotic for the people before the great misery begins." Worried that Bavaria might secede from the Reich, or that the communists might get the upper hand, Weber cautioned that if either of these should come to pass "the Americans, whether one likes it or not, would have to be called in to restore order." Weber had predicted as early as 1909 that the SPD might have difficulty controlling the German military; he also feared that a socialist government without bourgeois support would be isolated economically and might even prompt a military coup d'état.[25] Still, Weber was prepared to support certain steps toward economic reform in postwar Germany, including nationalizing mines and insurance companies and restricting the political influence of industrial magnates.

It is sometimes said that the ideals of objectivity or neutrality in the social sciences represent a borrowing of physical science methods into social thought. But certainly in the case of Weber (and Weber is not alone in this regard), this is only partially true. Weber rejected the *physique sociale* of Condorcet, along with the positivism of Auguste Comte, in favor of an "interpretive sociology" in which statistics played no part and for which the ideals of sympathy and understanding were the central methodological principles. Weber's thought shows an interesting development in this regard. In 1879, at the age of fifteen, the young Max Weber wrote a paper for his parents which Marianne Weber subsequently reproduced in her *Life of Weber*. Max wrote that "It is no more possible for a people to deviate completely from its initial orbit than this is possible for heavenly bodies, assuming that no external disturbances are present to modify the position of the stars." By the age of twenty-three he had apparently rejected such naturalistic notions. Writing to Emily Baumgarten in 1887, Weber argued that science and ethics were as different as positive and natural law—as different "as the law of the state that the murderer must be punished by death, and the law of nature that all bodies are attracted to one another."[26]

Throughout the rest of his career, Weber remained a critic of facile attempts to incorporate physicalist notions into social life. Weber was sharply critical of attempts to found a "science of value," whether that be based on Ostwald's "energetics," Gumplowicz's "racial worth," or Darwinian "fitness." In the early years of the twentieth century, Weber lambasted Wilhelm Ostwald for his claims to have discovered "The Energetic

Foundations of Cultural Science." The flaw in the work of an otherwise notable scientist, he argued, was in elevating abstractions specific to a particular discipline into absolutes supposedly valid for all spheres of life. It was not possible, Weber argued, to reduce all of human goals and intentions to special instances of some abstract ethical principle such as "energetic relations."[27]

Ostwald's attempt to derive moral principles from thermodynamics Weber considered to be a typical intellectual weakness of the day. Social Darwinians tended to err in similar fashion, as did their "equally objectionable" anti-Darwinian, pacifist counterparts. Each had an ethical formula, such as L. M. Hartmann's imperative to "act so that by your action social cohesiveness is preserved," Ostwald's "act so as to minimize entropy," or Petzoldt's "act so as to maximize stability." The central error of such systems was their attempt to incorporate physical science methods into the social sciences. In a review of an article submitted by a psychoanalyst for publication in the *Archiv*, Weber accused the author of having illegitimately dabbled in *Weltanschauung* and of writing therefore not as a "naturalist" but as a "moralist." Weber expressed his exasperation with scholars who take some technical advance and derive from it vast implications in the moral sphere:

> Until now, after *every* new scholarly or technical discovery, whether it be "meat extract" or the most abstract discoveries of the natural sciences, the inventor has felt himself called upon to discover new values and to be the reformer of "ethics," just as, for example, the inventors of modern photography feel called upon to be the reformers of painting. But we do not have to feel obliged to wash these apparently inescapable diapers in our *Archiv*.[28]

Weber rejected on similar grounds Lujo Brentano's attempt to link the Weber-Fechner law (stating that one's sensitivity to a particular impression diminishes as sensory load increases) with the economic principle of diminishing marginal utility—a connection not uncommon among economists of the day and one which Fechner himself intended in his *Das Psychophysisches Gesetz*.[29]

Weber not only rejected physical science models for the social sciences, he made it clear that his ultimate defense of value-neutrality rested on precisely this rejection. For Weber, the distinctive method of historical understanding *required* that the social scientist remain neutral. Weber is known today as a pioneer in the so-called *Verstehen* (sympathetic understanding) method in social theory—proposed as an alternative to statistical-analytical investigation. Knowledge of the social world, in this view, is best acquired through a certain sympathy with one's subject—whether this be a historical text, workers in a factory, or patterns of authority. "One needn't be Caesar to understand Caesar," Weber asserts, borrowing this expression from

Georg Simmel.[30] The method of *Verstehen* differs from natural science methods in that the goal is neither prediction nor control but rather a certain identification with one's object of study. In his *Economy and Society,* Weber claimed that through this method it was possible to accomplish something in the social sciences not possible in the physical sciences—namely, the subjective understanding of the behavior of individuals. This subjective understanding, constituting "the specific character of sociological knowledge," differs in certain essential respects from how we know in natural science. Understanding in the social sphere implies an effort to discover meaning and intent. This contrasts with, say, biology, where the goal is not to "understand" the behavior of cells but to discover functional relationships.[31]

At first glance, it might appear that the principle of sympathetic understanding contradicts the principle of neutrality. How can one sympathize without taking sides? In Weber's concept of sympathy there is a crucial distinction between *understanding* the intentions of an actor (a methodological position) and *accepting or supporting* those intentions (a political stance). Sympathy in this view does not imply support, the methodology does not imply a politics.

Weber's approach to the human sciences was based on Heinrich Rickert's neo-Kantian distinction between the natural and cultural sciences (*Natur- und Kulturwissenschaften*). The world, Rickert had proposed, is a unity, but the questions we pose to it are of very different kinds. Two central categories divide the world: "Natural products are those which grow freely from the earth; cultural products are those fruits of the earth that man himself has planted and nourished."[32] And the methodologies specific to each of these realms must be different, given the different characteristics of each. Social or cultural science cannot help but touch upon problems of values; Rickert agreed here with Dilthey that valuation (or intentionality) is one of the distinguishing aspects of what it means to be human. The social and historical sciences must therefore concern themselves with human intentions; the goal must be to create a certain sympathy for these intentions, seeing others as they see themselves. This does not mean that social scientists must make value judgments. Social science cannot avoid being "value-related" but it can and indeed must, if it is to be scientific, avoid involving itself in practical moral or political judgments.[33]

Rickert recognized that selectivity is necessary in all scientific work, that the part of nature to which scientists devote their attention, compared with that which they ignore, is vanishingly small. Knowledge is not *Abbildung* (accretion) but *Umbildung* (revision), and the end result is always a simplification. Rickert also recognized that the very *fact* of scientific research presumes that certain aspects of reality are interesting or typical or worth investigating. Weber followed Rickert in this view that scientific investigation always presumes the value of knowing what one sets about to investigate. He agreed with Rickert that "there is no science free of suppositions" (a popular slogan at this time), and that one's interest in a particular problem is always a product of one's values: "what becomes the object of research, and how far that research reaches into the causal chain—this is determined by the values of the individual researcher and the values that dominate his age."[34] According to Weber, this was true for both the natural and the social sciences. But it was especially true for the social sciences, for it was here that our interests are touched most closely.

Weber's conception of a distinctive methodology for the social sciences, the goal of which would be to understand and interpret human action in terms of its subjective meanings, differentiated him from many of his more reductionist colleagues and has won him a reputation as a critic of positivism. The method of *Verstehen* has become a pillar of modern sociological and historical methodology, especially in phenomenology, ethnomethodology, cultural anthropology, and historical sociology. It is important to recognize, however, that Weber shared with his empiricist-positivist colleagues a set of assumptions on the nature and limits of scientific discourse in at least three areas. First, the *Verstehen* method Weber proposed represented an attempt to develop an empirical methodology that would allow the description and analysis of social phenomena without resorting to theological, metaphysical, or moral or political judgment. Despite proposing distinctive methods for the social sciences, Weber's *Verstehen* nevertheless maintained, in common with the natural sciences, the goal of neutral and apolitical knowledge. Second, the kinds of generalizations achieved in such a science were not so different from those in the natural sciences. Weber's concept of "ideal types" (of legitimacy, authority, dominance, and so forth) was designed to allow the kind of abstractions in social science that had proved so successful in the physical sciences (as, for example, the use of models by Hertz, Helmholtz, or Mach). Though he objected vociferously to naturalist and positivist strains of social theory, Weber nevertheless agreed with those positivists who believed that the task of social theory was to interpret the world, not to change it.

And finally, Weber shared with the positivists an implicit assumption that methodology must be closely linked with subject matter, insofar as specific realms of discourse (physical versus social, for example) require methodologies tailored to those realms. A central assumption of Weber's and much of subsequent social science was that the social sciences touch more closely upon affairs of general human interest than do questions in the physical sciences, and that this is why it is harder to achieve reliable and nontrivial knowledge in sociology than in, say, physics. In the social sciences, so this argument goes, since humans (or human behavior) are both the objects of interest and the agents of investigation, subject and object cannot be easily isolated from one another. This is said to impair the objectivity of the investigator in a sense foreign to the physical sciences.

Auguste Comte, the founder of both positivism and sociology, constructed a similar argument as early as the 1830s (see Chapter 11), postulating a hierarchy of sciences according to both historical sequence of discovery (astronomy first, chemistry and biology next, sociology last), and epistemological difficulty (the social sciences are more difficult because our interests become confounded in them). The assumption behind this scheme is that we see most clearly that which is at a greatest distance, but also that it is more difficult to be objective when we focus on humans or their constructs than when we focus on other aspects of nature. The problem with this view is that it identifies too closely a specific realm or terrain with a particular mode of study. It is simply not true that anything that has to do with humans is inherently more difficult to study, or to study reliably, or even to gain consensus about than, say, barnacles or black holes. It depends on the topics investigated, the questions asked. If levels of funding are any index, physics is probably the least "disinterested" of all the sciences. Physics is certainly of far greater interest to governments than is sociology; one could easily argue that the "harder" sciences have become far more enmeshed in politics than the marginalized, "softer" sciences.

Those who argued at the end of the nineteenth century that the social sciences were inherently more difficult or subjective did so partly in order to broaden scientific discussion to include that which cannot be "counted, measured, or weighed." Those doing so were concerned about misunderstanding from two quarters: those within the scientific community who were too quick to draw physical analogies, likening the "orbits" of peoples to the motion of the planets, and those for whom the entire notion of "social science" was a contradiction in terms, given the capricious nature of human behavior. It was a mistake, however, to believe that reliable knowledge of human affairs is necessarily harder to obtain than knowledge of physical or chemical realities.

For Weber, methodological concerns such as these were only part of the equation. Weber's principle of value-neutrality depended upon a complex dialectic of fact and value, means and ends, and it is this dialectic that we must understand if we are to understand why he was so concerned to isolate the world of science from the world of politics and vice versa. The key text here is his 1919 "Science as a Vocation," published one year before his death at the age of fifty-seven.

Weber begins with a discussion of how science has changed the way we view the world. The modern world is "disenchanted" *(entzaubert)*, according to Weber, and science is largely to blame. Science has dispelled the myths of spirits and of morals; science has exposed the brute facts of the material world. But the progress of science does not mean that we understand the conditions of life any better than the Hottentot or the Indian—the average streetcar passenger knows little about what makes the engine go. What it does mean is that we *could* learn about such things in a quite matter-of-fact way should we so desire. It is in this sense that the world is disenchanted: there are no mysterious forces that rule the world. This is what has led us to separate science from meaning, science from values and politics. Plato claimed to have discovered eternal verities of ethics, science, and art in ideas of the Good, the True, and the Beautiful. Renaissance artists could still claim to be describing nature "in itself" in their drawings and sculptures. Even in the exact sciences of the seventeenth century, natural philosophers could claim to be honoring God through the study of his creations, as Swammerdam did, finding proof of God's providence "in the anatomy of the louse."

But what meaning, asks Weber, can there be in a science whose very rationality requires the abandonment of religion? Tolstoy, Weber tells us, has given the simplest answer: "Science is meaningless because it gives no answer to our question, the only question important for us: What shall we do and how should we live?"[35] Weber followed Rickert in his belief that science does presuppose certain values. Astronomy, physics, and biology all presuppose the value of knowing the principles by which nature works; science-based medicine presumes the value of life. But the source of values must lie outside science. Whether the life medicine saves is worth living, or whether the existence of the world described by science is good or not, these questions are not answered, nor even asked, by science. This is equally true for other areas of scholarly inquiry. Aesthetic analysis does not tell us whether there *should be* works of art; nor does jurisprudence tell us whether there *should be* law, or what justice is. Science is a cultural product: "The belief in the value of scientific truth is not derived from nature but is a product of particular cultures."[36] Values nevertheless remain external to the actual process of research—as motors driving the machine; shaping the form but not the substance; in its application but not its essence.

Weber's point—and here we come to the heart of it—is that the meaning or value of science must be sought outside science itself. Science is a profession, and the values of a profession must be sought outside that profession. It is not for the jurist to decide what is just, or which parts of the law are correct. It is not for the doctor to determine at what point a life is or is not worth saving. These are ethical questions that lie beyond the sphere of law and medicine. (Weber also states that the art critic examines the social, psychological, or historical origins of a work of art without judging whether that work belongs "to the world of heaven or the world of hell.")[37] It is in this same sense that values have no place in the academy.

Weber provided two very different reasons for this imperative of neutrality. There was first of all the question of academic freedom and censorship. Weber worried that scholars were not in fact free to express their opinions on matters of vital national interest, questions such as whether the policies of the monarchy in matters of war and diplomacy were correct ones. Indeed "the most decisive and important practical-political value questions of the day [1917]—precisely these questions are barred from the lecterns of German universities by virtue of the contemporary political situation." And thus "in view of the fact that those questions bearing most decisively on Germany's vital practical-political interests must be avoided, it is the duty of the man of science to remain silent not just on these, but also on value-questions which upon which he is so freely encouraged to expound."[38] If one cannot debate politics freely, then one should not debate politics at all.

Interestingly, as his own political consciousness evolved in the first two decades of the century, Weber's use of neutrality also changes, reflecting the changing political balance from monarchy to democratic socialism. In 1904 and even in 1909, neutrality for Weber was primarily an instrument to criticize socialism—that "movement without a *Weltanschauung*," that "dictatorship of the bureaucrat." By 1917, however, Weber had accommodated his views to the more moderate, parliamentary socialism he saw on the horizon. Neutrality now became a means of attacking the monarchy and its academic sycophants who tolerated discussion of values only insofar as it did not challenge state policy. This brings us to the second of Weber's major reasons for value-neutrality.

Weber's principle of *Wertfreiheit* was not something he invoked abstractly or at random. Nor was it simply a means of defending the autonomy of science against its critics. For Weber, there is also the curious sense that the isolation of values from science is the way things *ought* to be, given what we mean by morality in the first place. The point of "Science as a Vocation" is that science must free itself from values, but not out of any disrespect for values or politics—quite the opposite. Value-neutrality represents a critique of *scientism*—a desire to preserve a realm of values untouched by the authority of technical expertise. Science must be *value-free* to guarantee that values will remain *science-free*. Science must be separated from politics in order to free the products of "subjective will" from the straightjacket of objective determination. That is why politics "is out of place in the lecture hall." For while taking a political stance is one thing, passing it off as scholarly analysis is quite another. "The prophet and the demagogue do not belong together on the academic platform." Science is (and should remain) a technical activity dedicated to the elucidation of factual relationships and not "a series of omens and revelations issued from seers and prophets, or a collection of the reflections of wise men and philosophers on the *meaning* of the world." This, he says, "is an ineluctable *given* of our historical situation." For "wherever the man of science comes forth with his own values, full understanding of the facts ceases." The task of the teacher is "to share with students one's knowledge and scientific experience, and not to imprint upon them one's personal political views."[39]

Neutrality, in short, was to provide protection against those wanting to pass off political opinion as established science. Weber believed that values were to be resolved not by science but by struggle; he did not want to see "problems of earthshaking import, of great intellectual scope, the greatest problems that can move the human breast, reduced to objects of a technical field *(Fachdisziplin)* such as political economy."[40] Reductionistic "intellectualism" of this sort constituted one of the greatest obstacles to the pursuit of the religious life: "Redemption from the rationalism and intellectualism of science is the fundamental presupposition of living in union with the divine."[41]

In his 1917 essay on "The Meaning of 'Ethical Neutrality' in Sociology and Economics," Weber did recognize that it was possible to distinguish the question of whether science *is in fact* value-free from the question of whether it *should be* value-free. The former, he points out, is a question for historical and sociological investigation. To the latter question he gives a solid "yes"—values should not enter the classroom. He also adds to this, however, a certain qualification. If a scholar was going to make value claims in the classroom (or in his scientific work), then he should at least admit them. The greatest danger was not from those scholars whose politics were already clear, such as Treitschke and Mommsen, but rather from those who disguised their politics behind a cloak of science or empirical fact.

Weber wanted scholars to distinguish clearly when they are describing facts and when they are promoting a cause. But he also leaves his reader with the impression that the divorce of science from ethics is not a *political* choice, but rather a division etched in the nature of things themselves. (Ironically, Weber's defense of neutrality employs both political and ontological arguments: recall again his assertion that it is an "ineluctable *given* of our historical situation" that science should restrict itself to factual relationships.) Here we begin to see some of the politics of Weber's position—politics, ironically, of a natural order, of what can and cannot be achieved through human efforts, of what is and is not legitimate in university and national politics. For the unteachability of human values—and the dualism of fact and value at its root—he considered to emerge from the irreconcilable nature of all conflicts of value: "The advocacy of practical or ethical goals in the name of science is meaningless in principle, because the various value spheres of the world stand in irreconcilable conflict with one another." It was not for science to define "the public interest"; any attempt to do so must be doomed from the start, given the irrationality of the values upon which interests are based. The individual is the sole source and arbiter of his values; theory cannot (must not?) call this into question. Science may determine the facts, but when it comes to values a man must "follow what his God or demon tells him,"[42] keeping his science and his politics as separate as possible, not pretending to base the one upon the other, not peddling the one in the guise of the other.

Weber was a pessimist, and neutrality constituted what he saw as a courageous realism in the face of the impossibility of scientizing morals or moralizing the sciences. The neutrality of science, like the subjectivity of values, was the sad but inescapable consequence of the rift between the subjectively good and the objectively true, the realm of man and the realm of nature. This was a grim view, one that earned him the honor, in Max Scheler's terms, of being "the most perfect representative of the times"—and these were not the best of times.[43]

Yet here Weber has misunderstood the thrust of Tolstoy's critique of science. For the fact that science gives no answer to the greatest problems of life is, for Tolstoy, not cause for compromise but cause for alarm. If *our* science gives us no answers, it is not because answers are impossible but because our science is the wrong one. The science that is, is not the science that must be. In typical neo-Kantian fashion Weber has confused the empirically real and the historically necessary.[44] Listen again to Tolstoy:

> Men have to live. To live they must know how to live. And all men—whether ill or well—have always found this out and, in conformity with this knowledge, have lived and moved on, and this knowledge of how men should live has, since the days of Solomon, Moses, and Confucius, always been regarded as science, as the science of the sciences. It is only in our day that the science of how to live is not at all a science, and that only experimental science which begins with mathematics and ends with sociology, is the real science.[45]

Tolstoy's view of the proper scope of science was alien to Weber, as to most of the other champions of the ideal of value-free science.

Weber's Society for Sociology never managed to remain within the realm of purely neutral theory. Shortly after the 1910 meeting Weber, according to his wife, complained bitterly: "To Hell with the Society—for apart from the pretty meetings, nothing will come of it." By the second meeting of the society in 1912 Weber had stepped down from its governing board, complaining that

> I participated in the founding of this society, only because I hoped to find here a forum for value-free work and discussion . . . but at the Berlin meeting of 1912 every speaker, with only one exception, broke with the society's statutory guidelines [requiring that members abstain from making value judgments—R.P.], which is evidence to me of the

impossibility of implementing the principle. If only these gentlemen, not *one* of whom could avoid troubling me with their infinitely arbitrary, subjective "valuations," could have simply kept these to themselves. I've had it up to here with endlessly trying to implement like some Don Quixote the apparently impossible principle [of value-freedom—R.P.] and creating painful scenes.[46]

Neutral science may have failed in Weber's mind, but it succeeded in at least one sense. Generations of subsequent scholars would be taught that neutrality was an ideal toward which one should strive, even if unattainable in practice. (The young Viennese positivists, for example, turned to Weber's 1904 defense of value-neutrality for their own conception of science.)[47] By the time of Weber's death his principle of *Wertfreiheit* had caused a sensation, and not just in academe. It was discussed at meetings of political parties, and was even raised in the German parliament. Paul Honigsheim reported in 1921 that Weber's principle of value-freedom had come under scrutiny in the Berlin Investigative Committee exploring the causes for Germany's losing the war.[48] How could the ideal of value-neutrality acquire such notoriety? What was at stake?

The relation of science and value is a political problem. That is, any particular conception of their relations involves value judgments and cannot be neutrally determined. It is no less "political" to claim these are separate than it is to claim they are indissociably joined; the relation of science and politics is something we invent, more than it is something we discover. Historically, the ideal of neutral science and the corollary principle of subjective value are doctrines specific to a liberal world view, a world view sensitized to the dangers of censorship but also to threats to the autonomy of science posed by industry and military forces. Value-neutrality springs from the scientist's quest for autonomy; value-subjectivism springs from the moralist's fears that bureaucratized expertise will erode personal liberties.

Missing from most discussions of the origins and nature of value-free science is this connection between the principles of neutral science and subjective value, on the one hand, and the liberal theory of the state and economy, on the other. Yet the one cannot be understood apart from the other. Neutral science reflects a liberal vision of knowledge and the politics appropriate to its discovery; subjective value is central in the liberal ideal of the individualistic nature of ethical knowledge. Science in this view is objective—that which is true, in Weber's words, "even for the Chinese."[49] (Compare the assertion by Tönnies that sociologists should regard the phenomena of social life "as if they occurred on the moon.")[50] Values by contrast are personal and subjective, varying in time and place. Science represents the unerringly known, values the eternally unknowable; science moves in the public realm of demonstrable experience, while values reside in the private realm of arbitrary desire.

It is important to appreciate this dependence of the ideal of neutrality on a particular conception of an impermeable boundary separating science and values. The reasoning in this runs as follows. Science, narrowly conceived as an empirical collection of facts and theories, can tell us nothing about the moral world, what is good or bad. Ethical truth, narrowly conceived as personal preferences, tells us nothing about science or the natural world. Ethical truth must therefore come from outside the realm of natural or social science. Scientific truth in turn must lie outside the realm of ethics and politics—outside the realm of reflections on the nature of justice or the good life and how we ought to go about realizing these.

In the *Werturteilsstreit,* sociologists associated the realm of values with the unknowable and excluded values from science on this ground. Werner Sombart typified this train of thought in his discussion of the difficulties in measuring social well-being. It is not possible, Sombart argued, to determine in a scientific sense whether it is better to live in the city or the country, to drink or not to drink, to build a church or use the funds for other purposes. The farmer, the prohibitionist, and the faithful will arrive at one set of preferences; the city dweller, the drinker, and the atheist will arrive at another. Science cannot decide such things, he says, for they rest ultimately in the realm of subjective tastes: "I can say only the following: every value judgment is in the last analysis anchored in the personal world view of man. And personal world views always rest upon a metaphysical basis, which has to do with spheres which lie outside the empirical world, spheres whose depths the knowledge of man can never plumb."[51] Sombart vowed to exclude values from science "so long as there is no scientific proof whether blondes or brunettes are prettier"; Weber used this same metaphor to illustrate his position.[52]

It would be possible, of course, to say much more about the view of values implied in Weber's and Sombart's argument. It is a conception that pervades European social theory at the end of the nineteenth century, and it is one we find in subsequent social thinking as diverse as Vienna School positivism, the economic philosophy of Lord Robbins or the Chicago School, and Boasian anthropology. The structure of the argument is simple. A limitation in the power and task of science is derived from the supposedly "subjective" nature of moral judgments. Science is then required to limit itself to objective methods, for it has been declared impotent in the realm of morality.

Max Weber's ideal of "arbitrary and subjective value" emerges in social and economic theory of the late nineteenth century as obverse and reverse of the doctrine of neutral science. In fact, it is possible to argue that these two principles come to form the backbone of the dominant conception of the relations of science and ethics in twentieth-century Europe and America. Together, the dual ideals of value-free science and of science-free values have become central elements in the liberal conception of the place of science in society.

Part Three

The Legacy of Neutrality: Positivism and Its Critics

Physicalism knows no "depth," everything is on the "surface."
—OTTO NEURATH

All of science would be superfluous, if there were no difference between things as they appear and things as they actually are.
—KARL MARX

Once the rockets are up, who cares where they come down? "That's not my department," says Wernher von Braun.
—TOM LEHRER, 1965

11

Catholicism without Christianity

The value-neutrality celebrated by German sociologists in the early years of the century did not last—at least not in Germany. Weber died in 1920, and with him died the most vociferous and stalwart defender of value-free science. War, revolution, and starvation did not set well with the serenely passive view of science put forward by the sociologists. The 1920s sees the growth of movements to politicize the sciences on the basis of class, race, or nationality. Calls for "proletarian culture" (and proletarian science) or "Aryan science" (versus Jewish science) begin to chip away at the liberal ideal of science separate from, even when in the service of, political interests.

The question of values lay at the heart of what many in the 1920s perceived as a "crisis of science"—a crisis identified variously with technical specialization, philosophical relativism, unreflective materialism, mindless bureaucratization, or the decline of humanistic education. The phenomenologist Edmund Husserl criticized science that excluded human intentions in favor of a focus on "positive facts"; there was also the practical question: where was science in Europe's time of need? Siegfried Kracauer suggested that the crisis of science was a crisis of objectivity: if objectivity required the elimination of value judgments, then how could one hope for science that was more than a mass of disconnected facts? And if one allowed oneself to be guided by explicit value premises, how could one avoid the charge of subjectivity?[1]

Sociologists debated these questions at great length. Erich von Kahler traced the crisis of science to the Weberian ideal of value-neutrality, denouncing *Wertfreiheit* as an insult to classical ideals of the good life pursued through intellectual virtue. Weber, in his view, was simply rubber-stamping the substitution of technical expertise for classical ideals of scholarship. Science may have "disenchanted" the world, but it had also brought us a world that fetishized science. Weber's defenders (Arthur Salz, for example) argued that the much-praised education of "the whole man" was an elitist anachronism that had been superceded by modern scientific specialization.[2] Others responded to the crisis by making the social relations of science an object of scholarly study. Max Scheler began a series of studies on what he called "the sociology of knowledge" (see Chapter 15); Karl Mannheim followed Scheler in calling for sociology of knowledge to succeed the discipline of epistemology. Underlying these various endeavors were several questions: What are the politics of science? Are scientific struggles reducible to social struggles, or can intellectuals stand above the fray? If they can, does this mean they should?

Most of these debates were silenced—at least in Germany—by political events. The rise to power of German fascism in 1933 brought the collapse of most German sociological institutions. In the first year or two of fascist rule, many of the sociological institutions established in the early decades of the century were either destroyed or forced to emigrate. Marxist and/or Jewish publications were the first to go: the Frankfurt-based *Zeitschrift für Sozialforschung* moved its offices to Paris in 1933, resuming publication with Max Horkheimer as editor. A number of non-Marxist and non-Jewish institutions were also suppressed. The Forschungsinstitut für Sozialwissenschaft, founded April 1, 1919, closed its doors on March 31, 1934; with it, the *Kölner Vierteljahrshefte für Soziologie* ceased publication. The Society for Sociology was disbanded in 1934, and by 1938 two-thirds of Germany's sociology professors had been dismissed. By 1939 the entire field of sociology had been branded a "Jewish discipline" *(jüdisches Fach)*.[3]

Even the conservative Social Policy Association did not survive for long under the Nazis. In October 1933 Werner Sombart, president of the association since the autumn of 1932, moved to dismiss professors Lederer and von Mises from the governing council of the association in accordance with the *Gleichschaltungsgesetz* requiring that leaders of state-affiliated bodies be "of Aryan or related blood." On June 30, 1935, Sombart moved to dissolve the association altogether, defending this action with the argument that the governing council had been unable to establish satisfactory relations with the Nazi party. Germany's senior sociologist complained that the association had become a platform for social democratic "party pronouncements": the association was "a sick old man who long ago should have died." The time for discussion was past; in the new state, political will would be built not by "the detour of influencing public opinion" but by more direct means, in accord with the Nazi Führer principle *(Führer-*

prinzip). Sombart made his views on this quite clear: "Thank God it has come to this!" It was better to close the association than to watch it "vegetate."[4]

Nazi efforts to politicize the sciences are important in understanding the revival of the ideal of value-neutrality—especially in countries hostile to the Nazi movement. In the 1930s and 1940s, philosophers and scientists concerned about right- or left-wing efforts to politicize the sciences sought to distance themselves from these movements by proclaiming science value-free; as in the first decade of the century, the ideal of neutrality is put forward in response to concrete political movements. But this, as we shall see, is only part of our story. In the mid-twentieth century, neutrality is defended on a variety of other grounds as well. Positivist philosophers defend the neutrality of science as part of a larger attack on metaphysics; positivist sociologists defend the ideal of neutrality to establish social science on a methodological par with physics. Positivist lawyers defend the neutrality of legal science as part of an attack on "natural law"; positivist economists defend the neutrality of science in order to rescue it from reformist zealots. Positivist ethicists elaborate that great twentieth-century irony—"value-free ethics."

In the years between 1920 and 1960, logical positivism emerges to dominate leading centers of Austro-German, and later Anglo-American, philosophy of science. But positivism is a very loose term—one that has meant different things to different people. In his history of this movement, Leszek Kolakowski distinguishes six elements central to the doctrine of modern positivism: (1) an empiricist criterion of truth and verifiability (empiricism); (2) a related denial of the distinction between essence and appearance (phenomenalism); (3) a radical separation between the tasks of science, on the one hand, and those of morality, metaphysics, religion, art, and politics, on the other; (4) an assumption of the unity of science and of scientific method; (5) an assumption that science is the best way of knowing; (6) a dualism of fact and value, along with the associated doctrines that science is value-free (neutralism) and that values cannot be derived from facts (subjectivism).[5] Positivism thus presumes an ontology, a metaphysics, an ethics, and a theory of the growth of science. Science in the positivist view is a rational and cumulative enterprise; science grows through accretion of the new and replacement of the old. The "logic" of scientific growth consists in either the inductive generalization of laws from individual observations (inductive-statistical method), or the advancement of theory through conjectures and refutations (hypothetico-deductive method).

Descriptive taxonomies may be useful for introducing a movement, but they do little to explain its origins, persistence, or popularity. Positivism must be understood historically—that is, in light of circumstances prompting theorists to push their thinking in one direction rather than another. Movements emerge in response to situations, to climates of opinion or circumstance, and this is as true of mid-century positivism as it is of turn-of-the-century sociology.

The term *positivism* first appears in the 1831 *Doctrine de Saint-Simon, Exposition,* published in Paris by the disciples of the French utopian socialist and count. (The work also contains the first use of the term *individualisme*.) Saint-Simon's "positive" method was developed most elaborately by his student, Auguste Comte, in his *Cours de philosophie positive* (1830–1842). Comte intended his positive philosophy to contrast with the "negative" philosophies of skeptical empiricism of Berkeley and Hume and the critical philosophy of Kant; positivism was raised against Kant's claim that ours is "the age of criticism, and to criticism everything must submit."[6] Positivism for Comte was a blast against all things metaphysical or theological, following in the footsteps of a long line of critics—including Adam Smith, for whom metaphysics was "cobweb science," and David Hume, who ends his *Enquiries Concerning Human Understanding* with his infamous call to "commit . . . to the flames" books which cannot be shown to have an empirical base.

Positivism for Comte was a sweepingly optimistic, historical (Popper and Hayek call it "historicist") account of the movement of thought through three distinct stages or "theoretical conditions": the *theological,* the *metaphysical,* and the *positive.* In the theological or "fictitious" stage people seek the first and final causes of things, supposing these to be produced by the immediate action of supernatural beings. This most primitive stage of human intellect appeals to a supernatural being or beings, rather than to natural rights (as for the metaphysician) or observational experience (as for the positive scientist). The theological condition is characterized by the rule of armies and of clergy. In the metaphysical stage, by contrast, the goal is to discover the essence of phenomena by reason. The politics and thought of revolutionary France typify this stage. Here we find the rule of parliament, supported by rights and abstract ideals such as *liberté, egalité,* and *fraternité*—a system Comte once called "the mongrel and transitional system of metaphysicians and lawyers."[7] Finally, in the positive stage, one devotes oneself to the search for relationships through observation or experimentation. This is the culminating stage of human knowledge, the stage toward which all of human history has been advancing. The myths and fictions of previous stages are replaced by the positive results of observation and experience; the goal is predictive knowledge—in Comte's words: *savoir pour prévoir.*

Comte's historical cosmology presents both an ontogeny and a phylogeny of human intellectual development. "Each of us is aware that he was a theologian in his childhood, a metaphysician in his youth, and a natural philosopher in his manhood." Ontogeny thus recapitulates phylogeny—the history of the individual follows that of the species (see Table 4). Comte also postulates a hierarchy of the individual sciences, corresponding to the ease with which the various spheres of inquiry progress from one stage to another. Astronomy was the first to reach the positive stage, achieving this already in the ancient world with Ptolemy. Astronomy was followed by terrestrial physics and chemistry and finally physiology and *sociology* (Comte coins this term—he also uses the term *social physics*). The order of succession is from sciences that are more general, simple, and independent to those that are more individual, complicated, and dependent on other realms of knowledge. Natural knowledge, in contrast to social knowledge, is not only more simple, general, and abstract, it is also more remote from human experience.

Table 4 Auguste Comte's stages of intellectual development

Stage	Intellectual character	Political character
Positive (adult)	Discover relationships by observation Agency: invariable relations of succession and resemblance Highest principle: Law	Republicanism: rule of science, duty, and industry; peaceful
Metaphysical (youth)	Discover essence of phenomena by reason Agency: abstract forces, causes Highest principle: Nature	Parliamentary system: rule of rights and abstract ideals
Theological (infancy): Monotheism Polytheism Fetishism	Discover first and final causes Agency: supernatural being(s) Highest principle: God	Feudal, military system: rule of armies and clergy; warlike

The goal of such a system, for Comte, was to regenerate education: knowledge of the laws of thought would facilitate the progress of individual minds through the hierarchy of the sciences. Positive philosophy would also provide the basis from which to extricate society from the "anarchy" left in the wake of the French Revolution. Comte proposed a positivist calendar (where the months progress from "Moses" and "Homer" through "Gutenberg," "Descartes," and "Bichat"); a positivist library (150 volumes of poetry, science, and history); a positive system of worship (the so-called religion of humanity); a *Positivist Catechism*; and a

"Conspectus of Sociolatry" (a schedule of eighty-one annual festivals with "love as the principle, order as the basis, and progress as the end"). *Order and Progress* was the slogan of the movement; positivism would rescue the civilized nations from the revolutionary crisis into which prepositivist modes of thought had cast the world.[8]

Comte's faith that history requires the progressive triumph of science over superstition, order over chaos, has been both criticized and reaffirmed ever since he formulated his ideas in the middle of the nineteenth century. J. S. Mill, after breaking off an early romance with the system, accused Comte of trying to establish a despotism of society over the individual "surpassing anything contemplated in the political ideal of the most rigid disciplinarian among the ancient philosophers." Thomas Huxley labeled Comte's philosophy "Catholicism *minus* Christianity" (or, in a lesser-known refrain, an "incongruous mixture of bad science with eviscerated papistry"). In the twentieth century, Comte's positivism has come under attack as a species of *scientism,* defined by Habermas as "science's belief in itself—or the conviction that we can no longer understand science as one form of possible knowledge but rather must identify knowledge with science."[9]

Comte's, though, is only one among several positivisms. Like any other tradition, positivism is a diverse movement, with its dissidents and stalwarts, its ortho- and heterodoxies. There are positivist theories of law, positivist literary movements, and positivist theories of economics, sociology, and ethics. In the next three chapters we look at positivist views of science because it is within this movement that the ideal of value-neutrality becomes most clearly articulated. The final three chapters look at alternatives to value-neutralism, as we move from the social theory of science to the history of science criticism and the questions of freedom and social responsibility.

12

Logical Positivism

Auguste Comte's positivism represents both a political vision and a phylogeny of the individual sciences. This broader political animus contrasts with the narrower, more cautious version of positivism that arises in the late nineteenth and early twentieth century. The changing character of positivism reflects broader changes in science itself, as science undergoes a dual process of restriction and expansion. On the one hand, the methods and perspectives of science are applied to ever larger spheres of life (the origin of species and the elements of organic matter, structures of society and of the mind, the origins of religious texts and beliefs, and so forth). Yet scientists also become concerned to define more precisely what science is and what it is not. In association with these pressures to define the boundaries of science a new and narrower version of positivism emerges, one that defines the scientific method in essentially negative terms. Science is a method, *not* a realm of inquiry. Science is not metaphysics, or religion, or ethics, art, or politics. Positivists of the early twentieth century—with a few important exceptions—abandon Comte's view of science as a source of global political reform. Ernst Mach and his followers believed that science could help solve social problems, but their claims were more modest. The goal was now to discover methods for advancing the sciences, rather than to establish plans for reorganizing society. If Comte saw scientists as something like captains of industry, subsequent positivists saw them more as servants of industry.

The positivism that emerges in Germany and Austria in the second half of the nineteenth century is associated with the revival of Kant. Neo-

Kantian positivists were agnostic about the ultimate features of the world. Ernst Mach, Richard Avenarius, and Joseph Petzoldt all sought to establish a world view based only on knowledge presented to the senses and interpreted in the light of reason. Mach was one of the more radical in this regard. In his widely read *Science of Mechanics,* the Austrian physicist-philosopher postulated that reality does not exist apart from our knowledge of that reality; all that exists are *sensations.* Mach was an empiricist and an idealist, arguing that there is no difference between essence and appearance, between the real and the (in-principle) observable. If something cannot be known (that is, observed or inferred), it does not exist. Mach criticized the "conceptual monstrosities" of absolute space and time, arguing that these, being ultimately unobservable, can have no real existence. A priori truths, contra Kant, did not belong in science.[1]

For Mach, the central goal of science was to attain a certain "economy of thought" in explaining natural phenomena. A good theory explained as much as possible while assuming as little as possible. What this meant was that anything that could not be derived from sensory experience—all metaphysics, morals, politics, and religion—should be stripped from science. Critics and supporters alike agreed that this was provocative, to say the least. Herbert Feigl called Mach's philosophy, after Kant's, "the most radically anti-metaphysical philosophy of the nineteenth century."[2] V. I. Lenin wrote his *Materialism and Empirio-Criticism* as an attack on Mach's "anti-materialist empiricism." The French philosopher Guizot labeled those who built upon it "the semiotics police"; Karl Popper accused the positivists of having made innocent speech impossible.

Oftentimes in philosophical disputes it is the implications imagined to flow from a particular doctrine that determine whether one is for or against it. Mach's positivism was primarily concerned with physical rather than moral or political problems. The questions that interested him were those concerning the existence of atoms, the ether, and absolute space and time; indeed, in the field of physics, the positivism of Mach aided in the birth of two of the most profound achievements in the history of science: Einstein's special and general theories of relativity.[3]

Mach was not alone in using positivist principles to criticize physical theory. Another was Joseph Petzoldt, one of the leading popularizers of Mach's ideas and founding editor of Germany's short-lived positivist journal, the *Zeitschrift für positivistische Philosophie* (1913–1914). Petzoldt followed Comte and Mach in arguing for a science free of metaphysics, one that would separate the facts of nature from the additives of human thought. Metaphysics, in his view, had been like a veil separating the scientist from nature: covering and distorting it, though never obscuring it entirely. Metaphysical suppositions could be neither proved nor disproved: they were arbitrary fictions or fantasies, without foundation in human experience and thus beyond the pale of science. If meaningful statements could be made only about that which presents itself to the senses, then what could be said about the ultimate reality of atoms, or the ether, or anything else for that matter? Knowledge of reality is always provisional: one can speak of atoms or the ether at best in terms of convenient "fictions" introduced for purposes of calculation, prediction, or control. Positivist principles, according to Petzoldt, had allowed the triumph of relativity theory, electrodynamics, and thermodynamics over metaphysical ideas of absolute space, time, and energy.[4]

An interesting project would be to explore the different consequences of the rejection of metaphysics for the physical and social sciences, corresponding roughly to the separate legacies of Kant in these two spheres. For physical scientists, metaphysics was identified with assumptions concerning the reality of atoms, for example, or the existence of an ether mediating the propagation of light. Positivist prudence required that a Mach or a Kirchhoff or a Mayer reserve judgment on the reality of the wave nature of light; instead one may claim only that light behaves (among other things) "like a wave." The positivist or pragmatic conception of truth as what is reliable (Peirce, Dewey), or present to the senses (Mach), or convenient for purposes of explanation (Poincaré, Petzoldt) served to loosen the bonds of mechanistic philosophy and foster new approaches. Positivist philosophers (Moritz Schlick, for example) were in fact among the first to recognize the epistemological implications of relativity and quantum mechanics.

In the social sciences, however, the abandonment of metaphysics has, at least in one sense, contributed more to the solidification of inquiry than its liberation—in the sense that the restriction of social scientific study to the empirically given often leads one to ignore the historical character of social reality. This is the kernel of truth in Hegelian Marxist claims that empiricist social science is de facto conservative, or that the only constant in "human nature" is that human nature changes.

I do not want to enter into the long-standing dispute over whether social scientific methodology is or ought to be different from that in the physical sciences. One could probably argue that "essence and appearance" are no more difficult to disentangle in the social realm than in the physical or biological realms; that the physical or biological sciences profit as much from historical insight as do the social or political sciences.[5] Clear in any event is that the political resonance of empiricism, materialism, or any other methodology or ontology can be dramatically different in different historical periods. Viennese positivists in the 1920s brandished empiricism against the forces of conservative reaction: Otto Neurath, for example, claimed that metaphysics gave support to groups "close to the church with which today's ruling classes collaborate in considerable measure"; this was why official bodies were reluctant to encourage the teaching of "sociology on a materialist basis."[6] Positivist economists in Britain and the United States were at this same time using empiricism for more conservative ends. The exclusion of values or metaphysics from science can have very different political consequences, depending on the context of debate.

European positivists in the first and second decades of the twentieth century rejected metaphysics in order to remove impediments to the progress of physical knowledge. But value-neutrality is defended from yet another angle at this time, one that reached back to David Hume's suggestion that *ought* cannot be derived from *is*. In 1905 and then again in 1913, Henri Poincaré defended the immunity of science to moral critique by arguing that "Ethics and science have their own domains which touch but do not interpenetrate. The one shows us to what goal we should aspire, the other, given the goal, teaches us how to attain it. So they can never conflict since they can never meet. There can no more be immoral science than there can be scientific morals."[7] It was wrong to have imagined that science might place moral truth beyond all debate or question; equally wrong, however, was it to believe that science could be "immoral." Neither of these—scientistic nor romantic—was correct, and the reason was simple, indeed grammatical:

> If the premises of a syllogism are both in the indicative, the conclusion will equally be in the indicative. In order for the conclusion to be in the imperative, at least one of the premises would have to be in the imperative. Now the principles of science, the postulates of geometry, are and can only be in the indicative; experimental truths are also in this same mode, and at the foundations of science there is not, cannot be, anything else. The most subtle dialectician can juggle with these principles as he wishes, combine them, pile them up one on the other; all that he can derive from them will be in the indicative. He will never obtain a proposition which says: do this, or do not do that; that is to say a proposition which confirms or contradicts ethics.[8]

Poincaré's linguistic proof of the neutrality of science is one that logical positivists would make the centerpiece of philosophical discourse.

In 1922, a number of philosophers gathered around Moritz Schlick in Vienna in what Otto Neurath christened the "Vienna Circle";[9] six years later this same group established the *Verein Ernst Mach*, bringing into existence the most influential school of philosophy in the English-speaking world for nearly half a century. Among its early members were Rudolf Carnap, Otto Neurath, Friedrich Waismann, Herbert Feigl, Philipp Frank, Karl Menger, and Viktor Kraft. Ludwig Wittgenstein and Karl Popper were sometimes looked upon as fellow travelers, but they were not members. With an empiricist ontology borrowed from Locke, Hume, and Mach; a conception of moral discourse borrowed from Kant; a logical calculus borrowed from Bertrand Russell and Gottlob Frege; and a conception of the proper order of society borrowed variously from Bentham, Comte, Mill, and Marx; the group established what for several decades was hailed as the "received view" in Anglo-American philosophy.[10]

In 1929, on the occasion of Moritz Schlick's return from a visiting professorship at Stanford University, Neurath, Carnap, and Hans Hahn published a "manifesto" of the Vienna School under the title "The Scientific Conception of the World: The Vienna Circle," proclaiming what Herbert Feigl would later call "a philosophy to end all philosophies." The manifesto included among its principles an allegiance to epistemology as the central task of philosophy, to science as the single best way of knowing, and to the unity of science as a goal and methodological principle.[11]

The goal of the manifesto was a "unified" science, a science collectively achieved, empirically based, and expressed in propositions agreeable to all. Traditional philosophical problems were either to be unmasked as pseudo-problems or else transformed into empirical problems and thereby subjected to the judgment of experimental science. The positivist analyst was to bring into philosophy the method Galileo had introduced into physics, namely, "the substitution of piecemeal, detailed, and verifiable results for large untested generalities."[12] Logical analysis had revealed a sharp boundary between two kinds of statements: statements of the kind one finds in empirical science, and meaningless statements of the kind the metaphysician makes. "Analysis," the manifesto reads, reveals that the "feelings for life" expressed by the metaphysician "merely express a certain mood and spirit." Metaphysics was to be considered something between "mysticism and crossword puzzles."

The positivist rejection of metaphysics, combined with the view that statements are either "logical, empirical, or meaningless," meant that the task of philosophy was suddenly a much narrowed one. Philosophers were to explore the meanings of terms and the logical relations among propositions. Philosophy, in Carnap's terms, was to be "the logic of science."[13] Many long-standing philosophical problems—the existence of God, the meaning of life, the ultimate nature of truth, beauty, being, or time—were suddenly rendered vacuous, on a par with questions such as "Is seven a holy number?" or "Are the even numbers darker than the odd?" (Carnap's examples for so-called *Scheinfragen*). The so-called value-problem devolved into the question of whether propositions of value could be derived from propositions of fact; positivists proclaimed the neutrality of science and grounded their claim on Hume's radical separation between *ought* and *is*.

This is not to say that positivists did not concern themselves at all with historical questions or the social context of ideas. Comte was anything but apolitical; Mach left his chair in the history and theory of inductive science at Vienna when he was elected to membership in the upper house of the Austrian parliament. The logical positivists, too, did not eschew involvement in political or moral affairs; they simply isolated discussion of such affairs from the realm of science.

The case of Otto Neurath, first author of the Vienna Circle's manifesto, is a revealing one. In the years before the First World War, the young Austrian economist became interested in eugenics, translating (with his wife, Anna Schapire-Neurath) Francis Galton's *Hereditary Genius* for the first time into German.[14] His most important early work, however, was his analysis of the war economy. War economics, in his view, was a science with well-defined laws and principles which, like ballistics, are "independent of whether one is for or against the use of guns." From his studies of the Napoleonic and American civil wars, Neurath discovered certain inefficiencies in economies governed by free and unregulated markets. In the war economy, he observed, market mechanisms are often replaced by more efficient means of distribution according to need or function. Neurath saw the wartime economy, with its centrally planned and efficient methods of distribution, as a model for how a peacetime socialist economy might be organized.[15] Neurath sought to become, in his words, "a social engineer."

After the war, Neurath's interests in state-administered economic planning led him into service with the socialists of Bavaria. On January 23, 1918, he met with Bavarian Prime Minister Kurt Eisner, a leading member of the Independent Socialist party (USPD), to discuss the renovation of the economy. Eisner, as head of the coalition of parties that came to power with the proclamation of the Republic in November 1918 (SPD, USPD, and Bayrischer Bund, most prominently), invited Neurath to give a talk for the Munich Workers' Soviet (*Arbeiterrat*) on "The Road to Socialism—Social and Technical Perspectives." In January 1919, he was invited to draw up a plan for the socialization of the Bavarian economy and then in March, Neurath was appointed a member of the central committee of the Bavarian Soviet, serving also as minister of the Bavarian economy. One month later, on April 7, 1919, the Bavarian Soviet Republic was declared. The Republic was short-lived: on May 1, 1919, troops of the Social Democratic–led Reichswehr occupied Munich, unleashing a rash of bloody reprisals which crushed many people's hopes for the communist movement. The suppression of the Bavarian Republic was part of a larger counterrevolution throughout Germany: in January Rosa Luxemburg and Karl Liebknecht had been murdered by Reichswehr officers, and subsequent attempts to establish soviets in Bremen, Hamburg, and the Ruhr Valley were suppressed on orders of the new Social Democratic leadership. Neurath's positivist stance of neutrality served him well after the defeat: when brought before a court charging him with complicity at the highest levels of the Munich Räterrepublik, Otto Bauer was able to defend his colleague with the argument that he had served merely as a "technician" for the revolution.[16]

Peter Galison has recently shown that Neurath and several other Viennese positivists were also active in the modernist architectural movement to which Walter Gropius in 1919 had given the name Bauhaus. Neurath was convinced that new forms of architecture could help transform the life of the growing urban proletariat; the architectural Bauhaus, like the positivist Aufbau (construction, build-up), was to be accessible, international, free of ornamentation, functional, and cosmopolitan. In 1929, the same year as the Vienna Circle's manifesto, Neurath, Feigl, Carnap, and Hans Reichenbach all lectured at the Dessau Bauhaus, rubbing shoulders with the likes of Wassily Kandinsky and Paul Klee, Walter Gropius and Hannes Meyer. Neurath himself was the designer of Isotype—the simplified picture language that evolved into today's international signs for traffic and communication. His efforts in this area were part of his desire to transcend national and linguistic barriers—in his words, "Words divide, pictures unite."

Neurath's value-neutrality must be understood in this context. The positivists would sweep away the cobwebs of metaphysics the way the Bauhaus did away with decorative ornamentation. By his own admission Neurath was in fact *anti-philosophical*—equating philosophy with the "values," "systems," and "metaphysics" of an earlier and unscientific age. In Galison's terms, the positivists put forward an unphilosophical philosophy, along with their unaesthetic aesthetics and their apolitical politics.[17]

Neurath typified what I have called "neutral Marxism"—the value-neutral conception of Marxism as a science of society enshrined by the Second International largely under the influence of the late writings of Friedrich Engels. Others in the Vienna Circle were not Marxists by any stretch of the imagination. Moritz Schlick was an aristocrat who had done doctoral work in optics under Max Planck in Berlin. He was never sympathetic to the programmatic goals of Neurath and indeed was horrified to find on returning from Stanford that Hahn, Carnap, and Neurath had elevated the circle to a "movement." Ludwig Wittgenstein, one of the heroes of the early circle, was also a political conservative.

Neurath was not the only positivist to concern himself with broader social and political affairs—Carnap, for example, was a social democrat, as were several others. Still, if one surveys the writings of the positivists one is struck by their political naiveté. The journal of the movement, *Erkenntnis*, published continuously from 1930 to 1939, does not mention even once the words "Nazi" or "National Socialism" in all the years of its European publication—this, despite the fact that many writers for the journal (mostly Germans and Austrians) suffered personally and professionally from Nazi policies. In 1934, when the Nazi sympathizer Hugo Dingler attacked the journal as "formalistic" and "bolshevist," Hans Reichenbach could honestly respond that the journal had "nothing to do with politics."[18] Neurath despised Hitler and all that he stood for; his thoughts in this area, though, were largely confined to personal conversations and letters to his friends.

This brings us to the question of the Nazi response to positivism. When philosophers (and especially philosophers of science) consider the history of their discipline in the 1930s or '40s, attention is usually given to figures we would like to remember today. This is understandable: history is rarely written "for its own sake"—the purpose of history is often to highlight movements or ideas that may be of use to the present. Historical memory is highly selective. Traditions that are embarrassing are often neglected, unless of course the point is to illuminate the structure of evil or the march of progress.

In recent years, a great deal of attention has been focused on the questionable activities of certain high-profile philosophers during the Nazi era: the early anti-Semitic writings of Paul de Man have been exposed, as has the life and thought of Martin Heidegger, rector of the University of Freiburg for a time in 1933 and 1934. Interest in this area has generally focused on a few prominent men who managed for a time—after the war—to keep their involvement in Nazi philosophy, literature, or science out of public view. Unfortunately, we do not as yet have any satisfactory studies of the philosophical profession as a whole (rather than of specific leading individuals) under the Nazis.[19] Attention has tended to focus instead on a few "great and evil men"; Nazism is traced to Nietzsche's nihilism, Dilthey's *Lebensphilosophie,* Spengler's relativism, or the demon seeds of Fichte or Hegel.

One reason this topic has been ignored is that it has been viewed as essentially uninteresting. Philosophers have generally assumed that there could not have been anything worthy of the name "philosophy" under the Nazis. Indeed, as one philosopher in the 1960s put it, National Socialism "had hardly sufficient logical coherence to deserve the name of an ideology."[20] Even John Dewey in 1942 asserted that "one can hardly use the word *philosophy* in connection with Hitler's outgivings without putting quotation marks around it." Dewey cited Hitler's "contempt for intellectual measures and for science" except as a means of propaganda.[21]

However repulsive one finds Nazi thought, it is relevant to our discussion for three reasons: first, because the history of positivist thought is intimately tied up with Nazi persecutions; second, because the Nazis present one of the most consequential rejections of the ideal of value-free science in the twentieth century; and third, because it is commonly against the image of the Nazi perversion of science that the banner of neutrality is raised. One cannot understand the ideal of value-neutrality in twentieth-century science without understanding the insults posed to that ideal by Nazi science and philosophy.

With the triumph of Hitler in January 1933, the positivist ideal of neutrality comes under attack from philosophers eager to politicize the sciences, to "return" to *Weltanschauung,* to root the spirit of science once again in blood and earth, racial history, or a host of other causes. Central in the Nazi view of the world was that science is not, or should not be, value-free, but must instead reflect the values or *Weltanschauung* (world view) of the people producing it. For Hitler, absence of a unified *Weltanschauung* had been one of the causes of Germany's defeat in the First World War; restoration of that *Weltanschauung* was to be a sign of her recovery. Philosophers also celebrated the importance of world view. In Hermann Glockner's terms, the *germanische Weltanschauung* was to be the foundation of *deutsche Philosophie.*[22]

Weltanschauung becomes a concept in Nazi science as well as Nazi politics. Gerhard Wagner, for example, head of the Nazi Physicians' League and the most powerful doctor in all of Germany, announced in a speech before the Gesellschaft Deutscher Naturforscher und Ärzte in September 1934 that:

> The powerful spiritual transformation of the National Socialist revolution has not stopped outside the gates of science and the German universities—perhaps to the disappointment of some advocates of an "international" or even "trans-historical" science.
>
> There is no science without suppositions. And there is no longer any *German* science without the National Socialist *Weltanschauung* as its *first* presupposition. In all his thought and action the German scientist—and especially he—must always ask: am I being useful to my people?[23]

"No science without suppositions" (Nietzsche's phrase) was a common Nazi slogan—alongside the view that "the good of the whole comes before the good of the individual," or that Nazism was simply a matter of "applied biology" (Fritz Lenz's slogan of 1931 coinage).[24] Race, *Volk,* and nation defined the whole to which every part of society was to be subordinate; science was to become part of that whole as well. Sciences that could not or would not conform to Nazi ideals did not thrive.

The early years of the Nazi period witness the publication of dozens of books on what it will mean for science to be "volkisch," or German, or otherwise rooted in the ideals of the new Germany. Philosophers trumpet a dramatic reorientation of German science, boasting that the era of neutral or "value-free" science has finally come to an end. Fascist science theorists attack the ideal of neutrality as part of the "apathy" and "tolerance" of liberal democracy in the face of social and racial degeneration. Nazi philosophers criticize the specialization, bureaucratization, and fragmentation of science into separate disciplines, worrying that philosophy has become little more than one among several academic specialties (in Ernst Krieck's terms, a *Fach unter Fächern*). Walter Del-Negro, dozent at the University of Innsbruck, argues that philosophy had become "chaotic" and "impotent" with "little to offer the nation"; it was hardly surprising that the national socialist movement looked less to philosophy than to "history, geopolitics, racial theory, genetics, characterology, and ethnology" for intellectual inspiration.[25]

A number of philosophers did try to capitalize on the new racial currents sweeping the nation. Max Wundt, son of the eminent psychologist, Wilhelm, published an extraordinary book using Hans F. K. Günther's racial categories to correlate the ideas of Germany's leading philosophers with their skull shape and place of birth. Hegel was "nordic-dinaric"; Kant was "nordic with a hint of the dinaric"; Schopenhauer was somehow "nordic-westisch" in his youth, and "fälisch" in his maturity.[26]

More common was the argument that the Jews were to blame for the malaise of German philosophy. Walter Del-Negro, a party member since 1938—the year the Nazis annexed Austria—blamed the Jews not just for the turn toward liberalism and individualism but also for philosophical movements such as neo-Kantianism (Hermann Cohen's Marburger School), phenomenology (Edmund Husserl), neo-positivism (he cites "Carnap et al." though Carnap was not in fact Jewish), *Lebensphilosophie* (Dilthey—also not a Jew), and even pragmatism (it was "no accident that a Jew, Georg Simmel," supported the pragmatic doctrine of relativistic positivism in his early work). Del-Negro worried that the individualism popular in contemporary Anglo-American philosophy threatened to collapse objective reality into a morass of moral and historical relativism.[27]

Hermann Glockner, professor at Giessen, similarly criticized the "subjectivism, intuitionism and relativism" of liberal philosophy; Nazi philosophy was supposed to strive instead for objectivity and universality. Nazi philosophy was more than "mere acrobatics"; true philosophy was not formalist but filled with life; not playful but serious; not intellectual but humane. Philosophy was not a matter of personal whim but a scholarly expression of the *Weltanschauung* of a Volk.[28]

Given the widespread view among postwar Anglo-American social scientists that it was relativism of one sort or another that was to blame for the Nazi debacle, it is interesting to note that Nazi philosophers attacked *Relativismus*. Del-Negro traced historical relativism to Dilthey, but also to Max Weber's conception of values as dependent upon personal experience and the conditions of the time. (Spengler, too, according to Del-Negro, fell victim to a particularly virulent historical relativism.) Del-Negro recognized the value of that part of relativism that stated that science and world views are conditioned by the times; cultural and racial factors did shape science. But relativism was not to go so far as to challenge the very nature of science. Epistemological skepticism reflected the "inner insecurity of the age";

Del-Negro predicted that the Nazis would sweep away this unhealthy rot in German philosophy—the "skeptical relativism," the "exaggerated individualism," the "ill-founded mysticism." National Socialists would restore the wholeness to German philosophy, establishing what he called "the New Synthesis."[29]

Gerhard Lehmann, a philosopher at the University of Berlin, was another who attacked the ideal of value-free science, linking this, like Del-Negro, with the predominance of positivist relativism. In his 1943 *Deutsche Philosophie der Gegenwart,* the most comprehensive survey of German philosophy since Friedrich Ueberweg's 1923 *Grundriss der Geschichte der Philosophie,* Lehmann criticized the ideal of value-free science as part of the liberal ideal of science. Liberals had fragmented "man" into the economic man of neoclassical economics, the political man of Realpolitik, the *verstehende* man of Dilthey's and Weber's *Geisteswissenschaften.* Lehmann acknowledged Weber's analysis of authority, his distrust of the Treaty of Versailles, and his support for Germany's cause in the Great War. But he could not accept Weber's disavowal of any role for value judgments in science, his positivism, his view that the prophet, the philosopher, and the politician belong at church, market, and parliament but not at the university. Lehmann rejected Weber's view that science can set no goals for itself, that those goals must be set from the outside.[30]

Positivism for the Nazi philosopher was symptomatic of a larger philosophical degeneracy. For Del-Negro, positivism (including not just the Vienna School but also "allied movements" such as phenomenology and neo-Kantianism) was "the philosophy of the liberal and Marxist age"; positivism was "the historical correlate of industrialization and urbanization, of pure civilization and of intellectualism." Positivism had split the world into atoms—the ethical atoms of individuals guided only by utility; the ontological atoms of a world where "elements of sensation" define what is real; the political atoms of a nation fractured by competing interests. Del-Negro also criticized what he called "physical positivism" or "physical relativism" (quantum theory, for example—especially the so-called Copenhagen interpretation)—the view of many contemporary physicists that "only that exists which can be measured." Such was the view of "the Jews and their supporters" that absolute motion or simultaneity do not exist because means have not yet been found to measure them.[31]

One should not exaggerate the attention the Nazis gave to positivist philosophy. In one observer's words, the positivists were an "irritation," and little more.[32] The journal *Erkenntnis* continued publication in Germany (by Felix Meiner in Leipzig) long after the rise of the Nazis (until the summer of 1937), while many other journals with Jewish or socialist writers had been closed down or forced abroad. Positivists were careful to separate their politics from their scholarship. As already noted, *Erkenntnis* ignored politics and political movements altogether. The positivists discussed the foundations of physics and mathematics, quantum mechanics and relativity, protocol sentences and criteria of verification. But on the momentous *political* changes of the times—movements shaping both science and society—they were conspicuously silent.

I do not mean to suggest that the positivists did not suffer under the Nazis. Many were Jewish and at least several (notably Neurath and Carnap) were Marxists; neither was tolerated in the new regime. Positivism disappeared from Berlin in 1933, from Vienna in 1938, from Prague in 1936. After 1933, Jews found it difficult to publish in German journals, though the system was "leaky" and articles did occasionally see the light of day.[33] After 1938, with Hitler's annexation of Austria, the "sale of works by the Vienna Circle was forbidden, on political grounds, because among its members there were several Jews, and because activity of the Verein Ernst Mach was considered 'subversive.' "[34] Otto Neurath, Viktor Kraft, Friedrich Waismann, Herbert Feigl, Edgar Zilsel, Philipp Frank, and Karl Popper were all from Jewish backgrounds; positivism at least in Germany and Austria did not survive the rise of Nazism. Hans Hahn died a natural death in 1934. Neurath fled first to Holland (in 1934) and then to England (in 1940); Waismann fled to England in 1938. Felix Kaufmann fled to New York, where he found refuge in the New School for Social Research's "University in Exile." Hans Reichenbach fled to Istanbul (in 1933, with Richard von Mises) and then to America (in 1938), joining the faculty at the University of California. In 1936, Rudolf Carnap and his students Carl Hempel and Olaf Helmer left the German University of Prague, fearful of the growing fascist movement among both students and faculty. That same year Moritz Schlick was murdered by a deranged former student as he left the classroom at the University of Vienna; A. J. Ayer notes that local papers implied in their coverage of the incident that logical empiricists "deserved to be murdered by their pupils."[35] (Schlick was not Jewish.) Edgar Zilsel in 1938 fled to England and in 1939 to America, where he continued to work on the sociology of science until his 1944 suicide in a Berkeley bathroom.[36] Positivism thus shared with Gestalt psychology and Freudian psychoanalysis the distinction of having arrived on the shores of the United States and Britain as a consequence of the persecutions of Hitler's Germany.[37]

In America, positivist traditions merged with various strains of analytic and pragmatist thought cultivated since the early decades of the century. Reaction to the abuses of science under the Nazis (and in Soviet Russia) produced a strident feeling among many Anglo-American philosophers that science must remain value-free, that the politicization of science implies its destruction. Against the backdrop of Nazi science, it is not hard to understand the widespread postwar sympathy for the ideal of value-neutrality. Leaders in the pro-Nazi Deutsche Gesellschaft für Philosophie (Bruno

Bauch, for example) had argued that values were "objective"; Nazi scientists had sought to distinguish "valuable" and "worthless" citizens, lives "valuable to the community" and lives "not worth living." Philosophers seeking to ground a common ethics had spoken of race as "the ultimate principle of value";[38] racial hygienists had described the mentally or physically handicapped as "worthless incurables" *(unheilbar Wertlosen)*. Social theorists had classified Jews and gypsies as parasites, vermin, or bacilli; other groups were classed as genetically predisposed to crime, vagrancy, or anti-social behavior. When genocidal policies translated these words into action, one can understand the revulsion much of the world felt toward "value-laden" discourse. In perhaps no other period have scientists been so obsessed with calculating the "value" of an individual or a race, with discriminating between useful and useless, healthy and diseased, worthy and unworthy.

It is against this setting that postwar efforts to make science value-free must be understood. The so-called value-question becomes an object of endless methodological controversy in American social science soon after the war. The *Journal of Social Issues* devoted a full volume to the question in 1950; Ray Lepley organized a symposium which produced nearly five hundred pages devoted to "the place of value in a world of facts." It was still the centerpiece of discussion at the fifteenth Deutscher Soziologentag at Heidelberg in 1964, though Richard Rudner suggested as early as 1953 that arguments concerning science and values had already reached what e. e. cummings would have called "The Mystical Moment of Dullness."[39]

Different people had very different reasons for saying that science is, or ought to be, value-free. For those unwilling to acknowledge the complicity of scientists in Nazi crime, the ideal of value-neutrality served as a convenient cover for whatever invidious work may have been done in the name of anthropology or racial science. Value-neutrality allowed one to argue that genuine science could not have been implicated in the crimes of the period, despite substantial evidence to the contrary.[40] For anti-Nazi critics, by contrast, the tragedy of German science was in having allowed itself to become politicized; German scientists had failed to remain value-neutral, and it was this failure that was responsible for the excesses of the period.

Nazi politicization, though, was only one among several concerns in the immediate postwar period. There was the widely debated question of "the role of science in a democracy"; there was the question of "the responsibility of the scientist in a nuclear age." Especially among American physicists, the immensity of the power of atom gave an entirely new dimension to debates about the subordination of science to political, and specifically military, ends. *Business Week* warned in September 1946 that "the odds are getting better all the time that pure scientific research will become, permanently, a branch of the military establishment."[41] A *Bulletin of the Atomic*

Scientists editorial warned that the development of scientific methods of war had "chained science tighter than ever to the national state"; the postwar subordination of science to the military was "an evil thing."[42] Norbert Wiener wrote that the continuation in peacetime of measures taken during the war would lead to "the total irresponsibility of the scientist, and ultimately to the death of science." Philip Morrison cautioned that continued militarization would lead to increased secrecy, deterioration of teaching, and restrictions on travel, publication, and the personal background of scientists.[43] The editor of *Harper's* wrote that science had become politics and that it would henceforth be difficult for scientists to return to their laboratories and forget the outside world.[44]

Among social scientists, though, there were different concerns. Curiously, the most persistent question asked by social scientists in the decade following Auschwitz and Hiroshima was not *How should scientists be morally responsible?* but rather *Is social science possible?* This latter was the key question in George Lundberg's 1947 *Can Science Save Us?*; it was also the motivation behind J. A. Passmore's 1949 essay, "Can the Social Sciences Be Value-Free?" and Joseph Schumpeter's 1948 presidential address to the American Economics Association on "Science and Ideology."[45] The question was most commonly formulated as follows: physics and chemistry are objective sciences and have proven themselves a source of great power—power that may be used or abused. But can the social sciences achieve similar objectivity? Or is social science more like, as E. A. Hooton put it, "a Welsh rabbit—not really a rabbit at all!"

The most common answer to this question, by the 1940s and '50s, was that objectivity was possible in social science but far more difficult to achieve than in, say, physics or chemistry. Value-neutrality was needed precisely *because* objectivity was so difficult to attain. Robert Bierstedt, professor of sociology at the University of Illinois, took this position in a 1948 article for the *Bulletin* of the American Association of University Professors:

> Natural scientists are never a part of the problem they investigate. They begin their inquiries from a point of vantage which is wholly external to their data. The stars have no sentiments, the atoms no anxieties which have to be taken into account. Observation is objective with little effort on the part of the scientist to make it so. The binocular parallax may of course confound the observation of sidereal motions, the temperature of the chemist's body may introduce an extraneous variable into a precise experiment, and individual differences may produce constant errors . . . But these errors are known, their effects can be measured, and their influences can be subtracted from the total score.

Bierstedt was disturbed by the fact that while "physicists, financiers, and philosophers, publicists, politicians and poets" all had something to say about the atom bomb, sociologists had remained conspicuously silent.

More than this, though, he was bothered by the fact that people apparently *expected* sociologists to have something to say. Sociology had been confused with socialism, social reform, social work, or social prophecy. Like Weber, Bierstedt believed that the key to objectivity was value-neutrality. Sociology was "a science or it is nothing": "And in order to be a science it must diligently avoid all pronouncements of an ethical character . . . The sociologist, in company with his brother scientists, has taken seriously the famous remark of Jeremy Bentham, that the word 'ought' ought never to be used, except in saying it ought never to be used." Like Weber, Bierstedt conceded that values could not be excluded entirely from science—values enter into the choice of problems, the order in which they are solved, and the social use of the conclusions. (This Weberian view can probably be described as the orthodoxy of the age). Still, sociologists were not to be considered "saviors of society"; the boundary between sociology and social reform had to be carefully guarded. The pure science of sociology would ultimately prove of use but only after time. Meanwhile the sociologist, like other scientists, was to be "a fugitive from urgency."[46]

Joseph Schumpeter was another who believed that the central issue in the "values question" was whether social science was possible. Unlike Bierstedt, Schumpeter believed that advocacy did not necessarily compromise objectivity: as he put it, "advocacy does not imply lying." Like Bierstedt, though, Schumpeter believed that objectivity was a far easier thing to achieve in the physical than in the social sciences. In his 1948 presidential address to the American Economic Association Schumpeter postulated that, in the physical sciences, the influence of ideological bias does not extend beyond the choice of problems. This was because "logic, mathematics, physics and so on deal with experience that is largely invariant to the observer's social location and practically invariant to historical change: for capitalist and proletarian, a falling stone looks alike."[47]

Many of those who objected to value-neutrality in the postwar era did so on moral or political grounds. American pragmatists fought a long-standing battle with ethical relativists, methodological individualists, and Weberian positivists on the grounds that a "means-ends continuum" provided a more satisfying solution to the value problem than the irrationalist ethics and neutralist science coming into vogue. John Dewey hoped that the experimental method could serve as a model for how political disputes might be resolved. Adjudication of social differences by free exchange of views was for him "the very heart of political democracy"; this same method could be said to approximate "the method of effecting change by means of experimental inquiry and test: the scientific method." Pragmatists rejected the Weberian notion of values in eternal and unresolvable conflict: in 1949 Felix Kaufmann suggested that people disagree less about good versus evil than about which among various goods is better.[48]

For liberal pragmatists, the urgent moral problems of the postwar era left little time for value-neutrality. In 1950, at a symposium devoted to the question of values in social science, George Geiger of Antioch College suggested that "the alleged fact-value dualism has already been pierced by contemporary events," referring primarily to the fact that "these first five years of the Atom-Hydrogen Age have already made anachronistic the traditional compartmentalizing of human experience into what we do and what we ought to do." Even Robert Bierstedt conceded that "the atom bomb has no respect for ivory towers."[49]

Conservatives opposed value-neutrality for quite different reasons. For Leo Strauss at Chicago, the question at hand was not so much the *possibility* of a social science as the *objectivity* of ethical values. Strauss was a libertarian champion of natural rights and natural law for whom Weberian value relativism smacked of nihilistic positivism. (Weber indeed he characterized as "the greatest representative of social science positivism.") Strauss rejected Weber's assumption that the heterogeneity of fact and value precluded the possibility of an evaluative social science. Equally misguided was the view that there could never be knowledge of right and wrong. The Weberian categorical imperative boiled down to "thou shalt have preferences"—an "ought," Strauss suggested, intrinsically guaranteed by the "is."[50]

Strauss accused Weber of taking historicism to such an extreme that the values of every age appeared as genuine as any another. Such an assumption falsified reality, failing as it did to distinguish spiritual excellence from spiritual corruption: "Would one not laugh out of court a man who claimed to have written a sociology of art but who actually had written a sociology of trash?" Strauss pointed out that Weber himself was inconsistent, insofar as his own writings were filled with value judgments. Weber talks about "swindling," "mechanized petrifaction," "manly beauty," "works of art of the first rank," "specialists without spirit or vision and voluptuaries without heart"—as he must and should. Value-neutral social science was dishonest, given that values barred at the front door will usually reenter through the back. Strauss suggested that those who deny the possibility of objective evaluations must have difficulty understanding historical periods during which such a possibility has been widely acknowledged. The positivist would presumably have to regard the people of such an era as deluded, which is hardly a value-neutral attitude.

Strauss agreed with the pragmatists that science can and should be something broader—broad enough to include political philosophy and moral judgments. The value-neutralist was either naive or immoral: what could one expect from scholars (such as Lundberg) who argued that science should be as useful to fascists as to communists? Such a person left himself open to the suspicion that his activity "serves no other purpose than to increase his safety, his income, and his prestige, or that his competence as a social scientist is a skill which he is prepared to sell to the highest bidder." The value-neutralist was either an opportunist or a philistine conformist.[51]

Frankfurt School Marxists raised a different set of objections. Max Horkheimer in 1964 recalled his first contact with the ideal of value-freedom at a 1919 lecture of Weber's on the young Soviet Union. He had come to hear about the momentous changes of the day and instead was disappointed to hear only about "finely balanced definitions of the Russian system, shrewdly formulated ideal types . . . It was all so precise, so scientifically exact, so value-free that we all went sadly home . . . we thought that Max Weber must be ultraconservative." Horkheimer later recognized this was an over-hasty judgment—that Weber was as much a problem for conservatives as for leftists. Still, he argued, Weber should have realized that sociology cannot be completely divorced from philosophical obligations. Questions of values have to be raised, even *after* research topics have been chosen.[52]

Jürgen Habermas complained in similar fashion that the goal of Weberian social science was too narrowly focused on the production of technical or interpretive knowledge. Weber had recognized the value of the instrumental and the hermeneutic functions of knowledge, but not the critical. Weberian *Wertfreiheit* was a "scientific-political" presumption, analogous to "the political claim for protection of the decision-making authority from the competent expert's arrogance." Habermas followed Wolfgang Mommsen in reminding his American colleagues that, in Germany, the Weberian Caesar-like leader-democracy—the "leader with a machine"—must be judged with hindsight. One could not forget that among fascists there were those—like Carl Schmitt, the political theorist and SS man—who were "natural sons" of Weber.[53]

The single most probing critique of value-neutrality in the early postwar decades was Alvin Gouldner's 1961 essay, "Anti-Minotaur: The Myth of a Value-Free Sociology." Gouldner pointed out that value-freedom had become a hollow catechism, an excuse for no longer thinking seriously about social problems. Value-freedom had become a kind of "password" or token of professional respectability, the "gentleman's promise that boats will not be rocked." The concept was also clouded by a myriad of ambiguities:

> Does the belief in a value-free sociology mean that sociologists are or should be indifferent to the moral implications of their work? Does it mean that sociologists can and should make value judgments so long as they are careful to point out these are different from "merely" factual statements? Does it mean that sociologists cannot logically deduce values from facts? Does it mean that sociologists do not or should not have or express *feelings* for or against some of the things they study? Does it mean that sociologists may and should inform laymen about techniques useful in realizing their

own ends, if they are asked to do so, but that if they are not asked to do so they are to say nothing? Does it mean that sociologists should never take the initiative in asserting that some beliefs that laymen hold, such as the belief in the inherent inferiority of certain races, are false even when known to be contradicted by the facts of their discipline? Does it mean that social scientists should never speak out, or speak out only when invited, about the probable outcomes of a public course of action concerning which they are professionally knowledgeable? Does it mean that social scientists should never express values in their roles as teachers or in their roles as researchers, or in both? Does the belief in value-free sociology mean that sociologists, either as teachers or researchers, have a right to covertly and unwittingly express their values but have no right to do so overtly and deliberately?[54]

Gouldner was one of the few to realize the importance of understanding the historical context within which Weber had first proposed the ideal of value-neutrality. In Gouldner's words, Weber's intent was not so much to *amoralize* as to *depoliticize* the German university (Weber is the "Minotaur" in the account—half-man, half-beast, split between reason and faith, knowledge and feeling, head and heart). But the American university of 1961 was quite a different place from that of Germany in 1911. In America, the two major parties were quite similar to each other, much more so than the parties of Germany fifty years earlier. Many American academics had little interest in politics. The need of American universities, as of American society more generally, was thus very different: "the national need is to take the lid off, not to screw it on more tightly." Political will was needed to prevent nuclear war, but political will was difficult to mobilize so long as scholars were "tranquilized" by the doctrine of value-neutrality.

Gouldner recognizes that value-neutrality at one time served certain positive functions. Value-freedom allowed the emancipation of social science from politics (in Germany) and religion (in the United States); it permitted a partial escape from the parochialisms of personal background and the narrow immediacy of journalistic interests. Value-neutrality helped to maintain both the cohesion and the autonomy of the modern university, especially the autonomy of the newer social sciences. But value-freedom also had come to mean a willingness to sell one's skills to the highest bidder. Value-freedom allowed one to believe it was no worse to work to spread disease than to cure it; value-freedom meant that there was nothing wrong with doing market research for tobacco companies to help them sell cigarettes. The cash value of neutrality was to stifle social criticism—value-freedom translated into the commandment: "Thou shalt not commit a critical or negative value judgment—especially of one's own society." This imperative was dangerous in a world where, though there was not yet "a sociological atomic bomb," there was still the "brainwashing of prisoners of war and of housewives with their advertising-exacerbated compulsions."

If sociologists concerned themselves exclusively with the technical proficiency of their students, rejecting all responsibility for their moral sense, or lack thereof, "then we may someday be compelled to accept responsibility for having trained a generation willing to serve in a future Auschwitz."[55]

Gouldner does not ask why Weber's was the language incorporated into American value-neutralist rhetoric of the 1950s and 1960s. The fact that a distinctively Weberian idiom was adopted can be traced to several factors. There was a long-standing tradition of social scientists learning from Europe; the translation of Weber's methodological essays (including the 1904 "Objectivity" and the 1917 "Ethical Neutrality" essays) in 1946 and 1949 gave English-language academics a new and scientific-sounding idiom for these debates.[56] The popular use of the English expressions *value-free, value-freedom,* and *value-neutrality*—all translations from the German—date from this time. Cold War McCarthyism also fostered a general retreat from politics, and Weberian *Wertfreiheit* seemed to fit with calls for (or celebrations of) "the end of ideology." As we shall see, the methodological principle of *Wertfreiheit* also merged easily with the emotivist ideal of nondemonstrable values and the economic individualist principle that every mind "is inscrutable to every other mind." Weber himself could be conveniently interpreted as either the chief representative of the "distinctive methodology" school (and hence a bona fide "anti-positivist"), or the foremost advocate of value-neutrality (and hence an unassailable scientist). Weber also had the virtue of not being Karl Marx, and this was certainly an asset in the 1950s.

Postwar positivists articulated a very narrow view of science. Fearful of anything that might threaten the integrity or autonomy of science, positivist philosophers and social theorists embraced a narrow empiricism that made it difficult to moralize about social evils. Positivist philosophy ossified into a dogmatic and technical program, obsessed with logical formalism and the supposed clarification of terms. Philosophers emphasized the linguistic analysis of meanings, ignoring the larger social origins and consequences of scientific ideas. Philosophy was "the logic of science"; scientific theories were treated as deductions from systems of axioms. The task of the philosopher was to "rationally reconstruct" scientific theories by clarifying the relations of fact and theory, criteria of verification or confirmation. For both philosophers and social theorists, science was a body of knowledge and a well-defined set of methods. Positivist historians celebrated the works and lives of individual geniuses, the founders of present-day scientific theories; positivist social scientists embraced a value-free ideology that blunted impulses for social criticism.

There is another discipline, though, in which value-neutrality was especially consequential. We turn next to the field of economics, where value-neutrality has played a central role for more than a century.

13

Positive Economics

The Weberian ideal of *wertfreie Wissenschaft* arose as part of an attempt to defend the integrity of science against those who would use it to further some political agenda. Value-freedom was Weber's response to censorship: if certain values were to be excluded from the classroom, then all must be barred. Value-freedom was a liberal response to conservative domination of the German university and to radical demands (from outside the university) that "sociology that is not socialism, is nothing."

Weber and most of the participants in the *Werturteilsstreit* were trained as economists, and it has been in economics that the ideal of value-neutrality has received its most elaborate defense—perhaps because this is the social science that is proclaimed to be most closely allied to the physical sciences; perhaps because the science of "how one makes a living" is so obviously tied up with matters of vital human interest. But no doubt also because *value*—in the various senses of that term—is at the center of economic analysis. Why is a thing worth what it is? Is the value of a thing its price? There is also the methodological question: should economists, as scientists, make judgments concerning the structure of economic institutions or the distribution of the social product? Is value-free economics possible? If so, is it desirable?

Since the end of the nineteenth century economists have distinguished *normative* and *positive economics,* economics as an art and economics as a science, economics that makes value judgments and economics that does not. In John Neville Keynes's terms, "positive economics" (he coins this expression) deals with "what is," normative with "what ought to be." In

Milton Friedman's formulation, positive economics is "in principle independent of any particular ethical position or normative judgements." Economics is a neutral "tool kit" to which policymakers may turn, whatever their political values.[1]

Such ideas have never been without their critics. Critics have long pointed out that efforts to make economics a purely "positive science" are fraught with difficulties. Values are implicit in the starting points and ending points of inquiry; values structure the priorities of science and the assumptions guiding investigations. Ironically, those economists most emphatically proclaiming their neutrality have also been among those most actively involved in forming policy. (Friedman's Chicago School of economics was criticized for many years for advising several of the strong-arm rulers of South America, especially Chile's Augusto Pinochet.) Something is obviously awry if advocates of conservative, Chicago School economics can portray their work as "value-neutral." How has the ideal of neutrality come to occupy such a special place in economic theory? Why is it put forward so often and with such fervor—more so even than in, say, physics or chemistry?

The key event in the triumph of value-free economics is the "marginal revolution" at the end of the nineteenth century. In the marginal revolution classical is supplanted by neoclassical economics, political economy becomes economic analysis, and mathematics (especially the calculus) is established as the sine qua non of frontier economic science.[2] Price in the neoclassical view is traceable still to scarcity and desirability; the emphasis now, though, is on cost and utility at the margins. The revolution also sees the emergence of a new theory of value: the so-called *subjective theory*, the idea that the inherent worth of a thing stems not from its cost of production, or any inherent utility (usefulness), but rather from the personal desires of individuals to consume and to work.

Orthodox economists even today recognize the emergence of the principle of subjective value as the key shift in the rise of modern economics.[3] As is often pointed out, the subjective theory allowed a reorientation of the focus of economic analysis from cost of production to the desirability of consumption. From another point of view, however, the rise of this theory may be seen as an ideological weapon against both Marxian economics and the utilitarian ethic of redistribution. The subjective theory allowed value to be seen as a purely individual phenomenon, dependent on nothing but individual whim or personal preference.

As we shall see, the conservative implications of the subjective theory of value are further drawn out in the middle decades of the present century—especially in England and the United States. In the 1930s efforts to define, once again, an ethic of redistribution (this time based on A. C. Pigou's welfare theory, among other things) come under attack from positivist economists who find the attempt based on "metaphysical value judgments." The ideal of subjective value, as obverse and reverse of the principle of neutral science, becomes more than a methodological principle: especially in the wake of the Great Depression of the early 1930s it becomes a weapon in debates over how and whether the economic product should be redistributed. Here, as in the case of turn-of-the-century German sociology and mid-century positivism, the ideal of value-neutrality is proposed in response to concrete political movements. To appreciate this, we must look briefly at the history of theories of value and the character of political economy prior to the sea changes of the final decades of the nineteenth century.

Until sometime in the second half of the nineteenth century, theoretical debates in political economy centered on what Adam Smith had called the "Origins and Causes of the Wealth of Nations." Exchange value in the classical tradition was the product of nature, utility, and labor; prices were determined by the scarcity of an item and (especially) the effort required to produce it. Theory concentrated on the origins of prices and (after David Ricardo) the division of the social product among the various classes of society.

Concerned as they were with problems of the origins of wealth, political economists asked: why are some people poor and others rich? Why do some earn rents and others profits, while still others are paid wages? Marx, in particular, asked how it was that returns from the sale of a product (exchange value) are divided between the laborer and the capitalist. A key thesis of Marx (Engels called it one of his two great insights—the other being the materialist interpretation of history) was that workers are not in fact paid the full value of their labor; rather, the owners of labor power (capitalists) extract from the value of a commodity a "surplus value" which, when added to wages, forms the value of a commodity. The value of a commodity is determined by the labor required to produce it; and the ratio of surplus value to exchange value constitutes the *degree of exploitation*—which will vary according to how much below the "true value of labor" wages may be driven. The revolution predicted (and called for) by Marx was intended to return this surplus value to workers—or, more precisely, to eliminate surplus value altogether along with the capitalist class that expropriates it. Ownership of labor by capital would cease, and workers would finally earn the full value of their labor, reaping what they had sown.

The radical implications of this analysis were obvious. If the value of a good is the labor contained within it, why should a worker not receive in wages the equivalent in value of the labor he expends to make it? What does the owner of labor (the capitalist) contribute to production, and why should he deserve to extract anything at all from the process?

By the end of the century, economists had proposed an answer. The key concept in the marginal revolution of W. S. Jevons, Karl Menger, and Léon Walras was that the value of a commodity is "subjective"—that is, determined ultimately not by the labor it contains (or any other objective measure) but rather by how much people crave it (demand). In Archbishop Whately's words, pearls do not fetch a high price because men have dived for them; on the contrary, men dive for them because they fetch a high price.[4] Value is "subjective" in the sense that the price of a good may have little or nothing to do with either its use value or the work required to produce it; the key is simply its scarcity and its desirability. In Jevons' terms, "value depends entirely upon utility" (specifically, on the final degree of utility, or utility at the margin)—understood not as the use to which an object may be put but rather the pleasure it brings to the person who consumes it.[5]

In the context of economic theory, the doctrine of subjective value represented a response to political imperatives raised by both the Marxian theory of value and the utilitarian ethic of redistribution. Yet in many discussions of these problems this political dimension is lost, and the issue appears to rest on nothing more than the technical question of the factors relevant to the determination of prices. Marx's labor theory of value is assimilated to that of Ricardo and the entire classical corpus, misrepresented as the doctrine that the price of a good is determined by how long one must labor to make it.[6] In fact, the labor theory in Marx's sense represented an attempt to find a measure for the degree of exploitation of labor in a simplified system of classical capitalism, where workers sell their labor for a wage, capitalists do no work, and surplus value is extracted from the sale of commodities. The Marxian theory of value was tied to a larger theory of history—namely, that workers will not stand idly by as wages are driven down while machinery takes over increasing parts of the productive process (the so-called immiseration thesis and the principle of the "increasing organic composition of capital").

One should keep in mind that two questions were central in most premarginalist discussions of political economy: the technical problem of the origins of value, and the political or moral problem of how that value should be distributed. For utilitarians, a rather strong case for a more egalitarian distribution of wealth appeared to follow from the fact that the satisfaction one derives from a particular good depends on how much of that good one already has. In his 1848 *Principles of Political Economy,* John Stuart Mill had argued that if value is an objective good, then one must consider the social consequences of it going to one man rather than another. If the value of a good is the pleasure it brings to the person who consumes it, then who will benefit more from the consumption of, say, five pounds' worth of goods—a rich man or a poor man? The pleasure brought to the poor man will certainly exceed that brought to the rich man, Mill argued. The consequence of this line of reasoning was that distribution ought, morally speaking, to be as even as possible, for only then will the utilitarian ideal of the greatest good for the greatest number be achieved.[7]

Mill's conclusion depended upon both a utilitarian theory of value, according to which the social good consists in the greatest good for the greatest number, and the notion that one can *compare the pleasures* of one person with those of another. (Bentham once tried to quantify these pleasures as *utils*.) Conservatives in the marginalist tradition, by contrast, rejected the notion that the pleasures of individuals can be compared with one another in any quantitative fashion. W. S. Jevons, in the introduction to his 1871 *Theory of Political Economy* (a founding text in the marginal revolution), stated that nowhere had he attempted to compare "the amount of feeling in one mind with that in another"—indeed he saw no means by which such a comparison could be accomplished. He conceded that the sensitivity of one mind might, for all we know, "be a thousand times greater than that of another"; but he also denied that one could ever prove such an assertion. Every individual mind was "inscrutable to every other mind, and no common denominator of feeling seems to be possible." The weighing of motives was therefore always to be confined "to the bosom of the individual."[8]

The argument of the marginalists that one man's pleasure cannot be compared with another's depended upon a new conception of utility. Utility in an earlier sense had meant the use to which an object might be put: utility had a pragmatic implication, as in the Baconian, utilitarian ideal of science. For the marginalists, however, the term acquires a more restricted meaning. Utility becomes the *pleasure* derived by an individual from the consumption of a good or goods, whether or not that good be of any objective use. Utility becomes subjective in the sense that it is not an inherent property of a good: "the ore lying in the mine, the diamond escaping the eye of the searcher, the wheat lying unreaped, the fruit ungathered for want of consumers"—these things have no utility at all. Utility reflects nothing more than "the will or inclination" of an individual toward a good or an act.[9]

Jevons's famous marginalist insight followed from this definition—namely, that the utility (desirability) of an object depends on how much of the object one already has. Water is just water; but water is of far greater value to a man dying of thirst than to a man already sated. The incremental value of a commodity declines as one obtains more of it. Jevons formulated this as a law: "the degree of utility varies with the quantity of a commodity, and ultimately decreases as that quantity increases."[10] What matters most is the *final degree of utility*—the marginal utility—of a particular good: this is what ultimately determines the (exchange) value of a good.

The consequence of import to us, however, is that Jevons's marginalism (and specifically, his individualist/subjectivist conception of utility) represented an abandonment of the basis on which Mill's utilitarian ethics was built—namely, the comparability of wants. If the pleasures brought to two different individuals by the consumption of some good cannot be compared, then how can one say that a redistribution of wealth will bring about a greater good for a greater number? Who can say that the pleasure brought to the poor by a given increment of food is greater than that brought to the rich? Perhaps some people's tastes are more sensitive than others; perhaps the rich know better how to enjoy themselves than the poor (F. Y. Edgeworth suggested this in his 1881 *Mathematical Psychics*). If preferences are ultimately subjective, then how can anyone really ever know how preferences should be compared? The marginalist principle of subjective value was a brilliant refutation of Mill's argument for redistribution; it also served as a persuasive counterpoint to welfarist arguments that state support for pensions, health care, and so forth would serve the social good.

For economists consolidating the marginal revolution, subjective value theory implied that a whole host of questions formerly belonging to political economy could now be abandoned. By 1890 it is widely assumed among neoclassicists that questions of psychology, physiology, and sociology should not be part of economics. Neoclassicists considered questions of the structure of the economy or alternative forms of distribution to lie beyond the sphere of positive economics. Jevons confessed his political ignorance ("about politics, I confess myself in a fog . . . I prefer to leave *la haute politique* alone, as a subject which admits no scientific treatment"); V. K. Dmietriev wrote that "the inclusion of the social moment in the system of contemporary abstract-deductive theoretical economy is impossible in principle."[11] Orthodox economists worried less about the political order and less about alternative forms of distribution. This did not prevent them from advocating free trade or laissez-faire; indeed, free trade and laissez-faire came to represent the only political ideals consistent with value-free science and subjective value.

In evaluating the politics of the marginalist movement it is important to distinguish the origins of the theory from its impact. Marginalism was not initially conceived as a bourgeois response to Marx, though it was soon realized that the theory was admirably suited for the task. Eugen von Böhm-Bawerk's *Capital and Interest* (1884–1889) attacked the Marxist theory of value; a reviewer of the book in 1891 claimed that the Austrian marginalist had shown that "the labour theory is false, that the labourer in general does receive his whole product at its valuation at the time when he receives his wages," and that laborers therefore have no claim on the interest of capital.[12]

The most notorious neoclassical defender of capitalist economic order, though, was John B. Clark, whose "marginal productivity theory" stated unambiguously that the wages workers receive are in accordance with the marginal contribution labor makes to productivity. Clark's theory was clearly a response to the socialist concept of exploitation. In 1890 the Smith College professor of economics asked: "Is present society rooted in inequity and does it give to a few men the earnings of many? Is a robbery in which three quarters of the human family are victims perpetuated and legalized by the 'capitalistic' system? These things we shall know if we can find the forces that govern the rate of pay for labor."[13] In 1899, in his widely read *The Distribution of Wealth,* Clark made it clear that the rightness of the socialist conception of exploitation was at stake:

> The indictment that hangs over society is that of "exploiting labor." "Workmen" it is said, "are regularly robbed of what they produce. This is done within the forms of law, and by the natural working of competition." If this charge were proved, every right-minded man should become a socialist; and his zeal in transforming the industrial system would then measure and express his sense of justice.[14]

Fortunately, though, value-neutral analysis revealed that labor is not in fact exploited under capitalism. The system is "honest"; labor is like any other productive input insofar as wages are determined (under perfect competition) by the productive value of labor at the margin. Labor is not robbed. Free competition "tends to give to labor what labor creates, to capitalists what capital creates, and to *entrepreneurs* what the coordinating function creates"; this is a matter of "pure fact." From such pure facts Clark derives an unmistakable apologia: marginal productivity theory proves "the right of society to exist in its present form."[15]

Marginal productivity theory was widely (and correctly) interpreted as stating that workers "get what they deserve"; critics have more than once taken it as proof of the inherent conservatism of the neoclassical method. Neoclassicists were never entirely of one voice, though, on the political significance of marginalism. Some, like Clark, were outright defenders of a free-market status quo; others were more paternalist or reformist. Mark Blaug has suggested that, in many ways, classical was a more effective tool than neoclassical economics for defending wealth and privilege. What could be more despairing than Ricardo's "iron law of wages," according to which the natural price of labor tended toward that minimal sum required to allow workers "to subsist and to perpetuate their race, without either increase or diminution"?[16] Very few neoclassicists, though, were socialists. John Maloney points out that it is hard to name a single neoclassically minded socialist economist in England prior to 1914 (with the possible exception of Pigou). Alfred Marshall was never fond of Clark's marginal productivity theory; still, even he agreed with his non-marginalist colleagues (Joseph Nicholson, Henry Macleod, and Henry Sidgwick, for example) that one was either a "socialist" or an "economist"—but one could not be both.[17]

If marginalists almost to a man were anti-socialistic, they were divided over the question of whether governments ought to take steps to lessen the great inequalities of day—by progressive taxation and poor relief, for example. Many marginalists were still utilitarians, insofar as they upheld the Millsian notion that the poor will derive greater pleasure than the rich from the consumption of a particular good. Marginalism in fact could be interpreted as establishing an even firmer foundation for an ethic of redistribution, given its emphasis on the declining marginal satisfaction brought to consumers by consumption. Edwin Cannan early in his career followed Mill in suggesting that "roughly speaking, the more equally a given income is distributed, the further it will go in producing economic welfare."[18] Pigou made even more forceful arguments along this line (see below). Until the taboo against interpersonal comparisons of utility was enforced in the 1930s, marginalism could be interpreted either as conservative or reformist. As Maloney summarizes this period, the marginal revolution "added to the armoury of weapons for holding down class conflict," but it also "dispatched many of the self-interested excuses for preserving the status quo."[19]

The ideological openness of early neoclassicism was not to last. In the twentieth century, economists in the marginalist tradition articulate a very narrow view of science, a view hardened by the elegance of the neoclassical synthesis and buoyed by the prosperity of the First World's market-based economies. Economic treatises ritually cite the need to distinguish positive from normative economics; value-free science becomes the self-expressed goal of dozens of economics treatises from the 1920s through the 1950s. The ideal of neutrality strengthens both in response to efforts to politicize the sciences and as positivism (or logical empiricism) comes to serve as a common language for empirical science. Especially in the wake of the Great Depression, conservative economists utilize the ideal of value-neutral science as a weapon against efforts to replace laissez-faire by some form of social planning.

Lionel Robbins's 1932 *Nature and Significance of Economics* was probably the single most influential statement of value-neutrality in twentieth-century Anglo-American economic theory. It is significant not only because his formulation is clear ("economic analysis is *wertfrei* in the Weber sense"),[20] but also because he draws explicit links between the ideal of neutrality and the supposed need to deny "the scientific legitimacy of interpersonal comparisons of utility." As occupant of the prestigious Chair of Economics at the University of London (1929–1961), as chairman of the *Financial Times* (from 1961), and as a Life Peer ("Baron Robbins of Clare Market"), Robbins was a man to whom one listened.

What does it mean for science to be value-free in Robbins's sense? For one thing, it means that the economist must consider, without prejudice, anything and everything that falls within the category of "human behavior as a relationship between ends and scarce means which have alternative uses." Economics is therefore "entirely neutral between ends; that is, in so far as the achievement of *any* end is dependent on scarce means, it is germane to the preoccupations of the economist." The ends may be noble or base, material or immaterial. Fluctuations in the price of hired love are as explicable as fluctuations in the price of hired rhetoric; the "pig-philosophy" (Carlyle's epithet, cited by Robbins) must be all-embracing.

But there is a more important sense in which science, for Robbins, must be value-free. Like many earlier marginalists, Robbins was concerned to eliminate value judgments in order to stymie socialist or utilitarian efforts to redistribute income. The issue in question, again, was whether interpersonal comparisons of utility are legitimate. More sharply even than the nineteenth-century marginalists, Robbins rejected any and all attempts to compare "the different satisfactions of different individuals." The problem with all such efforts was that they invariably involve "judgments of value rather than judgments of fact"; and value judgments, in his view, were "beyond the scope of positive science."[21]

In Chapter VI of his *Essay* Robbins develops this argument to its fullest, directing a broadside attack on the utilitarian ideal that a more egalitarian distribution of wealth or income might bring us closer to "the greatest good for the greatest number." He begins by noting that "It is sometimes thought that certain developments in modern Economic Theory furnish *by themselves* a set of norms capable of providing a basis for political practice. The Law of Diminishing Marginal Utility is held to provide a criterion of all forms of political and social activity affecting distribution. Anything conducive to greater equality, which does not adversely affect production, is said to be justified by this law; anything conducive to inequality, condemned." Robbins suggests (with "great diffidence") that such arguments are "entirely unwarranted by any doctrine of scientific economics"; indeed, outside of England, "they have very largely ceased to hold sway."

Robbins states in very clear terms the primary argument he is rejecting: "The Law of Diminishing Marginal Utility implies that the more one has of anything the less one values additional units thereof. Therefore, it is said, the more real income one has, the less one values additional units of income. Therefore the marginal utility of a rich man's income is less than the marginal utility of a poor man's income. Therefore, if transfers are made, and these transfers do not appreciably affect production, total utility will be increased. Therefore, such transfers are 'economically justified.'" The fallacy in such an argument, he claims, is that it makes certain assumptions which, whether true or false, "can never be verified by observation or introspection." The question of the comparability of different individual experiences is ultimately a metaphysical one: comparisons of utility fall "outside the scope of any positive science." Comparisons of this sort involve "an element of conventional valuation," are "essentially normative," and therefore have "no place in pure science."[22]

Ironically, Robbins appealed to egalitarian sentiments to bolster his ultimately anti-egalitarian argument. How, he asks, could one ever test the hypothesis that a rich person receives less pleasure at the margin than a poor person? If someone from another civilization were to claim that his caste or his race was more sensitive to satisfactions than an individual from an inferior caste, how could we refute him? "We could not say that he was wrong in any objective sense, any more than we could show that we were right." Such conclusions might be justified in terms of convenience, but they are unverifiable and therefore have no place in science. Nor is it even possible to say what the "interests" of a group really are. Subjective value theory allows Robbins to argue that efforts to explain the evolution of political forms in terms of economic interests are doomed to fail, because the interests of political groups are no more capable of objective definition than the wants of individuals. "The concept of interest involved in all these explanations is not objective, but subjective. It is a function of what people believe and feel." We must be humble in our efforts to predict long-term human behavior; human interests are "ultimate data" which no amount of analysis can further explain.

Robbins suggests, as a consequence, that many of the claims made in the name of applied economics are lacking in scientific foundation. The fact of diminishing marginal utility, in particular, "does not justify the inference that transferences from the rich to the poor will increase total satisfaction." Nor does it tell us "that a graduated income tax is less injurious to the social dividend than a non-graduated poll tax." However one might judge the social utility of such measures, conclusions of this kind cannot be derived from "the positive assumptions of pure theory." Historical connections between the development of economic theory and utilitarianism are purely accidental; the two have no logical connection.

Robbins pushes this argument one step further. Suppose it *were* the case that the conventional assumptions of utilitarianism are true, that we could in fact measure the satisfactions obtained by two different individuals from a given good. Suppose even that we could show that certain policies would therefore augment social utility—would this mean that we *ought* to implement those policies? Not at all. "For such an inference would beg the whole question whether the increase of satisfaction in this sense was socially obligatory. And there is nothing within the body of economic generalizations, even thus enlarged by the inclusion of elements of conventional valuation, which affords any means of deciding this question. Propositions involving 'ought' are on an entirely different plane from propositions involving 'is.' "[23]

Robbins's attack on the utilitarian ethic of redistribution provided a powerful argument against liberal economic reforms. It is no great mystery who Robbins had in mind in making his criticisms. Though he does not mention him by name, A. C. Pigou's mildly socialist welfare economics (the so-called material welfare school) was clearly one of his primary targets. In 1912, the successor to Alfred Marshall's chair of political economy at Cambridge had published his *Wealth and Welfare* in which he argued even more forcefully than Mill that the economic welfare of a community was likely to be augmented "by anything that, leaving other things unaltered, renders the distribution of the national dividend less unequal." For proof, Pigou appealed to the principle of diminishing marginal utility, from which he argued that a transference of wealth from rich to poor would enable "more intense wants to be satisfied at the expense of less intense wants," thereby increasing the aggregate sum of the community's satisfactions. (Pigou suggested as one possible measure of inequality the mean square deviation from the mean income.) Pigou's ideas found sympathizers among both liberals and socialists of the Fabian variety; his book, which appeared in several editions through 1932, helped to keep alive the notion that an analysis of the impact of inequality on welfare could be a legitimate part of economic science.[24]

Critics of Pigou's "welfare economics," however, were able to turn to Pareto, Robbins, and recent positivist literature to rule out of court any such considerations. In fact, an entirely new branch of theory—a self-styled "new welfare economics"—grew out of the insight that it was impossible to aggregate the utilities of individuals (to compose a "social utility function") without weighting the utilities of various individuals in some fashion. Drawing from Pareto as well as from Robbins, economists reasserted the principle that satisfactions may be ranked but not quantified; the focus of analysis shifts from cardinal to ordinal utility.

After Robbins, the key figure in this transition to a "new welfare economics" was J. R. Hicks. In an article in the 1939 *Journal of Economics,* the Cambridge don claimed that it was not in fact beyond the scope of economics "to lay down principles of economic policy, to say what policies are likely to be conducive to social welfare, and what policies are likely to lead to waste and impoverishment." The purpose of economics was to explain and to predict; but economists were not in the "dreadful" position of having to avoid altogether questions of economic policy. There was a danger in allowing "economic positivism" to lead one to shirk such issues—a danger that might very well lead to the "euthanasia of our science." Hicks was enough of a positivist, however, to believe that interpersonal comparisons of utility were illegitimate; indeed, his self-appointed task was to cleanse welfare economics of the value judgments that had sullied the science since Pigou, to render it "immune from positivist criticism."[25] The claims allowable in the new welfare economics were in fact drastically narrowed from those of Pigou and earlier utilitarians.

The key insight for the pioneers of the new welfare economics of the 1930s was that one could still make certain modest claims about overall economic efficiency (welfare) without making interpersonal comparisons of utility. However one weighted the value of five pounds to a rich man or a poor man, it was certainly true that a policy which made at least one person better off, without making anyone worse off, could be considered an improvement. Welfare economists followed Pareto in defining an optimally organized economy as one where "every individual is as well off as he can be made, subject to the condition that no reorganization permitted shall make any individual worse off."[26]

The consequence of this approach, however, was to remove entirely the question of inequality from calculations of economic welfare. An economy is "optimal" in Hicks's (and Pareto's) sense if the marginal rate of substitution between any two commodities is the same for everyone who consumes them both and for everyone who produces them both. "Welfare" in this sense is simply a measure of aggregate economic efficiency, shown to be equivalent to the conditions required for markets to clear and for prices to be stable. An economy may be "optimal" according to the new welfare economics even if half the population is fantastically rich and the other half starving. All that is required is that the situation be such that a change will not make anyone better off without making someone else (even the very wealthy) worse off. Questions of distributive justice are ruled out of court, given that these are questions on which there can be no "identity of interest" and thus no "generally acceptable principle."[27]

From a point of view internal to the neoclassical theory, the shift from cardinal to ordinal utility represented a simplification of technical assumptions, one that allowed the translation of Marshall's utility theory into Edgeworth's and Pareto's indifference curves. "Marginal utility" becomes "marginal rate of substitution," and hidden value judgments are expunged. From a political or philosophical point of view, however, the shift allowed one to eliminate whatever vestiges remained of the utilitarian notion that a more egalitarian distribution of income is preferable from a welfare-theoretical point of view. J. R. Hicks made this clear in 1939: "If one is a utilitarian in philosophy, one has a perfect right to be a utilitarian in one's economics. But if one is not (and few people are utilitarians nowadays), one also has the right to an economics free of utilitarian assumptions."[28]

I do not want to leave the impression that value-free economics and its associated ideological baggage was unopposed. Indeed it is hard to name an orthodoxy more embattled, refuted, and despised—or one more entrenched, homogeneous, and powerful. If classical economics was reviled as the "dismal science" or "pig-philosophy" (Carlyle's coinages), if the classical economist was "a one-eyed flat fish with the side on which there was an eye always in the mud" (Ruskin's words), neoclassical economics became the object of even sharper derision. Marxists blamed the theory for ignoring productive relations (reducing workers, management, and firms to equally abstract "consumers" and "producers"); institutionalists complained that the doctrine was ahistorical or "predarwinian" (Thorstein Veblen's term). Neoclassicists have been accused of ignoring the power of producers over consumers (through advertising, for example); they have been faulted for their narrow focus on the short run, the static, the individualistic. Critics have charged neoclassicism with unrealistic assumptions (perfect competition, complete information, rationality); the "queen of the social sciences" has been charged with either ignoring or violating the cumulative wisdom of psychology, sociology, and anthropology. The style of inquiry has also been the subject of scorn. Joseph Nicholson early in the present century complained that the Marshallians and the Pigouese had mathematized the truths of economics to the point of making them "unintelligible or vapid." Radicals have long branded neoclassical economics a "bourgeois cul-de-sac," a kind of exclusivist, Orwellian Newspeak preventing genuine economic discussion from taking place. John Maloney notes that, even a hundred years ago, the failure of most economists to address questions of distribution, housing, legal justice, and so forth so outraged the general public that, for many, to profess economics was to deny ethics.[29]

Apart from questions of relevance and realism, one of the most persistent charges against neoclassicism has been that it serves as an elaborate apology for free-market capitalism. Marginalism is accused of being an elaborate tautology designed to show that what is must be—Joan Robinson thus asserts that "the whole point of *utility* was to justify *laisser faire*." Such suspicions surfaced occasionally even from within the neoclassical camp. Philip Wicksteed in 1914 recognized that a great deal of economic literature was "little better than apologetics, welcomed by those whose consciences need a soothing syrup." Gunnar Myrdal in 1930 published a book-length attack on the "uncompromising laissez-faire doctrine" put forward by Gustav Cassel and others in the 1920s—especially its pretensions to value-neutrality. Myrdal argued, by contrast, that "without value judgements, the whole notion of a social conduct of economic affairs is meaningless." What he objected to were political pronouncements—whether Marxist or laissez-faire—put forward in the name of science. When in 1954 he updated his critique, he ridiculed the new welfare economics as "not even new wine for the old bottles."[30]

In America, institutional economists were among the most persistent critics of neoclassical orthodoxy. Thorstein Veblen, in a long-standing battle with apologists such as Clark, defended a more anthropologically oriented economics that would chronicle the evolution and diversity of human tools and institutions.[31] The goal of economic science was to explain economic institutions, not to glorify the equilibration of prices. According to Veblen, "predarwinian" economists (including most of his twentieth-century colleagues) had imputed habits of business calculation to the population more generally, fetishizing market institutions as if economic activity were exclusively a matter of buying and selling. "Business enterprise" he contrasted with "the machine process"—making money versus making goods. The latter was the main force advancing civilization, the former the greatest force retarding it.[32] Veblen stressed the need to understand nonrational aspects of human behavior—conspicuous consumption and waste, adherence to religious or economic dogmas, resistance to innovation, and so forth. Humans were as much creatures of instinct and habit as of rational calculation. His cynicism concerning the triumph of "imbecile institutions" set him apart from most of his colleagues, for whom the analysis of market equilibria was the be-all and end-all of economic theory. Veblen was probably the most widely read American economist in the early decades of the twentieth century, yet by rejecting price theory he eventually forfeited his right—in the eyes of the neoclassicists—even to call himself an economist.

Wesley Mitchell followed somewhat cautiously in Veblen's footsteps. He lectured at Columbia and the New School for Social Research from 1913 to 1937, emphasizing always that economics must remain "primarily a discussion of public policies."[33] True to Veblen's teachings, Mitchell argued that excessive devotion to technical acrobatics could mask the larger institutional structures economists must appreciate to understand cause and effect. As founding director of the National Bureau of Economic Research (1920–1945), Mitchell initiated a broad range of empirical and statistical studies (of business cycles, for example) to remedy the abstract formalism of orthodox theory.

After Veblen, though, the most persistent institutionalist critic of neoclassical orthodoxy was the maverick Texan economist, Clarence Ayres. Ayres, like Mitchell, was a student of Veblen, but also of the pragmatist John Dewey, from whom the heterodox economist gained a more philosophical approach to economic criticism. In 1935 Ayres debated Frank Knight in the pages of the journal *Ethics,* challenging the Chicago economist to admit that neoclassical theory, far from being "neutral" on questions of politics, was actually a celebration of the market and market institutions.[34] Knight, one of the architects of Chicago School economics, conceded that the "social equilibrium of price theory" was morally ideal only to the extent that "the distribution of resources among individuals is itself ethical."[35] This was a substantial concession, one which many of Knight's less reflective students would not be prepared to acknowledge.

Ayres's criticism went even further, though. Ayres followed Howard Becker in arguing that the neoclassicist world view presumed a static, "natural order" conception of human nature. Neoclassicists had failed to recognize market institutions as historical phenomena. More than this, they failed to recognize that there were productive and unproductive goods, ways of making a living that promote the "life-process" and others that are injurious. The abstraction of all human interaction into prices obscured these larger, technological and institutional issues—issues that would broaden economic theory to include inquiries into questions of technological development and social policy. Ayres sought to develop a value theory where value was more than price, more even than the worth of a product; science, too, was something much broader than most economists were willing to grant.[36]

How are we to explain the popularity of the ideal of value-freedom in mid-twentieth-century economic theory? Critics such as Veblen or Ayres were peripheral to the profession; many economists today will probably never have read either of them (many will probably never even have heard of Ayres). It would be easy to imagine that value-neutrality was simply part of a larger effort to streamline and purify economic science; this might be the kind of explanation provided by someone who believes that as a science grows, so does its reliance on abstract mathematics or its distance from practical problems. This was the interpretation of Samuelson, for example, in his *Foundations of Economic Analysis*;[37] similar arguments were put forward by German social theorists in the *Werturteilsstreit*. Many economists today would probably see the question in these terms.

One might also imagine that the triumph of neutrality stemmed from the influence of Vienna School positivism. Yet Robbins himself rejected any such imputation, and links between his work (or that of Hicks) and logical positivism of the Viennese sort are actually quite difficult to establish. Behaviorism, in Robbins's view, was an inadequate foundation for economics, as was Gustav Cassel's strict *Ausschaltung der Wertlehre*. When Robbins was accused of being a positivist (by R. W. Souter), he could honestly respond that he intended nothing of the sort, that it was from no less a figure than Max Weber that he had learned the ideal of value-neutrality. Robbins frequently cited Weber in support of the idea that science ought to be value-free.[38]

Terence Hutchison was one of the few prewar economists with clear links to Viennese logical positivism. Hutchison lectured at the University of Bonn from 1935 to 1938, returning to the United States armed with the tools and ideals of logical positivism gathered from what remained of the Vienna Circle. In his 1938 *Significance and Basic Postulates of Economic*

Theory, he introduced the vocabulary of the Vienna School to American debates on the place of values in science. Like others in the positivist camp, Hutchison devoted a great deal of attention to the so-called demarcation problem—namely, how one was to distinguish properly between "science and pseudo-science." Hutchison maintained that economics must exclude propositions that could not be put to empirical test—whether those be expressions of "ethical uplift or persuasion, political propaganda, poetic emotion, psychological 'association', or metaphysical 'intuition' or speculation." There had to be "some effective barrier" for excluding such expressions from economic science.[39]

In the middle of his chapter on "Science and Pseudo-Science," Hutchison made it clear that the fascist perversions he had witnessed in Bonn were among the enemies his positivist method had been designed to combat:

> The most sinister phenomenon of recent decades for the true scientist, and indeed for Western civilization as a whole, may be said to be the growth of Pseudo-Sciences no longer confined to hole-and-corner cranks or passive popular superstitions, but organised in comprehensive militant and persecuting mass-creeds, attempting simply to justify crude prejudice and the lust for power. There is, however, one criterion by which the scientist can keep his results pure from the contamination of pseudo-science[:] " . . . the appeal to fact."

The positivist "appeal to fact" would allow one to distinguish between true science and propaganda; it would provide a barrier against the rising tide of pseudo-science. Warning against the dangers of dividing science into Nordic and Jewish, proletarian and bourgeois, Hutchison cited Pareto's admonishment to remain "absolutely" in the logico-experimental field, refusing to depart from it "under any inducement whatsoever."[40]

Hutchison was not, however, one of those swept up in the "ordinalist revolution." He did not believe that interpersonal comparisons of utility were illegitimate; indeed he devoted the final pages of his treatise to arguing that the old-style welfarist arguments for redistribution were verifiable in terms of the sense people attach to words in ordinary usage. It was simply not true to say that such comparisons were outside the realm of testable experience. Nor did he believe those who claimed (Ludwig von Mises, for example) that economic science had demonstrated that "a liberal, capitalist, *laissez-faire* economic policy leads to maximum returns for the community or to greater returns than any collectively planned economic policy." Hutchison opposed the a priori defense von Mises offered of economic liberalism against socialist planning, especially his view that science had succeeded in showing that every conceivable substitute for the capitalist social order was inherently "contradictory and nonsensical."[41]

Hutchison, I should repeat, was somewhat unusual in his explicit use of Vienna School vocabulary to scrutinize the science of economics. Other traditions—operationalism, for example—were moving to restrict the focus of inquiry to the empirically verifiable, and economists borrowed freely from these traditions.[42] However one traces the intellectual lineages of this period, the origins of value-neutrality in economics cannot be sought only in the influence of this or that school of thought. The filiation of ideas is important, but it does little to explain why particular doctrines or methods become popular at certain times, why others disappear. As in the case of the *Werturteilsstreit,* the value-neutrality of mid-century economics was advocated not in the abstract but in response to specific political movements: socialist or welfare economics, fascist triumph in Europe, economic collapse at home. For conservatives such as Robbins, neutrality served to defend a separate and private sphere of values (and property!) against the encroachment of science or so-called rational planning; for progressive liberals like Hutchison, value-neutrality provided a bulwark against efforts to politicize science in the name of race or class. Value-neutrality in both cases provided a means of demarcating science from pseudo-science, sense from supposed nonsense.

John Maloney has made a persuasive case that the rise of value-neutrality can be understood as part and parcel of the professionalization of economics. The fact that economics was professionalized under an exclusively neoclassical banner helped to cement the links between neutrality and marginalism. By 1880 there was a strong correlation between enthusiasm for the marginal revolution and the belief that science must be value-free. Marginalism and positivism indeed "fed on one another," and the reasons were at least partly strategic. In W. J. Ashley's mind, neutrality was the economist's best hope for retaining both scientific rigor and political influence. Edwin Cannan made a similar argument in 1904. Maloney suggests that neutrality may have become popular among professional economists for the simple reason that economists were likely to be chosen from among those people "who either approve of existing society or trust existing politicians to try and do something about it."[43]

Maloney, apparently following Schumpeter, distinguishes between ethical and ideological neutrality.[44] It is easy to avoid "value judgments," if all that is meant by this is that one avoids statements to the effect that x is good or y ought to be done. This is a simple matter of grammar, and it was easy for economists to proclaim their interests purely "technical" (or theoretical). In Robbins's terms, the difference between an economic and a technical problem was simply that whereas in a technical problem the end is one and the means are many, an economic problem is one where both the ends and the means are many.[45]

Maloney points out that statements may be ideologically charged even if ethically neutral—even, that is, if they contain no specific moral imperative. The example he gives is one from Cannan: "Among well-to-do people there is not one person in a hundred who will not work more rather than less in consequence of an increase in taxation." The statement does not explicitly advocate a policy, but it is hard to deny that it is ideologically charged, even if it is empirically verifiable. There are many such statements in the neoclassical corpus. (Indeed, Maloney says, the decision to restrict the scope of economics to value-neutrality may itself be an ideological decision, insofar as it is based "less on satisfactory evidence than on an unfulfilled wish to believe.") Ideology enters economics insofar as specific theories may be more or less tightly bound to particular world views—so much so that in elaborating a theory one ends up servicing a world view, regardless of what one thinks of the world view in question. Economists use theory, but theory also uses economists. Such was the case, Maloney suggests, with the marginal revolution. By the beginning of the twentieth century it is difficult to be an economist without adopting marginalist principles. The power of professionalization was (and is) such that economists, once trained as marginalists, find it difficult to be anything else: "automatic adherence to the neoclassical paradigm of anyone who could handle it is the most disquieting aspect of the professionalisation of economics."[46]

Ideology lies in where one situates the burden of proof, which in turn depends on the starting point of inquiry, which depends on the tools available for analysis. The ideology of the neoclassical system lies in its implicit (sometimes explicit) acceptance of the market economy as the natural economy. Marginalists were, and generally remain, partisans of the morality of the invisible hand—the libertarian notion that governments govern best when governing least and that free markets produce a maximal satisfaction of consumer wants. Even if one wants to defend a mixed economy (capitalist and state-interventionist), the neoclassicist must start with a competitive market, perfect information, and the rest, and then relax these assumptions to incorporate perturbations. Where one starts, and where one relaxes, says a great deal about the ideology in question. With the market as the norm and nonmarket "intervention" (the very word presumes a market norm) the deviation, uncertainties will weigh in favor of free-market solutions. As Maloney notes, the neoclassical paradigm is best at uncovering reasons why government intervention is *not* needed in a particular case:

> The neoclassical telescope primarily helps the economist to see why superficially undesirable features of capitalism do *not* lead to a loss of welfare, or why the State's remedy may be worse than the disease. Incomes policies, minimum wage laws, anti-pollution laws (rather than taxes), laws to halt the "brain drain", are all "common sense measures" whose rationale is reduced by neoclassical analysis. The nature of the neoclassical paradigm fits it best

to the task of showing up cases where, contrary to "common sense", government intervention is not desirable.[47]

This, then, is the problem with the philosophical orthodoxy of neoclassical economics—the view that holds that values enter into the choice of problems but not the substance of the results. The problem with this view is that it ignores the fact that people live by their tools. If economic tools are sharp for the market and dull for planning, it is hardly surprising that the tools have tended to favor certain policies over others.

In the science of economics, value-neutrality is commonly defended as part of what it means for a discipline to become scientific. It is also a tool for defining the proper bounds of scientific discourse. Economists have used this tool in different ways at different times, but its primary social function, especially in the twentieth century, has been to discourage certain structural and ideological questions—concerning alternative systems of producing or distributing wealth, for example, or how innovations should be encouraged or managed—from becoming the objects of theoretical discourse. Neutrality in the tradition of positive economics is not just a defense of the right of the discipline to call itself a science; it is an instrument for the creation and defense of economic and political order.

14

Emotivist Ethics

The purpose of this book is to explore the origins of the ideal of value-free science. So far we have concentrated on positivist views of science—but positivism is also a theory of values and we may ask: what is the positivist conception of *values*? We've already seen how for Weber, value-free science was proposed in order to guarantee "science-free values"; we have also seen how, for neoclassical economists, the ideal of value-freedom played an important part in the defense of the free-market economy. Here we look at the writings of the British analytic philosopher Alfred J. Ayer, the foremost exponent of a specifically positivist, or *emotivist,* ethics and the greatest champion of that twentieth-century irony: value-free moral philosophy.

Positivism, I should repeat, is a theory of values as well as of science.[1] More specifically, it is a theory of the relation of facts and values. Critics of positivism, who tend to concentrate either on the theory of science or the theory of values, have given insufficient attention to both sides of the divorce. So convincing has the positivist separation of facts and values become that critics usually accept this part of the positivist terrain even as they criticize other parts. Where is the discussion of ethics in Kuhn, or Hanson, or Feyerabend, or the more recent critics of positivist science theory, such as Barnes, Bloor, or Latour? Where is the discussion of science or science policy in Hudson, or Urmson, or Railton? The leading questions in postwar philosophy of science have been those concerning the relation of observation and experiment; problems of science and society, science and ethics, have remained on the margins.

For Weber, as we have seen, values were "that about which men can ultimately only fight." The positivist tradition relied upon a similar conception of ethics, a conception that went under the name of the "emotive theory" of ethics. The emotivist theory was first set forth in I. A. Richards and C. K. Ogden's 1923 *The Meaning of Meaning,* a book in which the authors suggested that *good* was "a unique, unanalysable concept" referring, literally, to nothing: a statement that this or that object or action is "good" represents nothing more than "an emotive sign expressing our attitude" toward that object or action, perhaps eliciting similar attitudes in other persons.[2] Ogden was the inventor of Basic English, the 850-word vocabulary that was to serve as a common introductory medium of expression for the English language; Richards was a literary critic known primarily for having launched the internalist, antihistorical movement known as "new criticism."

The most powerful exposition of the emotivist theory, however, was that presented in chapter 6 of A. J. Ayer's *Language, Truth and Logic,* a work that caused a sensation when it first appeared in January, 1936. Ayer had studied briefly in Vienna (1932–1933), after which he found himself in close agreement with "those who compose the 'Viennese Circle', under the leadership of Moritz Schlick, and are commonly known as logical positivists."[3] His conception of philosophy was both extraordinarily narrow and extraordinarily bold: narrow, because philosophy was to restrict itself to the clarification of the propositions of science "by exhibiting their logical relationships, and by defining the symbols which occur in them"; yet bold, because the young philosopher (not yet twenty-five when he finished the book) was proposing a scheme that was advertised as putting an end, once and for all, to "conflicting philosophical parties or 'schools.'"

The means by which this was to be achieved were simple. Philosophy, in his view, should concern itself only with logical questions (empirical questions were the province of the individual sciences). Propositions were to be distinguished as one of three kinds: logical, empirical, or metaphysical. The metaphysical were to be dismissed, the empirical were to be investigated by science, and the logical (or analytic) were to be analyzed to determine whether they were necessarily true or necessarily false. Philosophy was to become "the logic of science," and ethics (or what remained of it) was to become a subbranch of that logic.

Ayer turned to ethics in order, as he put it, to maintain "the general consistency" of his view—specifically, to defend his view that all synthetic propositions are empirical hypotheses. (Aesthetics and ethics appeared to pose a problem for his "radical empiricism.") Following Hume, Ayer divided all genuine propositions into two classes: questions of *logic* (Hume's "relations of ideas") and questions of *empirical fact.* All other propositions (metaphysical ones, for example) were not genuine—indeed they were, in his terms, literally nonsensical.

Moral judgments, in particular, fit within neither of these categories and therefore were not "genuine." Ayer divided statements of traditional ethical philosophy into four types: definitions of ethical terms, descriptions of moral behavior, "exhortations to moral virtue," and ethical judgments proper. Only the first of these (in his view) had a place in moral philosophy; the others either belonged to other disciplines or had no meaning whatsoever. Descriptions of moral behavior belonged to psychology or sociology; exhortations to virtue were not propositions at all but simply "ejaculations or commands which are designed to provoke the reader to action of a certain sort." Nor did ethical judgments themselves belong in ethical philosophy, given that they are neither "definitions nor comments upon definitions."[4]

This is the irony of Ayer's positivist ethics: "A strictly philosophical treatise on ethics should therefore make no ethical pronouncements." Ethical philosophy was to restrict itself to the analysis and definition of ethical terms; positivist ethics, like positivist science, was to be value-free.[5]

Ayer was concerned that his rejection of any empirical foundation for ethics might leave the path open to absolutist or intuitionist ethics. Rejecting both absolutism and naturalism as untestable, and therefore meaningless, the alternative he proposed was *emotivism*. The key idea in emotivism is that judgments of the value of a thing are expressions of emotion which can be neither true nor false. An ethical judgment in a sentence adds nothing to the content of that sentence: the statement "You acted wrongly in stealing that money" says nothing more than that "You stole that money." The purely evaluative statement "stealing is wrong" does not contradict the statement "stealing is right," because neither statement has empirical content. Neither is a genuine proposition in the sense Ayer has defined, namely, that a statement must be analytic or synthetic. Ethical terms express or arouse emotion, but they do not convey meaning. The sentences of "a moralizer" therefore cannot be judged true or false; indeed, they have "no objective validity whatsoever."[6]

Ayer's text was remarkable for its influence. From 1936 to 1939, the British journal *Mind* published ten articles or discussion notes devoted to the book (Wittgenstein's *Philosophical Investigations*, by contrast, received only one such discussion in its first four years after publication). By 1986 the book had gone through twenty-seven printings and at least fourteen translations—including Catalan, Mahrathi, and Italian Braille. It has never gone out of print; the American Dover paperback alone by 1977 had sold more than 270,000 copies. The editors of a recent volume of essays on the book state rather presumptuously that "only a few" British philosophers could claim never to have fallen under its spell.[7]

Ayer's emotivism was, however, as much a symptom as a cause; the strong analytic bent of British moral theory had been building since the early part of the century. In 1903 G. E. Moore of Trinity College, Cambridge, had published his *Principia Ethica*, a book that aroused nearly as much controversy as Ayer's, purporting as it did to present (following Kant) a "Prolegomena to any future Ethics that can possibly pretend to be scientific."[8] The thesis of the book, one of the founding documents in the British analytic school, was that the idea of "good" is simple and indefinable—that is, that there is nothing common to things or actions which makes them good or evil. The good of a thing or action is something we intuit; we do not derive it from other qualities of things.

Moore is chiefly known today for his uncompromising attack on "naturalistic ethics" and especially on what he called the *naturalistic fallacy*: the effort to identify "good" with something other than itself or, as more commonly expressed, the effort to derive ought from is, values from facts. Those guilty of committing the naturalistic fallacy included Jeremy Bentham, the hedonist (reducing good to pleasure); J. S. Mill, the utilitarian (reducing good to sentiments of the majority); the devoutly religious (reducing good to the will of God); Marxists (who assume that the good can be reduced to allegiance to ultimate historical victors); and pragmatists (who want to reduce good to consequences). Such efforts were fallacious because—as Hume was later said to have shown[9]—you cannot derive good from something it is not. Good is not a natural property of a thing; something is good or evil *irreducibly*, that is, we cannot claim that it is good because it shares a particular property with something else (pleasure, the will of God, or evolutionary success, for example). Ethical propositions can be neither proven nor disproven; good is not to be judged by its consequences, as is the case for utilitarians or pragmatists.

Moore was immensely popular among those who gathered under the flag of the Bloomsbury circle. Abraham Edel describes how, as a student at Oxford in the late 1920s, controversies raged about the relation of "the good" and "the right"; these were questions "over which deep passions were roused and on which a man's whole welfare in this cosmos seemed to depend."[10] Moore's apostles followed him like a guru: after reading the *Principia*, Lytton Strachey wrote to Moore that he had "wrecked and shattered all writers on Ethics from Aristotle and Christ to Herbert Spencer and Mr Bradley." Moore was supposed to have been the first to apply the scientific method to moral reasoning: "henceforth, who will be able to tell lies one thousand times as easily as before? The truth, there can be no doubt, is really upon the march. I date from Oct. 1903 the beginning of the Age of Reason."[11]

Insofar as the analytic philosopher's "age of reason" can be said to have begun with Moore, it culminated in the two decades following his 1921–1947 editorship of *Mind*. By the time Ayer published his *Language, Truth and Logic* in 1936, others in the positivist/analytic camp had begun to argue that ethical terms have only emotive significance. Wittgenstein in his 1921 *Tractatus* had asserted that "if there is any value that does have value, it must lie outside the whole sphere of what happens and is the case." Moral and scientific discourse were to be sharply distinguished; and ethical judgments were literally meaningless. Schlick and Carnap put forward similar views. Schlick suggested that "the sense of every proposition concerning the value of an object" consisted merely in the fact that the object, or the idea of it, "produces a feeling of pleasure or pain in some feeling subject"; Carnap had stated that ethical propositions have no "theoretical meaning."[12] Similar arguments were put forth by Charles L. Stevenson in 1937 and 1944,[13] by which time emotivism had become one of the leading moral theories of the analytic camp. The *Philosopher's Index* for the period 1940–1966 lists more than a hundred articles on the topic. Books, interestingly, were less common: as G. J. Warnock later put it, many felt by the 1950s that the age of philosophical books had come to an end, that philosophy would henceforth advance through piecemeal studies of a less systematic character. (Wittgenstein is said to have told C. Lewy about this time that his philosophy couldn't be "continued," it could only be "applied.")[14]

As influential as Ayer's theory was, it was also subject to strong criticism. *Time* magazine blamed the book for the rise of fascism;[15] the Jesuit M. C. D'Arcy criticized the book in the June 1936 issue of *Criterion*. E. W. F. Tomlin's review in the September issue of *Scrutiny* was one of the most savage. Tomlin claimed that he had never read a book revealing such a combination of "immaturity, loose thinking and wholly unwarranted cocksureness." Ayer's separation of emotive and scientific propositions was dangerously close to faculty psychology; the chapter on emotivism was "surely the most puerile piece of casuistry that a philosopher has ever put forward in the name of reason." Tomlin also criticized the young Oxford don for his implicit assumption that, for twenty-five centuries, philosophers had been talking nonsense. Ayer himself eventually conceded that his moral theory had been "much too crude."[16]

Analytically minded historians of moral philosophy have tended to view emotivism as a critical response to G. E. Moore's nonnaturalism. Moore saw "good" as something that, though nonnatural, was nevertheless objectively real and knowable by intuition; the emotivists, by contrast, saw moral claims as literally meaningless. Emotivists rejected both subjectivist and objectivist accounts of value, preferring instead to see moral claims as meaningless outbursts of emotion, the way a person grunts when uncomfortable or sighs when experiencing pleasure.[17] From a larger view, however, it is possible to argue that emotivists and nonnaturalists such as Moore were actually in agreement on a number of issues. Both rejected the derivability of values from facts; both took the morality of the judge to be more important than the morality of the agent. Both saw the key to moral philosophy in the analysis of moral language, and both conceived of the moral philosopher as a *spectator* to moral practice.

This same basic point of view could still be found in the 1980s. The distinction was typically made between *moralism* and *moral philosophy*. W. D. Hudson made this distinction on the very first page of his 1983 *Modern Moral Philosophy,* siding with the latter against the former, arguing that the task of moral philosophy must be to analyze the logic of moral discourse, not to render moral judgments or teach moral values:

> This book is not about what people ought to do. It is about what people are doing when they *talk* about what they ought to do. Moral philosophy, as I understand it, must not be confused with moralizing. A moralist is someone who uses moral language in what may be called a first-order way. He, qua moralist, engages in reflection, argument, or discussion about what is morally right or wrong, good or evil. He talks about what people ought to do . . . a moralist, as such, is interested in ethics; and a moral philosopher, as such, in metaethics.[18]

"Metaethics" for the analytic philosopher means value-free ethics; discourse about ethics, but not ethical discourse. The task of moral philosophy in this view is to clarify the logic of moral judgments—not to say what is right or wrong, but to say what is meant when others say what is right or wrong.

Analytic metaethics was, of course, a very different project from classical conceptions of morality. Socrates asked "What is justice?" not "What is the language of justice?" Aristotle objected to Plato's purely contemplative approach, but still he asked for philosophers to inquire into how one goes about achieving the good life in a reflective manner. Aristotle distinguished between theoretical and productive sciences: ethics was to be a productive science in this sense—a science practiced not simply for its own sake (as he says was true for astronomy and geometry) but in order *to produce something.* The purpose of inquiry into ethics was to produce the good, as the goal of medicine was to produce health. Modern moral philosophy—at least in the sense of Hudson and many of his analytic colleagues—rose above the question of whether there is or even should be justice. Ethics was "the logical study of the language of morals" (Hare);[19] philosophers were to limit themselves to discussing "moral reasoning" because we can never really know what is morally good. We can only know what people say is good.

As early as the 1940s, even some within the analytic tradition had rejected such a narrow conception of philosophy. Stuart Hampshire criticized the emotive theory, along with the larger Kantian notion of an unbridgeable separation between moral and factual judgments on which it was based. Hampshire pointed out that Hume himself had never denied that moral judgments are based on arguments about matters of fact—he had only showed that such arguments are not compelling in the sense that deductive arguments are. Distinguishing between the morality of the agent and the morality of the judge, Hampshire suggested that though analytic philosophers had become obsessed with the latter, the former was actually the more genuine and fruitful sphere for moral inquiry. Moral philosophy should be about how to deliberate about policy, and not restricted to the logic and language of moral sentences.[20] Others in the analytic camp cautioned that a purely metaethical approach to morality left much to be desired: P. H. Nowell-Smith, reviewing R. M. Hare's *Language of Morals*, noted that "nothing that we discover about the nature of moral judgments [from a theory such as Hare's] entails that it is wrong to put all Jews in gaschambers."[21] Metaethics, in other words, tells us very little, if what we are interested in is finding out what is right or wrong.

American pragmatists rejected in even stronger terms the idea of ethics as an autonomous discipline. John Dewey did worry that unchecked moralism was one of the primary hindrances to fruitful ethical reflection; still, he opposed the positivist view that ethical statements were unverifiable and therefore meaningless.[22] Using the example of medicine, he pointed out that there is always a continuum of ends and means: there is never an end that is not also a means to some other end; there is no means for which we cannot ask, what are its consequences for human well-being? Ethics in Dewey's view involved a larger analysis of problems of political theory and practice, and a host of other problems more commonly thought to be the province of other disciplines. Indeed many of Dewey's students abandoned philosophy in the 1940s, '50s, and '60s and turned instead to education, economics, psychology, political science, or sociology, leaving philosophy to those content to lose themselves in linguistic formalism.

In the late 1960s, the changing political climate fostered dissatisfaction, especially among younger philosophers, with the docile metaethical stance of most analytic philosophy. Philosophers began to worry more about cruelty and punishment, medical ethics, civil rights and liberties, and questions of justice and distribution.[23] Historical, legal, and medical questions began to enter philosophical discourse in a more assertive form, and questions as diverse as animal rights and the origins of the Holocaust began to fill philosophical journals. In the fall of 1971 the journal *Philosophy and Public Affairs* was founded with the explicit claim that it was possible for philosophers to take and defend moral positions on questions such as abortion, civil disobedience, political oppression, and so forth. That same year John Rawls published his influential *Theory of Justice*, raising questions of distribution into the center of mainstream philosophy.[24]

The fact-value question began to fade from memory, as philosophers recognized that it no longer mattered whether values could be derived from facts. Moral questions were real questions, regardless of whether they took on this or that particular logical structure. Analytic philosophers continued to analyze (J. L. Mackie suggested, for example, that value judgments are "ontologically queer"—neither real nor purely ideal), but their arguments appeared increasingly anachronistic. In 1974 the president of the American Philosophical Association claimed to have solved the ought-is dilemma; by this time, though, fewer and fewer people could still be found who thought that it really mattered.[25]

15

Social Theory of Science

In 1967 John Passmore wrote in the *Encyclopedia of Philosophy* that logical positivism "is dead, or as dead as any philosophy ever becomes."[1] One can debate various reasons for the decline, and even whether positivism did in fact die in the sixties. American pragmatists had never accepted the fact-value dualism; according to the British journal *Mind,* logical positivism had died as early as 1940—at least in England.[2] Whatever date one chooses, the salient fact is that philosophers became increasingly dissatisfied with the highly abstract, linguistic orientation of analytic philosophy, a philosophy which to many concerned itself excessively with the logical structure of science and insufficiently with the mechanisms by which science changes and with the more important issues of the social origins and effects of science.

The positivists did not ignore entirely the problem of change. Hans Reichenbach, following Hume and Mill, had argued that scientific change occurs through induction; induction was in fact "the key to the difference between science and poetry"—a difference which, for the Berlin or Vienna positivist, was no small matter.[3] Scientific change, though, was never a central focus of positivist work. Karl Popper in 1959 pointed out that the positivists and analytic philosophers "by their method of constructing miniature model languages" tended to miss "the most exciting problems of the theory of knowledge—those connected with its advancement."[4] Popper in fact devoted much of his life's work to correcting this oversight. In his 1934 *Logik der Forschung* (translated as *The Logic of Scientific Discovery*) Popper argued that science advanced through a two-stage process of discovery and justification. The English title of his work, he later explained, was a misnomer; for there is no "logic" of discovery in the strict sense of this term. Discovery is not a rational process but an intuitive, creative process of the imagination. Science is rational not in the production of its theories but in the means by which they are tested. Imagination is the source of theory, experience is the test. Popper rejected the Baconian inductivist idea of perceptions as " 'grapes, ripe and in season which have to be patiently and industriously gathered,' from which, if pressed, the pure wine of knowledge will flow."[5] Knowledge was not like a stack of bricks piled on a level surface—knowledge deepens as it grows; knowledge is more like a house built on a swamp, supported on posts which must be driven ever deeper into the shifting ground.

Popper has disavowed attempts that have been made to label him a "positivist." He was not a member of the Vienna Circle, and emphasized in many of his writings his differences with positivist doctrines (Carnap's principle of verification, for example). Popper indeed considers himself not only to have broken with the Vienna Circle but to have been the one who "killed it."[6] He nonetheless shares with Vienna School positivists the *internalist* conception of science, according to which the progress of science is motivated from within, independent of (or in spite of) larger historical processes. His goal remains to explain the logical relations of scientific propositions, albeit not fixed but in motion. There is no attempt to explore the origins of science in broader social processes.[7]

Popper, despite his own claims, was certainly not the one to have "killed" positivism. If anyone could make such a claim it would be Thomas Kuhn, whose 1962 *Structure of Scientific Revolutions* has become the most widely read book on the history of science in this century (with sales of more than 750,000 plus sixteen foreign-language editions as of 1990). Kuhn more than any other recent author has facilitated a richer understanding of the process of scientific change. In the 1950s, positivist faith in the increasing generality of science had led either to a disregard for its history or to an exclusive concern for those parts of history which could be seen as preparation for the present. The past, in this view, was an impoverished form of the present; present theories contained previous ones like Chinese boxes, the grandest of which is always added last. Einstein's relativity reduced Newton's mechanics to a special case, molecular genetics subsumed Mendelian principles, and so forth and so on, happily and forever after.

Kuhn argued, by contrast, that the key to understanding scientific change is not the logical form of propositions but rather the shift in the broader "paradigm" with which theories are invariably associated. Experience is still the test of theory, but the facts of experience are theory-dependent: one man's duck is another's rabbit, crystalline spheres in one view are rotating planets in another. (Kuhn drew from the German Gestalt psychologists, who recognized that visual patterns are perceived immediately as wholes and not as isolated parts.) Facts are never perceived in the abstract, but rather in the context of a larger set of theories and facts, tools and methods, beliefs and values. One sees (at least to some degree) what one wants to see, what one thinks it *possible* to see. In China, where cosmological traditions envisioned a changing universe, sunspots and new stars were recorded as part of the natural and expected course of events. In the West, such violations of the immutable crystalline spheres often went unnoticed, at least until the Copernican revolution destroyed the spheres and ushered in a new conception of the universe.[8]

Scientific revolutions, in this view, emerge according to a two-step process. In the overwhelming bulk of scientific work, scientists do not question the basic tenets of their discipline. They solve minor puzzles, problems that are brought forth in the course of what Kuhn calls "normal science." The paradigm is stable in the face of difficulties, for theories are seldom tested against nature directly but rather indirectly, as complexes of theories and assumptions any one of which might be fiddled with to save the phenomena. Eventually, as paradigms are elaborated with greater and greater subtlety and sophistication, anomalies begin to arise in the capacity of the paradigm to explain and interpret the phenomena it is designed to cover. When conflicts between paradigm and interpreted fact become too great, and when an alternative is sufficiently elaborated, the diehards of the old paradigm are pushed aside by the champions of the new. A new stage of "revolutionary science" begins to unfold in which facts change their meaning. The "morning star" becomes the "evening star," the loss of phlogiston becomes the absorption of oxygen. Yet a theory is never abandoned without another to replace it: "the decision to reject one paradigm is always simultaneously the decision to accept another."[9] Scientists have an investment in their theories and their tools. Even in the face of great anomalies, until a new paradigm emerges epicycles will be added to orbits, exceptions added to rules. Scientists will try, in Ptolemy's words, to "save the phenomena."

Both Kuhn's and Popper's theories of scientific change have served to modify, albeit in different ways, the logical positivist conception of science. Yet despite some obvious differences, each also shares certain basic features with the older movement. Science for both Kuhn and Popper emerges, as it were, from its own internal dynamic. For Kuhn, progress in science is motivated by "the challenge of the puzzle": "A man may be attracted to science for all sorts of reasons. Among them are the desire to be useful, the excitement of exploring new territory, the hope of finding order, and the drive to test established knowledge." Kuhn himself points out that his essay says nothing about "the role of technological advance or of external social, economic, and intellectual conditions in the development of the sciences"; he also notes that though such influences cannot be denied and should be explored, the inclusion of such forces would not alter in any major way the main thesis of his essay.[10]

Popper's position on this question is subtle. On the one hand, he maintains with Michael Polanyi that it is the social nature of science that is responsible for its objectivity: "ironically enough, objectivity is closely bound up with the *social aspect of scientific method,* with the fact that science and scientific objectivity do not (and cannot) result from the attempts of an individual scientist to be 'objective,' but from the *friendly-hostile co-operation of many scientists.* Scientific objectivity can be described as the inter-subjectivity of scientific method."[11] This is what Popper means by the "public character of scientific method"; it is an aspect of science he says is almost entirely neglected by sociologists of knowledge. Science in one sense, then, has social origins. But scientific truth, once achieved, no longer bears the marks of those origins. Science, in other words, is social in its methods but not in its results. Science is (ideally) an autonomous institution, independent of political interests. Popper supports a liberal political model for the development of science, according to which, as in other spheres of life, government which "governs least, governs best." In this, he follows closely the political philosophy of his libertarian friend, Friedrich von Hayek. Science suffers in the measure that it becomes mixed up with politics; there is no room in Popper's model for the idea that genuine science may be constructed to serve particular social interests or may be so designed as to aid in the maintenance of a particular social order. Nor is there any important sense in which the truths of science are contextual: "If an assertion is true, it is true for ever."[12]

For both Kuhn and Popper, the focus of theory is on the structure of science rather than on the conditions of its production, the uses to which it is put, or the ideologies it might serve. For both, science is a given, and the task of theory is to explain the logic or morphology of its growth. Both are "internalist" in the sense that insights into the origins, structure, and priorities of science are sought internal to the community of scientists. Both assume a fundamental unity of science (Kuhn's book, one should recall, was published as part of the positivists' *International Encyclopedia of Unified Science*); both assume the goal of science theory to be the construction of a single, abstract model of (modern) science "in general," valid for all times and places. Though Popper has discussed the problem of distinctive methodologies for the social sciences and shares a concern to struggle against the pursuit of politics in the name of science, his efforts are designed to exclude politics from science; neither he nor Kuhn explore how the structure and priorities of science may be shaped by larger social forces.

* * *

The tremendous success of Kuhn's book can be attributed, first, to the extent to which the philosophical thought of postwar England and America had taken on a highly abstract and ahistorical character and, second, to the growing search for "alternatives" in the sciences that emerged in the course of the 1960s. The idea that there have been revolutions in the history of science was not, after all, a novel one, nor was the notion that experience must be guided by theory. Popper in 1949 had criticized what he called the "bucket theory of knowledge"—the empiricist notion that the mind is an empty and receptive vessel into which experience freely flows. (The mind is rather like a searchlight, illuminating only that onto which its beam is cast and focused.) Milton Friedman in 1953 had recognized that "a theory is the way we perceive 'facts,' and we cannot perceive 'facts' without a theory"; a century earlier Auguste Comte had observed that "facts cannot be observed without the guidance of some theory." Even the supposed arch-inductivist Francis Bacon had recognized that the gathering of facts is meaningless without theories to give them meaning and direction.[13]

The idea that knowledge is "social" is therefore older than one might imagine—this is true regardless of what aspect of the idea one focuses on. If it is that men working together produce more than men alone, one need only refer to Sprat's seventeenth-century *History of the Royal Society,* where we find a clear expression of this view. If it is the idea that knowledge may differ radically from place to place or time to time, one need only consult Pascal's *Pensées,* where we hear that what is true on one side of the Pyrenees may be false on the other. If it is the principle that theory takes roots in practice, then one need only recall Thomas Hobbes' reference to "need, the mother of all inventions."[14]

It is not until the 1920s and 1930s, however, that one finds a sustained interest in the social origins and expression of knowledge—partly as a result of efforts to politicize the sciences, partly as a result of concerns growing out of the subordination of science to industry, warfare, or one or another of the various racial or economic ideologies of the day. The 1920s and 1930s is what might be called the "classical" period in social theory of science. In Austria, Otto Bauer, Franz Borkenau, and Henryk Grossmann explored the intellectual changes associated with the shift from feudalism to capitalism; Edgar Zilsel explored the social origins of the ideas of genius, physical law, romanticism, progress, and experimental method. In Poland, Ludwik Fleck explored the effects of world view on medical research; in the Soviet Union, Nikolai Bukharin and Boris Hessen applied Marxian principles to the growth of knowledge. In Britain, Benjamin Farrington described the political climate surrounding the rise of Greek science; in America, Lewis Mumford traced the intertwined growth of technics and civilization. E. A. Burtt described the "metaphysical foundations" of modern science, and Robert Merton explored the interaction of religion and economy in early modern English science.[15]

The term *sociology of knowledge* was coined in 1921, in an article by Max Scheler in the *Kölner Vierteljahrshefte für Sozialwissenschaften* entitled "The positivist philosophy of knowledge and the tasks of a sociology of knowledge." Scheler proposed a "pure sociology of knowledge" *(reine Erkenntnissoziologie)* to explore, among other things, "the degree to which the different sciences are culturally bound." Surprising perhaps to today's eyes is the fact that Scheler saw the sociology of knowledge growing out of positivism. Traditional epistemology in his view had ignored the most interesting questions—only the positive philosophy of Comte and Spencer had attempted to combine *Erkenntnistheorie* with sociological statics and dynamics.

But Scheler, sometimes known as "the Catholic Nietzsche," was also critical of the positivist vision. Comte, in his view, had put forward an excessively narrow image of science: it was not enough to say that prediction was the goal of science *(savoir pour prévoir)* or that science renounced any and all efforts to discover "the essence of things." Scheler rejected the three-stage model of Comte and his followers (Mill, Spencer, Mach, and Avenarius): the positivists were right to have distinguished theological, metaphysical, and positive forms of thought, but they were wrong to have seen these as sequential stages along a single hierarchy. The religious, the metaphysical, and the positive are, Scheler says, actually varieties of human thought that persist in all cultures. Religion is not a primitive form of science, but a mode of thought orthogonal to—and not necessarily in competition with—the metaphysical and the positive. Progress is not from one to the other; rather, each grows in its own way. (Scheler suggests, for example, that among East Asian cultures the metaphysical form is typically dominant over both religious and scientific traditions.) Positivism according to Scheler is at root *eurocentric:* the positivists misinterpreted the crisis of religion in contemporary bourgeois societies as a natural and inevitable decline of religion and metaphysics throughout the world.[16]

Scheler's phenomenological classification of the forms of knowledge contrasts with the Marx-inspired historical materialist view, by far and away the most prolific source for early social theory of science. In its simplest formulation, historical materialism states that the means by which people make a living (economy) has much to do with the lives people lead in that economy (society). Applied to science, it is the doctrine that science is the product not of individual genius but of social needs, and that these needs arise from how a people makes its living. Thus in an influential essay of 1921, the Soviet sociologist and *Pravda* editor N. I. Bukharin postulated that *astronomy* arose from the need to find one's bearings in desert plains; *measurement of time* from the importance of seasons in agriculture; *physics* from techniques of material production and warfare; *chemistry* from the mining or manufacture of metals, glass, enamels, and paints; *botany* from the use of medicinal plants; *zoology* from efforts to domesticate animals; *geography and ethnography* from the development of trade and colonial warfare; *anatomy, physiology, and pathology* from practical medicine; *geometry* from the measurement of the earth and the contents of vessels; *arithmetic* from figuring inventories and balancing accounts; and *political economy* from the needs of merchants and their governing nation-states.[17]

One does not, of course, have to be a Marxist to trace changes in ideas to changes in the ways we make a living. Aristotle attributed the rise of geometry to the leisure time of Egyptian priests; Herodotus, to the need to remeasure land after the annual floods of the Nile. Bacon described how men are beholden to the nightingale for music, to the pot lid that flew off for artillery, and to the wild goat that heals itself for surgery.[18] In the period of which we are speaking, though, Marxist sociology of knowledge had the advantage of being allied with optimistic, forward-looking liberation movements—movements purporting to be either grounded in science, or predicted by science, or best able to utilize a science freed from feudal or capitalistic fetters. Social theory of science was often activist and rarely apolitical. Among historians of science, one of the most influential authors in this area was the Soviet physicist turned historian, Boris Hessen; we shall consider his contribution in some detail.

In 1931, a large delegation of Soviet scholars arrived in London for the Second International Congress on the History of Science. From the day of their arrival (they were the only delegation to arrive by plane) to the day of their departure (an extra session had to be planned to allow the presentation of the large number of Soviet papers), the delegation created a sensation, not least of all for their call for a new theory of science based on Marxian materialism and their promises for science under the new Soviet nation. Their papers were to shape the thought of an entire generation of British leftist intellectuals, prominent among them J. D. Bernal, Lancelot Hogben, Joseph Needham, and J. B. S. Haldane. Two papers—those of Boris Hessen and Nikolai Bukharin—were to cause the greatest sensation.

In "The Social and Economic Roots of Newton's *Principia*," Hessen put forth what at that time was a radical thesis—namely, that "the formation of ideas has to be explained by reference to material practice."[19] It was not just that scientific ideas have important applications; this was hardly controversial by this time and did not play an important part in Hessen's paper. The thesis that aroused controversy was not that science is useful, but that the usefulness of science is important in its origins. Hessen argued in particular that Newton's *magnum opus* was not merely an abstract treatise on mathematical physics, but rather a response to concrete technical and economic problems facing seventeenth-century British capitalism—especially problems of transport, industry, and military production.

By the end of the fourteenth century, Hessen argued, water transport had become the primary mode of transport in Europe. An average-size vessel could carry some 600 tons of cargo, 300 times that of the largest two-wheeled cart drawn by oxen. A trip from Constantinople to Venice took three times as long by land as by sea. With no means to measure longitude, ships were forced to hug the shores. The development of merchant capital thus posed the following problems: determining position on the open sea; determining the movements of the tides (for harboring ever-larger vessels); constructing internal waterways and canals; building ships with greater stability and maneuverability. Each of these, in turn, presumed certain bodies of knowledge. Increasing the reliability of vessels, their tonnage, and speed required a knowledge of hydrostatics and dynamics—laws, in other words, governing the floating of bodies in liquids and the movement of bodies in a resistant medium. Navigating on the open sea required knowledge of the movement of the heavens, a reliable chronometer, and a reliable compass.

By the seventeenth century, Hessen continued, exploitation of the mines had become Europe's leading industrial problem. Copper, tin, and iron were required for gunnery and defense, gold and silver for money, and coal for fuel (especially with growing firewood shortages). As surface mines began to be exhausted, mining ventures were faced with problems of raising ore, ventilating shafts, and pumping water from ever increasing depths. The problem of pumping water especially puzzled engineers. With the newly developed air pumps, water could be pumped vertically up to thirty-one feet but never more than this, regardless of the size of the tubes or design of the pump—so long as it was based on the suction of air. (Further pumping simply drew a vacuum.) It was in the face of such difficulties, Hessen argues, that Torricelli and Pascal developed the theory of pressure, attempting to weigh the air and prove the power of the vacuum. Scientists *explained* the vacuum, but only after industry (and in particular, mining) had *created* one.

Hessen argued in a similar vein that military technology posed challenges to the physicists of the day. Problems of ballistics required calculations of trajectories and of optimal range under a variety of circumstances. Tartaglia in 1537 first proved that an object propelled at a forty-five-degree angle yields optimal ballistic range. Galileo, who began his *Mathematical Demonstrations* with praise for the rich materials offered by the Florentine arsenal to the student of nature, was the first to describe the parabolic trajectory of a missile.

Each of these, Hessen argued—problems of industry, transport, and military production—were fundamentally problems of mechanics. And each of these was addressed in Newton's *Principia*. The problem of the trajectory of projectiles was solved in a general way through the calculus; the problem of the movement of bodies through a resistant medium was solved in the general theory of force and resistance. Problems of hydrostatics and hydrodynamics and the problem of the vacuum were advanced through the theory of force and pressure and through the law of gravity. Hessen maintained that Newton's own interests coincided with practical problems of the time. Hessen described a letter Newton wrote to his friend Francis Aston in 1669, advising him in his visit to the continent

> diligently to study the mechanism of steering and the methods of navigating ships; attentively to survey all the fortresses he should happen to find, their method of construction, their powers of resistance, their advantages in defence, and in general to acquaint himself with war organisation. To study the natural riches of the country, especially the metals and their production and purification. To study the methods of obtaining metals from ores. To discover whether it was a fact that in Hungary, Slovakia and Bohemia close to the town of Eila or in the Bohemian mountains not far from Silesia there was a river with waters containing gold, also to ascertain whether the methods of obtaining gold from gold-bearing rivers by amalgamating with mercury remained a secret, or whether it was now generally known. In Holland a factory for polishing glass had recently been established; he must go and see it. He must learn how the Dutch protected their vessels from rot during their voyages to India. He must discover whether pendulum clocks were of any use in determining longitude during distant ocean expeditions.[20]

Hessen's thesis that science emerges in response to particular economic and technical problems was warmly greeted by left-leaning British scientists and historians, not surprising given the context of the times. Britain was in the throes of a depression which many interpreted as the consequence of a lack of rational scientific management. The image of science united with politics under socialism was an appealing one, one that many of the early Vienna Circle positivists (for example) found attractive. Despite radical differences in their theories of science, Hessen and his British converts (J. D. Bernal, Joseph Needham, J. G. Crowther, Lancelot Hogben, Hyman Levy, and J. B. S. Haldane) on the one hand and the Viennese positivists on the other both agreed on the importance of science in solving social problems. Bukharin in his 1931 address cited Neurath in support of his hopes for organizing great scientific institutions, coordinated through economic planning into "one vast association of scientific workers."[21] Bukharin differed little from the positivists in speaking of the social function of science as (1) to increase our knowledge of the external world; (2) to invent and perfect technical processes; and (3) to overcome the forces opposed to technical advancement.

The Russian example of the unity of science and politics enchanted a generation of left-wing British intellectuals—so much so that by the late

1930s a strong scientific optimism, combined with a sense of the possibility of planning science ("Bernalism"), had come to represent the orthodox line in not only the British Communist party but among a wide range of liberals as well. Herbert J. Muller's 1936 *Out of the Night* promised the creation of a new race of men fostered by science and technology; H. G. Wells's 1933 *Shape of Things to Come* (and the film by W. C. Menzies based on this book) forecast a world where enlightened men of science would save the world from fascism. A flood of scientific utopias promised a brave new world powered by science-based technology.[22]

But the euphoria was not without dissenting voices. Following the tradition of E. M. Forster's *The Machine Stops* (1912), a series of dystopias began to appear in the 1920s and '30s, led by films such as Fritz Lang's *Metropolis* (1926) and (on a lighter note) Charlie Chaplin's *Modern Times* (1936) and culminating in Aldous Huxley's 1932 book, *Brave New World*—all of which stood the technocratic vision on its head. Conservatives meanwhile regrouped in the late '30s and early '40s to face what they saw as a unified threat of Soviet communism, Nazi tyranny, and the British planned science movement. In 1941 a group of conservative scientists and philosophers (eventually to include Michael Polanyi, John R. Baker, Karl Popper, Friedrich von Hayek, and Ludwig von Mises, among others) formed the Committee for Freedom in Science (later the Congress for Cultural Freedom) dedicated to the preservation of pure science.[23]

Most traumatic for the British socialists, however, were events in Soviet Russia itself. In 1948, at a session of the Academy of Agricultural Sciences, T. D. Lysenko's theory of the inheritance of acquired characteristics was endorsed as the official Soviet state biology. The decision was part of an effort to radically transform Soviet agriculture, bypassing the tedious procedures of traditional breeding to improve plant varieties and animal stocks.[24] Mendelian genetics was denounced, as were a number of other sciences (eventually relativity theory, quantum mechanics, and certain aspects of resonance chemistry), all as part of an effort to supplant "bourgeois" by "proletarian" science. The suppression of genetics shocked a generation of British radical scientists who had cherished Soviet society as a model for both socialist politics and support for science. Progressive British scientists were suddenly forced to choose between their science and their politics—a choice they never dreamed they would have to make.[25]

Capping the problem was the fate of Hessen himself. As Loren Graham has shown, the British radicals had understood Hessen's thesis on the social relations of Newton's *Principia*, but they had never really understood the social context of Hessen himself. Hessen's paper was not just an abstract treatise on the nature of science but part of an ongoing struggle in the Soviet Union to define the place of science in the new regime. Boris Hessen was one of the leading Soviet defenders of Einstein at a time when relativity, along with Mendelian genetics, was beginning to come under criticism as a species of "bourgeois idealism." In the years before his London paper Hessen had come under attack in Soviet journals as a "Machist" and "right deviationist"; in the fall of 1930, at a conference on the state of Soviet philosophy, Hessen had been denounced as a "metaphysicist of the worst sort" for his interpretation of relativity in terms similar to those of Sir Arthur Stanley Eddington. In his 1931 London paper, Hessen had hoped to prove that science, though indisputably a product of certain social and economic formations, was nevertheless able to transcend those origins and become useful in other contexts. If Newton's great *Principia* could be shown to have emerged in the course of particular struggles without this impugning its validity, might not the same be said for Einstein's theories?[26] Must the origins of a theory forever soil its use? Cannot a theory be *true*, regardless of its class origins?

Hessen was not to win his struggle. In the early 1930s he was arrested and imprisoned, and eventually died in the wave of purges that culminated in the late 1930s. By the middle of World War II all but two members of the Soviet delegation to the 1931 history of science congress (Kol'man and Zavadovsky) had perished in Stalin's prisons. Subsequent Soviet histories of the life and times of Newton (for example, the publication celebrating the three-hundredth anniversary of Newton's birth) ignored Hessen altogether.[27]

In the Anglo-American world, Hessen's theory has come under attack for being too "externalist," for attributing too much influence to material factors in the rise of science.[28] One of the earliest attempts to modify Hessen's account to include the role of religious elements was Robert Merton's *Science, Technology and Society in Seventeenth-Century England*. Here Merton combined the approaches of Hessen and Max Weber to produce a work that has become a classic in the sociology of science.

In 1904 Max Weber published a series of essays on "The Protestant Ethic and the Spirit of Capitalism," arguing that capitalism had emerged in northern Europe largely as a result of the rise of Puritan Calvinism and its ethic of salvation by grace, together with the belief that evidence of grace could be seen through works. Weber suggested the link of Puritanism and science but made no moves to explore it.[29] Thirty-four years later Robert Merton published his *Science, Technology and Society in Seventeenth-Century England*, applying Weber's thesis on the origins of capitalism to the origins of modern science.[30]

The thrust of Merton's thesis, following Weber, is that science is a social process, that the origins of its vitality and prestige must be sought in the "cultural values" of the age in which it arises and in the economic and military problems posed to it for solution. Science in the seventeenth century, for example, served a variety of "utilities":

- the religious utility of exhibiting the wisdom of divine handiwork;
- the economic and technological utility of enabling mines to be worked at increasing depths;
- the economic and technological utility of helping mariners to sail safely to ever more far-off places, in quest of adventure and trade;
- the military utility of providing for ever more efficient and inexpensive ways of killing the enemy;
- the self-development utility of providing a form of mental discipline (much as the study of Latin or even mathematics sometimes continues to be justified today); and
- the nationalistic utility of enlarging and deepening the collective self-esteem of Englishmen as they advanced their claims to priority of discovery and invention.[31]

Science, in other words, before it was widely accepted as a value in its own right had to be justified in terms of other social values. In seventeenth-century England, Puritanism shared with the emerging natural philosophy a set of common values that fostered the mutual appreciation of these separate institutions. The Puritan emphasis on the doctrine of salvation through grace and on proving evidence of grace through works provided a spur to science; conversely, it was in science that Puritanism found one of its key supports, as English Puritans found evidence of the glory of God in the study of his works. Following the work of Dorothy Stimson, Merton demonstrates that many of the early members of the Royal Society of London were Puritan Calvinists, that many saw in the production of scientific works proof of personal salvation. Merton also adduces evidence to prove that science was allied with Puritanism in the religious-political reforms of Charles and Cromwell.

Merton's thesis on the role of the Puritan ethic in the rise of modern science has given rise to an enormous critical literature. There is the problem of whether it is really fair to link "Puritan" values with science; there is the problem of Galileo's Catholic Italy, of science in the Netherlands, of Islamic science, Chinese science, and the numerous contributions of other heretics of various shades. Some have argued that it was not the specifically *Puritan* nature of English society that allowed science to flourish, but rather the tolerant, "latitudinarian" nature of the English Church and state in the days of the Glorious Revolution.[32] Still, from a larger perspective, the historical import of Merton's thesis lies less in the details of his claims for Puritanism or economic necessity than in the fact that, contrary to most scholarship at this time, factors widely believed to lie outside the practice of science are given a central role in its origins and flourishing.

In America, Merton's work remained for a dozen years or so one of very few landmarks in the young field of "sociology of knowledge." Sociologists tended to ignore science, preferring instead to concern themselves "with the juvenile delinquent, the hobo and saleslady, the professional thief and the professional beggar."[33] In his foreword to Bernard Barber's 1952 *Science and the Social Order,* Merton predicted that sociologists would turn seriously to the sociology of science only when science was widely regarded as a source of social problems. Merton himself, however, had continued work in this area, outlining what are now referred to as the "norms of science." In 1942, in his now classic "Note on Science and Democracy," Merton distinguished four such norms:

- universalism: "that truth claims are to be subjected to *preestablished impersonal criteria*" consonant with observation and previously confirmed knowledge.
- communism: that the "substantive findings of science are a product of social collaboration and are assigned to the community [as] a common heritage."
- disinterestedness: that the scientist searches for truth for its own sake, apart from the interests of class, status, nation, or economic or other rewards.
- organized skepticism: that judgment should be suspended until the facts are at hand and until beliefs have been scrutinized "in terms of empirical and logical criteria."[34]

The importance of Merton's paper for the history of the ideal of neutrality is twofold. On the one hand, it represents one of the first attempts to analyze the social functions of the ideal of "pure science." At the same time, it represents an important defense of the ideal of neutrality, a defense which, as he himself pointed out, emerged in response to threats to the autonomy of science.

In his 1938 article on "Science and the Social Order," Merton had explained that the ideal of "pure science" serves a dual function in modern society. On the one hand, the exaltation of pure science represents "a defense against the invasion of norms that limit directions of potential advance and threaten the stability of scientific research as a valued social activity." The ideal of purity is instilled early on in the scientist; science "must not suffer itself to become the handmaiden of theology or economy or state," for as soon as compliance with religious doctrine, political demands, economic utility, or any other "extra-scientific" criteria substitutes for purely intellectual criteria, science "becomes subject to the direct control of other institutional agencies and its place in society becomes increasingly uncertain." The ideal of pure science also serves certain psychological functions. Failure to ignore the broader context of one's research increases the possibility of bias and of error. The need to withdraw from such "externalities" is rooted in human psychology—in the fact that "discriminative behavior" requires a certain degree of interest, "yet if there is too much [interest] the behavior will cease to be discriminative."[35]

The ideal of pure science has also furnished a basis for revolt against science—a revolt that, Merton suggests, is found in nearly every society where science has reached a high stage of development:

> Since the scientist does not or cannot control the direction in which his discoveries are applied, he becomes the subject of reproach and of more violent reactions insofar as these applications are disapproved by the agents of authority or by pressure groups. The antipathy toward the technological products is projected toward science itself. Thus, when newly discovered gases or explosives are applied as military instruments, chemistry as a whole is censured by those whose humanitarian sentiments are outraged. Science is held largely responsible for endowing those engines of human destruction which, it is said, may plunge our civilization into everlasting night and confusion.

Merton explained such criticism in terms of what he called an "imperious immediacy of interest," an exaggerated concern for the short-term utility of science, a concern that is ultimately irrational insofar as it defeats important elements in the broader social scale of values. Such criticism "confuses truth and social utility." Equally confused was the view that science is necessarily for the good of humankind. The notion that the progress of science will prove beneficial in the long run may provide a rationale for scientific research "but it is manifestly not a statement of fact." Both of these views—the critical and the optimistic—involve a confusion of truth and social utility "characteristic in the nonlogical penumbra of science."[36]

The social function of Merton's own interest in "norms of science" is not difficult to see. Merton set out to elucidate the norms of science not as an abstract intellectual exercise but as part of a reaffirmation of the values of Western liberal democracy, values threatened by the rise of European totalitarian regimes at the time of his writing. Merton makes this clear in the first few lines of his 1942 article—originally titled not "The Normative Structure of Science" as it appears in subsequent reprints but rather "Note on Science and Democracy"—and later, in 1949, "Science and Technology in a Democratic Order."[37]

Merton begins his 1942 essay by warning against the "contagions of anti-intellectualism" threatening to become epidemic, then observes that "incipient and actual attacks upon the integrity of science have led *scientists to recognize their dependence on particular types of social structure*." A tower of ivory, he continues, "becomes untenable when its walls are under prolonged assault." In the face of this assault it is the task of democratic societies to reaffirm the values compatible with genuine science. Science develops in various social structures—but which of these "provide an institutional context for the fullest measure of development"? It is in an attempt to answer this question that Merton elaborates the four institutional imperatives comprising the "ethos of science"—universalism, communism, disinterestedness, and organized skepticism.[38]

The norms of science Merton elaborates are inverted images of the values advanced in the name of Nazi "Aryan" science and, to a lesser degree, Soviet proletarian science. In his defense of universalism Merton attacks the Nazi alignment of science along racial lines and the Soviet attempt to direct science along "proletarian" lines. Such attempts must fail: "The Haber process cannot be invalidated by a Nuremberg decree nor can an Anglophobe repeal the law of gravitation." Merton points out that "purity" has been invoked to require that scientists blind themselves to practical impacts of their work. He cites George Lundberg's assertion that it is "not the business of a chemist who invents a high explosive to be influenced in his task by considerations as to whether his product will be used to blow up cathedrals or to build tunnels through the mountains."[39]

In recent years criticism has been leveled against Merton's sociology on the grounds that it concerns itself purely with the structure of scientific institutions and insufficiently with the effects of structures, interests, and ideologies on knowledge itself. Norman Storer, in his edition of Merton's essays, observed that the Columbia sociologist concerned himself with science "as a social institution rather than as a type of knowledge."[40] (Merton's analysis of Nazi science, for example, was concerned almost exclusively with the negative effects of the Nazi state on the progress of science—ignoring the role of science in the development of Nazi ideology and the reliance of that ideology upon certain forms of science, including the ideal of "pure science.")[41] Critics have argued that the "norms" of science are idealizations—that parochialism, self-aggrandizement, and conservatism are as common in science as in other pursuits. Many of his most vocal critics have argued that the real task of sociology must be to examine how scientific ideas are "socially constructed"—how military and medical priorities, or patterns of exclusion of women and minorities, or rules for the maintenance of disciplinary boundaries, have structured the logic of scientific practice.

Here too, though, as we shall see, the question of value-neutrality has remained a vexatious one. Those who announce that science is "socially constructed" often do so from a neutralist vantage point that rivals the narrowest positivist empiricism of the 1950s.

16

Realism versus Moralism

Sociologists in recent years have dedicated themselves to exposing various myths or idealizations in positivist/internalist accounts of science—that science is inherently progressive or automatically beneficent, that science grows by slow and steady steps, that observation is primary to theory, or that science is, like nature for Aristotle or God for the Persians, the only "self-moved mover." Sociologists have brought a new *realism* to the study of science, recognizing that science is "what scientists do" and not just what they say they do. The import of this view has been enormous. It is now widely recognized that science serves certain interests, that science is rarely neutral insofar as it touches the vital affairs of humanity—our health, our status, our wealth and power, our security, our happiness—that science is not neutral in regard to these things but participates in their fulfillment or frustration.

Historians have also rejected the image of scientists as "autonomous god-like creatures acting in a world of unconditioned freedom" (Herbert Butterfield's parody), and have sought instead to explore the origins of science in terms of its social context. The development of science, in this view, is not a self-sustaining filiation of disembodied ideas but a social process with material and social prerequisites. We appreciate why geologists know more about oil-bearing strata than any other strata, why virologists studies viruses that attack the tobacco plant. We know that thermodynamics was stimulated by problems in the development of the steam engine and armaments industry; that Pasteur's discovery of microorganisms was aided by progress in the fermentation of wine and beer; that research into genetics was fostered by concerns to improve agriculture and the human racial constitution. We know that the development of electromagnetic theory was spurred by hopes for a source of power that could cover large areas; we also know that scientists for centuries have distorted the female body, at the same time that women have been systematically excluded from the practice of science. We have begun to see the emergence of studies of popular and alternative movements in the sciences, of the place of women, blacks, and other groups marginalized in the sciences, of science in Third World or non-Western cultures. Paleoastronomy and the "anthropology of science" have become recognized disciplines; participant-observation in scientific laboratories has been added to the stock of sociological methods. We have studies of Big Science, proletarian science, and Nazi science; there are studies of science in Gupta India, medieval Islam, Cultural Revolutionary China, and in the progressive or technocratic West. There is much evidence, in other words, to show that the scientist, no less than other mortals, is a *zoon politikon,* sharing the hopes and aspirations, frailties and foibles, of other men and women. There are many slogans to fill this bill: that objectivity is "just one more form of persuasion" (John Peters' attribute to C. Wright Mills), that science is "politics by other means" (Bruno Latour).[1] Realism, in short, has provided a healthy corrective to idealized textbook visions of science.

But realism, I would argue, is not enough. In the face of unprecedented environmental destruction and the militarization of science, the relations of science and society are not epistemological or historical niceties but pivotal issues in the well-being of humans on the planet. The point, in other words, is not to chronicle our madness but to escape it. Neutral sociological "realism" in this case may blind itself to a deeper issue—that science is at least part of the problem, and that alternatives must be sought in the theory and practice of science itself.

I would not take issue with the realist thesis that structural-functionalist sociology of science for many years exaggerated the autonomy of science, focusing excessively on the (often idealized) mechanisms by which scientific institutions maintain their distance from other parts of society. I also agree that it is wrong to take the success of an experiment as a sufficient explanation for its having been carried out;[2] so would I also agree that sociologists since Mannheim have tended wrongly to seek an autonomous sphere of intellectual work, free from social influence. I would tend to agree rather with Schumpeter that it is wrong to allow such a "fire escape"; one has to take the heat. Joseph Ben-David was wrong in assuming that, though it is fine to pursue a sociology of "blind-alleys and wrong pathways," there is little possibility of a "sociology of the conceptual and theoretical contents of science."[3]

So, too, would I support Steven Shapin's criticism of what he calls the "coercive model" of the sociology of knowledge, according to which sociological explanation consists in claims that "all (or most) individuals in a specified social situation will believe in a specified intellectual position." Sociological explanation should not be set against the contention that scientific knowledge is empirically grounded in sensory input from natural reality.[4] The problem, though, with what I shall be calling "realist" sociology of knowledge (including most prominently, and ironically, many of those who call themselves "social constructivists")[5] is not that it is insufficiently empirical (as Shapin implies, in his discussion of his critics) but that it is too radically empiricist. Shapin, like other realists, rejects "privileging specific formulations of reality." His ultimate goal is "writing more sensitive history"; this means abandoning the "patently normative attitude towards rationality" which appears to inform the coercive model's view of the determination of the social.[6] But: there is an alternative to both the coercive and the realist views (ignoring altogether the Mertonian, "description of norms" approach)—namely, the activist or *moralist* view.

What do I mean in calling for "moralistic" science theory? What I mean is really quite simple—namely, that attention needs to be paid to the concrete forms of suffering and injustice in the world, and that this is as much a challenge for the science theorist as for anyone else. Danger and failure can become the objects of scholarship, as much as glory and success have been. Where are the studies of science's role in producing poverty, disease, or political oppression? Where are the studies of the militarization of science, or even of science's part in keeping the peace? Where are the studies of how science has worked to exacerbate racism, sexism, or colonialism?[7]

Moralism does not mean that the canons of objectivity and accuracy should be relaxed. Advocacy need not compromise objectivity. Objectivity is a matter of method; advocacy is a matter of commitment. The moralist has to assume that facts and values are distinct in at least one sense—that the real is often not the good. Eric Hobsbawm has pointed out that new social trends are often first perceived by critics, because it is in their interest to be aware of novelties. Sandra Harding makes a similar point in her suggestion that criticism is often a precondition for objectivity (see Chapter 17). Both of these notions contrast sharply with what I will be calling the "realist" thesis—namely, that it is the task of science theory to analyze but not to criticize, to explain but not to complain.

One of the most influential "realist" texts is now nearly two decades old. In 1974, Barry Barnes published his *Scientific Knowledge and Sociological Theory*, announcing what he called a new, "strong program" for the sociology of science and reasserting the virtues of value-neutrality in science theory. The program has been an influential one, capturing the imagination of dozens of students in the new field of "science studies." The program is by no means homogeneous, but there is a strain of self-conscious "agnosticism" about what is true or good in science that pervades many of the writings in this vein.[8] Barnes is typical in his antimoralism, and so we examine his book at length.

The program Barnes announces is a broad one, namely, "to understand and explain beliefs about nature and their variation." It is a program he qualifies, however, with a methodological objection to existing research: "The investigation of science as a phenomenon has been incompletely differentiated from its justification." As a result our present understanding of science has become "hopelessly conflated" with concerns about what science ought to be. Barnes contrasts this "evaluative" approach with his own "sociological approach," which is to "take science as it finds it"—that is, to consider scientific knowledge as a set of accepted beliefs and to ignore as much as possible any assumptions of what are correct beliefs. Barnes notes that this is a goal which, strictly speaking, never can be achieved in practice; he apologizes in advance for whatever "harmless evaluations" or elements of advocacy or criticism may slip into his book.[9]

Value-neutrality is not an incidental aspect of Barnes's approach. He elaborates a principle he calls "the sociological equivalence of belief systems," according to which all beliefs must be judged of equal worth. "Many academic theories about beliefs divide beliefs about nature into 'true' and 'false' categories." The former, he says, are presented as unproblematic in the sense that they directly derive from an awareness of reality. The latter, "false," categories are in this view to be accounted for by "biasing and distorting factors." Philosophers are supposed to be responsible for establishing criteria to determine "the truth or falsity of beliefs"; sociologists and psychologists are given the task of accounting for the falsehoods the philosophers expose by unearthing causes of bias.

The alternative approach Barnes suggests is an "instrumentalist" or radical relativist view, in which the meaning of ideas is entirely dependent upon the uses to which they are put. He rejects the notion that ideologies are necessarily "false, distorted, or otherwise adversely influenced by social factors," contrasting this with his own perspective according to which we may say that *all* thought is ideological. It is meaningless to talk of particular beliefs as "intrinsically ideological," for ideology is a contextual affair. The ideology of functionalism, for example, is not conservative in the abstract but only according to the particular uses to which it is put. Beliefs must be seen as tools adaptable to different uses in different circumstances.[10] The view that "all men are created equal" may be conservative in one context (as, for example, in a repressive meritocracy) and liberal or progressive in another (if it is used to attack aristocratic privilege, for example).

Thus far, Barnes has said nothing new or radical—he acknowledges a debt to Veblen and to Durkheim, though he ignores others in the German tradition. But his Edinburghian relativism is more than "methodological" (in the sense that Melville Herskovits used this term).[11] It is *ontological:* that is, there is no sense in which we may say that any particular belief is truer or better than any other. His point is not simply that we should consider beliefs in the context of their cultural meaning, but that ultimately no belief can be considered better or more rational than any other, no belief can be considered closer to reality than any other.[12] It is not just that "any belief may be made to serve any particular interest," but that there are no "privileged" views concerning truth or value.

Barnes links his rejection of "privileged views" to a demand to eschew all judgments of value in regard to science. The error of previous philosophical systems has been their tendency to judge other belief systems according to standards set by modern science. Barnes attacks both Clifford Geertz and Talcott Parsons for their residual "evaluative elements," for assuming that social forces operate in one sphere (nonscience) and not in the other (science). He criticizes those anthropologists who "rely on the truth of science" to identify, for example, the inefficacy of magic; he argues that it is on this and no other basis that anthropologists justify their claims that "ceremonies do not cause the rain that follows, or that chanting does not aid the growth of corn."

Barnes objects to the view he says has been shared by most sociologists since Merton, that science somehow stands apart from other forms of culture as singularly pure and undistorted, a view associated with the conception of scientific institutions as autonomous. Barnes's task is to "exorcise" this view; his alternative: a sociology that "applies to true and false beliefs alike," one that sees true beliefs no less than false as the product of social forces. This is the Edinburgh School's *strong program,* according to which all forms of knowledge must be considered as "causal" as any other. This he calls variously *naturalism* or *instrumentalism:* a standpoint according to which the science theorist is not interested in "whether beliefs are true, but to show they are natural."[13]

The problem with such a view is that it is abstract—ironically, in ways that are similar to the "intrinsic" conception of ideology he wishes to transcend. If one adopts Barnes's principle of the "equivalence of belief systems," how does one distinguish between oppression and liberation, truth and lies? This would not be so important were the "strong program" simply a sociological method or a proclamation of personal epistemological impotence. But the program makes stronger claims—Barry Barnes claims to have solved "what have been regarded as the epistemological problems of the sociology of knowledge."

Despite his posture as attacking positivism, Barnes remains a staunch defender of the ideal of value-free science, if admittedly in a somewhat different sense from other versions of this ideal. Advocacy in his view compromises objectivity. Barnes is not entirely unaware of the social functions of neutrality: at the end of his book he recognizes that "since science is expected to be descriptive, evaluations within it may *masquerade* as facts or neutral statements." Indeed the "not entirely undeserved reputation for disinterest" scientists enjoy may give legitimation to values implicit within their knowledge claims. He dismisses this general problem, however, with the remark that, "although significant, [it] is not of great importance." It is rather in "semi-popular writings and in marginally-scientific areas" that overt evaluations can become "insidious"; he mentions pop ethology and ecology in this regard. (One wonders on what basis Barnes considers evaluations insidious, given his belief that all belief systems are sociologically equivalent.) Science may also be evaluative, he notes, by virtue of its terminology. But this he suggests should not be pushed too hard, for the "statements in physiology and pathology textbooks may legitimately be held to be factual or genuinely descriptive." We do not "seriously discriminate, exploit, or rank order in terms of heart conditions or in-grown toenails"; most of the terminology of biological science is similarly "innocuous and lacking in social function." The evaluative component of science can therefore safely be treated as "significant but not of great importance, and as *eliminable*."[14] Barnes defends this point of view on the grounds that this is how scientists themselves conceive of the relation of science and values—and in his conception of sociological method, it is always the positions of "the actors themselves" which must be the starting point (and endpoint?) of inquiry.

But which actors? Scientists have never spoken with a single voice on the question of neutrality versus activism. Barnes's conception of the sanctity of the historical past and the sociological equivalence of beliefs leads him to reject all judgments in terms other than those of some supposedly homogeneous "actors themselves." He therefore asserts that it is "incongruous" to expose the writings of Francis Galton and the eugenicists at the end of the nineteenth century as "racist" (his quotes), given that these ideas "fitted naturally and securely in the taken for granted world of the time."[15] One might wonder whether criticism of Nazi eugenics as racist is equally "incongruous," given that Nazi ideas also "fitted naturally and securely in the taken for granted world of the time." (Of course, those who suffered under the Nazis might have different views about how natural such ideas were.) To assume the incongruity of critique is to accept the servility of scholarship, the idea that might makes right, the erroneous idea that any particular time has only one prevailing view.

Barnes's neutralism is continuous with historical efforts on the part of sociologists to professionalize under the banner of science, a theme we have encountered in earlier chapters. The strong program preserves a traditional canon of liberal, scientific sociology—namely, the view that the objects and events of social analysis must be considered, in Tönnies's words, as if they occurred "on the moon." Barnes and his positivist and professional predecessors are united in their conception of the task of theory as passive explanation, not as transformative critique—but this makes his appeal to Veblen's instrumentalism very partial and puzzling. Barnes completely ignores Veblen's biting analysis of invidious and noninvidious judgments. Veblen was a social critic, a thorn in the side of robber barons and business enterprise. It would be hard to find a more stark contrast with Barnes's own sociological "equivalence principle."[16]

There is one final point I should make concerning the problem of ideology in science, a problem which deserves further study. It has become common in social theory of science to criticize the ideal of neutral science and to argue instead that all science is value-laden in the same sense that there is no fact independent of theory, no neutral observation language, and so forth.[17] This position may be seen in many of the sociological writings on science for the last two decades or so. There is one qualification I would suggest to this trend, one that has to do with the effects of ideologies on the practice of science.

It is possible to argue that all science is "interested" (practiced in the service of interests), yet one should not underestimate the capacity of scholars to lose themselves in rather uninteresting trivia and detail. The "uselessness" of science, in other words, is not just a myth but can also be a reality, especially under social circumstances where irrelevance is encouraged. It is important to distinguish two senses of neutrality in this regard: neutrality as an apology or myth, used to disguise interests served by science (in the form of false consciousness or outright deception); and neutrality as a style of scientific practice—a way of doing science. The value of this distinction lies in recognizing that ideologies are not just illusions but have concrete effects and serve particular functions. Neutrality (for example) serves to preserve the autonomy of science, insulating science from critique. It serves as well as a mask, disguising whatever interests may lie behind the origins and maintenance of research priorities. The ideal of value-neutrality may have consequences for the kinds of sciences actually pursued: a science that does not question the ends it serves (or potential sources of bias) may miss certain fruitful areas of research that do not serve those interests. In other words, science practiced under the banner of neutrality does not escape unscathed. A price is paid, not just by science but by society as a whole. This is because theory can shape the world as it is shaped by it. And in a world where the quality of life depends at least in part on the quality of our science, this is a serious consideration.

The problem of ideologies in science, in other words, is not simply a problem of "exposing" myths; the problem is also to explore the extent to which the concrete practice of present-day science might be different. Sociologists, insofar as they restrict themselves to the neutral description of science in its social context, leave this key legacy of positivism—the legacy of detached indifference—unchallenged. There are two sides, then, to the new sociological realism as applied to science theory. On the one hand, realism serves the function of exposé, and in this sense may serve to challenge and enlighten. On the other hand, raised to an epistemological level, the radical relativist sociology of knowledge fails to engage the question of alternative scientific priorities in a constructive manner. The science theorist explores what is, but not what might or should be.

It is in this sense, then, that I would advocate a renewed and politicized "moralism" to complement the realism that has emerged in recent science theory. The problem with neutral empiricist realism is that it confuses the tasks of theory with its prerequisites. The task of theory is to judge and to act, not to prove one cannot help but judge. The point is to *enter* the fray, not just to chart its boundaries. Science must not be let off the hook—we cannot rest comfortably in the view that science needn't become "political" since it already is by definition.

Theory, however sociological, is too excellent a bystander; genuine theory of science must be *political* theory, because the objects of social theory always exist in unfinished form. The task of theory is a creative, or at least a catalytic, one; it is not enough to "discover" that values play a role in science. Ideology exposed in theory does not disappear in practice: science cannot be declared moral, but must be made so.[18] The problem of science theory in this political or moral sense is not a sociological or epistemological nicety, but a question of how and to what extent science might be practiced in a more "socially responsible" manner. This is the spirit of science criticism—to which we now turn.

17

Critiques of Science

If the ideology of scientific neutrality represents, among other things, a response to efforts to criticize science, then it is important to understand these critiques if we are to understand the origins of the ideal of neutrality. Science criticism, as I shall be using the term, presupposes a rejection of the neutrality of science. Criticism in this sense is directed against the moral or political implications of particular scientific practices; it may or may not be directed against the truth claims of particular theories.

The critique of science (or of technology) is as old as science itself and, like science, has changed substantially over the course of time. Ancient Greek philosophers recognized the difference between "means" and "ends" and with this, the idea that tools might be used for alternative uses, for good or for evil. Plato in the *Republic* notes that the virtue of a tool is its proper use, but that those well-trained in its use are usually also expert in its abuse. Socrates recognizes that those most capable of healing are also those most capable of hurting, and that those most competent to speak truthfully about an issue are also most competent to lie.[1] Ancient mythologies often warn against the dangers of the unrestricted pursuit of knowledge: Prometheus is punished for bringing man fire from the gods; Icarus challenged the gods' possession of the skies and died for his presumption. Eating from the tree of knowledge (of good and evil) drove Adam and Eve from the Garden.

Among the earliest forms of "science criticism" (if one can call it that) are critiques of logic. Aristotle took pains to expose the fallacies of logic, distinguishing genuine and spurious forms of reasoning. Plato's *Gorgias* is a lesson on the dangers of formal reason freed from substantive wisdom; his

Parmenides is a lesson on the caution demanded of deep and searching thought. The idea of "sophistry" represents the dangers of reason led astray; nowhere is this more humorously expressed than in Aristophanes' comedy, *The Clouds* (420 B.C.), where by ridiculous logic the son of Strepsiades is able to justify beating up his father. The comedy also contains a delightful lampoon of scholars puzzled by whether gnats hum "through their mouth, or backwards, through their tails." Strepsiades celebrates upon hearing the answer from the "master of the gnat": "So then the rump is trumpet to the gnats! O happy, happy in your entrail learning!"[2]

Christian philosophers of the Middle Ages added to this the concern that knowledge must always be pursued in a prudent and restrained fashion. The medieval historian Robert Lopez notes that the Jesuit *Enciclopedia Cattolica*, borrowing from Saint Thomas and Saint Jerome, described three attitudes toward the acquisition of knowledge: too little interest in knowledge was *culpable ignorance*, a vice; a prudent interest was *studiousness*, a virtue; an excessive craving for knowledge was *curiosity*, a sin. "Sinful curiosity" included seeking knowledge for pride or some evil purpose, seeking knowledge with forbidden methods (such as necromancy or divination), and prying into God's secrets—mysteries of the faith, the end of the world, and the hidden intentions of the Lord. Saint Augustine, in Book X of his *Confessions*, had similarly deplored that "futile curiosity that masquerades under the name of science and learning," comparing the investigation of the secrets of nature to the fascination we find for freaks and prodigies at the circus. Sensory experience was a "lust of the eyes" that one must diligently guard against to preserve piety.[3]

With the rise of Renaissance humanism one begins to see criticisms of the narrow or trivial character of much natural inquiry. Science might categorize nature and discover some nicety, but what does it do to uplift the human soul? Petrarch, often regarded as the founder of modern humanism, complained in the fourteenth century that the Aristotelian schoolmen

> know many things about animals, birds and fish: how many hairs the lion carries in his mane, how many feathers the falcon carries in his tail, and with how many coils the sea serpent surrounds the foundering ship at sea. They know how the elephant gives birth, and that she can live to be two or three hundred years old; that the crocodile alone, of all the animals, can move his upper jaw—all of which surely in large part is false, and even were it true, would not help one lead a happy life. For I ask you, what use is it to know the nature of animals, birds, fish and snakes, if one does not know, or even disregards, the nature of man—his goals, his origins, and his destiny?[4]

Petrarch expresses a contempt for science that would remain in humanistic writings for centuries to follow, the same kind of critique that Cyrano de Bergerac would raise in his mockery of devices to send men to the moon or that Thomas Shadwell brought forth in his spoof on *The Virtuoso,* who presumed to know everything on the basis of science.

It was in a similar, if more light-hearted, spirit that Jonathan Swift described Gulliver's extraordinary travels to the Academy of Lagado in Laputa. In a thinly veiled jab at the Royal Society of London, Swift describes projects for "extracting sunbeams out of cucumbers, which were to be put into vials hermetically sealed, and let out to warm the air in raw inclement summers"; another academician was busily seeking a way "to reduce human excrement to its original food, by separating the several parts, removing the tincture which it receives from the gall, making the odour exhale, and scumming off the saliva." Still others were working to calcine ice into gunpowder or to determine the malleability of fire. An architect had contrived a new method for building houses "by beginning at the roof and working downwards to the foundation"; there was also a man born blind who, together with several apprentices also in his condition, was trying to mix colors by feel and by smell.

Swift describes his encounter with the Academy's "universal artist," a man employed thirty years seeking the improvement of human life:

> He had two large rooms full of wonderful curiosities, and fifty men at work. Some were condensing air into a dry tangible substance, by extracting the nitre, and letting the aqueous or fluid particles percolate; others softening marble for pillows and pincushions; others petrifying the hoofs of a living horse to preserve them from foundering. The artist himself was at that time busy upon two great designs; the first, to sow land with chaff, wherein he affirmed the true seminal virtue to be contained, as he demonstrated by several experiments which I was not skilful enough to comprehend. The other was, by a certain composition of gums, minerals, and vegetables outwardly applied to prevent the growth of wool upon two young lambs; and he hoped in a reasonable time to propagate the breed of naked sheep all over the kingdom.[5]

The image of the natural philosopher in Swift is that of the ridiculous bungler or misguided fool. In Petrarch, the critique is directed against those who concentrate on technical details while ignoring problems of the moral life, losing the forest for the trees. This is the humanist critique of science, and it is one that continues into the eighteenth and nineteenth centuries. John Locke, for example, warned of knowledge that might comprehend the workings of a clock yet misread the time: "He that was sharp-sighted enough to see the configuration of the minute particles of the spring of a clock, and observe on what its elasticity depends, would discover something very admirable; but if eyes so framed would not at a distance see what o'clock it was, their owner would not be benefited by their acuteness." Goethe in his *Faust* mocked those "educated gentlemen" for whom meaning resides only in that which can be calculated, weighed, or coined.

Thomas Carlyle complained about scholars for whom the creation of the world was "little more mysterious than the cooking of a dumpling"; the chapter on "Pure Reason" of his 1833 *Sartor Resartus* presents a diatribe against the "logic-choppers, and treble-pipe scoffers, and professed Enemies to Wonder; who, in these days, so numerously patrol as night-constables about the Mechanics' Institute of Science." Carlyle was a conservative romantic who believed there was very much wrong with the modern world. He was as anti-science as he was pro-slavery—witness his blast against social science ("the Dismal Science") in his essay on "The Nigger Question."[6]

One can distinguish this humanist critique from the clerical critique of science as subversive or contravening morality. Repression of science by ecclesiastical authorities is what one most commonly imagines when one thinks of sixteenth- and seventeenth-century criticisms of science. The stories here are familiar. In 1543, Nicolaus Copernicus published his *De revolutionibus,* arguing that it is the earth that moves while the sun stands still. Copernicus disrupted the harmony of the universe: heliocentrism made it difficult to interpret the Bible in a literal fashion. Only by virtue of a diplomatic introduction by Andreas Osiander, claiming that the treatise presented only a model, and not necessarily a picture of reality, was Copernicus saved.

Copernicus suffered little more than having his treatise placed on the papal index of prohibited books, the *Index Librorum Prohibitorum.* Colleagues of his were not to get off so lightly. Giordano Bruno, taking Copernicus seriously, argued that if our earth is a planet then there must be other worlds, possibly with other inhabitants like us. The Catholic Church would not stand for such heresy, and in 1600 Bruno was burned at the stake at Campo dei Fiori, Rome. Galileo's work, published in the vernacular, also proclaimed support for Copernicus; his collected works, together with those of Spinoza, were placed on the *Index* until the early nineteenth century. Luther and Calvin both condemned Copernicus' heliocentrism; both rejected Harvey's discovery of the circulation of the blood.[7]

Until quite recently most critics of science—whether humanist or clerical—tended to focus on untoward consequences for morals, wisdom, or religious belief rather than on health or environmental consequences. Until the twentieth century, in fact, one finds relatively little reference to the dangerous physical or ecological effects of science, partly because, despite Baconian hopes, the results of "high science" (the principles discovered by Galileo, Newton, or Harvey, for example) were slow to penetrate the practical trades and arts. At the time of Newton, sailors were still using Ptolemaic methods to determine position at sea. Gunners used artillery tables drawn up from practical experience, not mathematical calculations based on mechanical principles. Defects in the casting and boring of gun barrels or the preparation of explosive charges, combined with uncertainties in battlefield logistics, meant that the degree of accuracy provided by mathematical calculation was of little value. I. Bernard Cohen argues that Benjamin Franklin's invention of the lightning rod is the first dramatic example of the conscious application of scientific principles to practical life. As a consequence, he suggests, the fear and distrust of science for its effects that we are accustomed to in the present century—ecological damage, nuclear catastrophe, and so forth—do not figure importantly in writings on science prior to the nineteenth century.[8]

The case is otherwise with respect to technologies, or sciences more closely connected to practical life. In the *Phaedrus*, Plato has the Egyptian king Thamus express his fear that the invention of writing will soon "implant forgetfulness in the souls of men" by allowing them to write down that which, in previous times, would have been committed to memory. Criticisms of weapons of war has a history just as old. The early Hebrews banished iron from holy places, regarding it as a base and unclean substance, given its use in war. At the Lateran Synod of 1097, Pope Urban II banned the crossbow as an inhuman weapon; the papal ban was reaffirmed in 1139, when Pope Innocent II further denounced the "deadly and God-detested art of slingers and archers." Conrad III in Germany banned the crossbow from his kingdom; European leaders were no doubt worried by the fact that the weapon enabled a relatively unskilled yeoman to pierce the armor of his lord. In the fifteenth century, Leonardo da Vinci refused to publish plans for a submarine, because he feared it could be used as a weapon. And in the seventeenth century, Robert Boyle kept secret a poison he had developed for fear of its abuse. Francis Bacon himself bemoaned the tragedies that had issued from the invention of gunpowder—as did William Shakespeare, who penned several lines about "villainous saltpetre." Indeed the very first reference to gunpowder anywhere, a ninth-century Chinese Taoist treatise, instructs alchemists *against* the mixing of saltpeter, charcoal, and sulfur for fear of explosions.[9]

None of these particular warnings seems to have met with much success. The Teutonic knights used the crossbow to rout their Estonian enemies; the perfection of poisons, submarines, and explosives has continued apace into our own times. Technologies of war have been defended as technologies of peace, as when Alfred Nobel justified his invention of dynamite with the argument that such terrible powers of destruction would make war unthinkable.

Criticism of the arts of war is old, but so is criticism of the impact of innovations on employment. Throughout history, we have examples of laborers rejecting the introduction of machines that threaten their livelihood. In 1479 workers in Danzig demanded that the government drown the inventor of a loom that could weave several pieces of cloth at the same time. "Machine breaking" culminates early in the nineteenth century, when workers in the lace and hosiery trades (and later in the woolen and cotton industry) rioted to protest working conditions. Employers tried to read into machine breaking a kind of romantic, anti-technological impulse, though it most often occurred in the context of violent strikes or work stoppages where workers simply destroyed the closest things at hand, including the machines with which they worked. "Luddites" were sometimes also motivated by concerns that the new machines were producing a shoddy product, bringing the entire industry into disrepute. As Eric Hobsbawm has shown, the destruction of local mills and machinery by early industrial and farm workers formed the social background to the modern industrial strike.[10]

Attitudes toward innovations, one must keep in mind, are not abstract but depend upon what people think they stand to lose or gain. In 1673 a bill was introduced into British parliament to ban stage coach travel within about fifty miles of London, on the grounds that the new form of transport was threatening the livelihood not just of watermen (those who carried the mails up and down the Thames) but also of "cloth-workers, drapers, taylors, saddlers, tanners, curriers, shoemakers, spurriers, lorayners, felt-makers" and a host of other trades dependent upon the old riding industry (the stage was not nearly so hard on one's clothing!). Girdlers, sword-cutlers, gunsmiths, and trunk-makers were also said to have suffered, as had the owners of roadside inns and the breeders and trainers of horses.[11] The authors of this legislation worried about the inconveniences imposed by the new transport (rising early to meet fixed schedules, waiting when the equipment broke down, not being able to choose where or when one stopped or started), but their primary focus was on the loss of livelihood.

There are other examples from more recent times. In the nineteenth century farmers cultivating the madder plant, the traditional source of the dye known as "Turkey red," protested the introduction of synthetic substitutes on the grounds that these would hurt their livelihood. The French army dyed the trousers of its soldiers with natural dyes from the madder plant long after alizarin dyes from coal were available, in order to maintain employment for thousands of madder farmers in Provence. In our own century, one of the main criticisms of the high-yielding varieties of rice and wheat developed in the 1940s and '50s has been that the new varieties are inherently biased against smaller farmers, who are unable to purchase the necessary pesticides, fertilizers, and heavy equipment required to exploit the new strains. Many of the world's poorer farmers have actually been impoverished by such "advances": Mexico City today is the largest city in the world partly because the "green revolution" forced smaller farmers to leave the land and seek work in the city.[12] Concrete fears thus often underlie abstract calls for limits to (or redirection of) technical or scientific progress.

As one might imagine, concerns about the negative impact of science on society are often voiced in times of economic crisis. In the 1930s, many blamed the Great Depression on the failure of scientists and engineers to consider the human impact of their work. L. Magruder Passano, a professor of mathematics at MIT, wrote to *Science* magazine in 1935 suggesting that the chief aim of scientific research was "to enable those who already receive an undue share of the wealth produced by industry and research, to appropriate a share still larger." Harvey Cushing, a professor of surgery at Yale, complained that "the average man" had begun to feel that "scientific research and the labor-saving inventions which grow out of it are chiefly responsible for the hard times and unemployment and uneven distribution of property" made so apparent in the Depression. When Sir Josiah Stamp and others (most notably the Bishop of Ripon) called for a moratorium on science and invention, what they were really worried about was the displacement of men by machines—one of the primary causes, in their view, of the Great Depression.[13] One reason Theodore Koppanyi in the 1940s opposed the founding of the National Science Foundation was his concern that there was already enough science being done. His solution appears to be unique in the history of science criticism, recommending as he did that the government follow the example of the Agricultural Adjustment Administration and pay people *not* to do research![14]

The concerns of Koppanyi and the Bishop of Ripon sound curiously anachronistic today. More familiar are the kinds of doubts raised by postwar atomic scientists worried that they had unleashed "a monster" into the world. After the detonation of atomic bombs over Hiroshima and Nagasaki, project leader J. R. Oppenheimer summed up the doubts of several of his colleagues when he stated that scientists "had come to know sin." Scientists organized to try and halt the rapidly growing arms race; in 1945 the *Bulletin of the Atomic Scientists* began to mark time until "midnight," when final nuclear war would begin. The sense of shame felt by physical scientists spread to other fields: psychologists began to question the ethics of using subliminal techniques to sell products (by inserting into movies 1/3000-second messages exhorting viewers to "Eat Popcorn" or "Drink Coca-Cola"); Arthur Bachrach drew an explicit parallel between the moral responsibilities faced by psychologists armed with such techniques and those of physicists who built the bomb.[15]

Physics is not the only area for which World War II represents a turning point in the ethics of science. Human experimentation became a postwar concern especially in the wake of the 1946–47 Nuremberg trials, where German physicians accused of crimes against humanity were shown to have performed cruel and often fatal experiments on prisoners in Nazi concentration camps. The Nuremberg Code that emerged from the physicians' trial was supposed to codify a set of standards for the ethical treatment of human subjects, including informed consent, the use of humans only as a last resort, and the avoidance of any experiment that might bring about the death or permanent injury of the subject.

In subsequent years, however, the code would be violated time and again, even in the country of its authors. In 1949, for example, officials at the Hanford Nuclear Reservation in southeast Washington released some 27,000 curies of radioactive xenon-133 and iodine-131 into the atmosphere as part of an experiment to determine dispersal patterns for radioactive isotopes. Details of the so-called Green Run remain classified, but it is suspected that the experiment (in which 8,000 square miles were contaminated with more than one thousand times the radiation released at Three Mile Island) was part of an effort to discover the location of Soviet plutonium plants. Residents of local communities were not informed until 1986, when an environmentalist group obtained documentation on the experiments from the Department of Energy under the Freedom of Information Act.[16]

There are other examples. In the 1960s the U.S. Army and Central Intelligence Agency gave LSD to several unsuspecting persons as part of a project to explore the effects of the drug; at least one person died from trauma caused by the experience. In the notorious Tuskegee Syphilis Experiment, 400 black men from Alabama suffering from the advanced stages of syphilis were allowed to go untreated for as long as forty years (1932 –1972) as part of a U.S. Public Health Service experiment to chart the course of the disease. It was not until 1972 that the experiment was halted, after a reporter for the Associated Press broke the story. There are the infamous Willowbrook experiments, where retarded children were deliberately infected with hepatitis in order to test the effects of an experimental vaccine; there are the secret radiation experiments conducted in the 1950s by the Atomic Energy Commission to determine the health effects of ingested uranium.[17] There are the secret germ-warfare tests the U.S. Army conducted by spraying bacteria into New York subways and populated areas around San Francisco—experiments described by one critic as "science gone mad."[18]

These and other examples were in blatant violation of the principles of the Nuremberg Code. Ironically, the code never has been recognized as having legal standing. When a soldier victim of an LSD study sued the U.S. Army, the case went all the way to the Supreme Court. In 1987, in the case of *Stanley vs. the United States,* the court voted 5 to 4 against the soldier; the decision made it clear that the government could not be held legally or even financially responsible to the standards articulated in the Nuremberg Code. Disturbing also is the fact that many Americans were apparently willing even to help administer such experiments if ordered to do so by scientific authorities. The psychologist Stanley Milgram in a series of pathbreaking studies in the early 1960s demonstrated the extremes to which people would go in administering pain to experimental subjects on orders of scientists urging that "the experiment must go on."[19]

The 1960s was a watershed in science criticism. As many orthodox academic philosophers remained content to formalize the laws of science, critics began to question the logic of scientific "rationality" itself, condemning what they saw as reason propelled by its own inner logic, or the ideal of science as the best way of knowing, or the use of science as an instrument of social control. Herbert Marcuse's *One-Dimensional Man* excoriated the society in which peace meant readiness for war; Rachel Carson's *Silent Spring* warned that chemical agriculture was killing butterflies and birds. Some began to question whether tobacco and chemical executives really *belonged* on the boards of national cancer research institutions. Others asked if it was right and proper that Madison Avenue advertising should be the second largest employer of Ph.D. psychologists, the C.I.A. the largest employer of Ph.D. historians, the National Security Agency the world's largest employer of mathematicians.

Some have tended to lump together all opponents of science under the general rubric of "antiscience," or those who simply fear any tampering with nature.[20] (Jeremy Rifkin's work has often been cast in this light—sometimes fairly, sometimes not.) But not all of those who criticize science are "antiscience" in principle, or prefer the Stone Age to the present, or deny that "life itself would be impossible without chemicals" (the slogan of an Allied Chemical media campaign in 1981–82). In recent years, the romantic critique of science has been superceded by more complex critiques based not on longings for an earlier and simpler age but rather on broadening participation in science or redirecting science toward environmental responsibility, international cooperation, or other social goals. Criticism in some cases has led to the establishment of entirely new academic disciplines: environmental toxicology, restoration ecology, technological assessment, bioethics, conservation biology, ecological economics, environmental ethics, geophysiology, and so forth.

In recent years the critique of science has been especially pronounced in four areas: medicine, agriculture, biological determinism, and the militarization of science. These areas are not neatly separate from one another, because many new problems (pesticide pollution, for example) and new techniques (plasmid cloning, for example) have eroded the boundaries that once separated, say, agriculture and medicine. The same gene-insertion techniques used to produce human insulin (to treat diabetes) are being used today to produce bovine growth hormone (for use in agriculture) and biowarfare agents (for use by the military). A biotechnology firm may produce pharmaceuticals and genetically altered seeds along with bacteria that eat up oil spills. A military agency may fund research on global warming, high-definition TV, or the environmental effects of nuclear war. Critics of militarization may base their critique on environmentalist or feminist principles; the same groups protesting biological determinism may also object to the commercialization of university laboratories or the environmental hazards brought about by petrochemical agriculture.

Medicine has become a field of endless controversy—not just because nearly everyone's interests are at stake, but also because advances in the science in question may be only tenuously linked to advances in human welfare. Concerns about human experimentation, rights to refuse treatment (or to die), the use of fetal tissue or animals for research, and new forms of transplants or fertilization and birthing have all raised doubts about what it means for medical science to advance. One broad set of concerns has to do with the nature and scale of medical intervention. Have physicians overestimated the value of "heroic" medical techniques? When does the prolongation of life no longer serve the best interests of a patient? Do patients have the right to choose the type of treatment they want—what about "alternative" cancer treatments, for example, many of which are now illegal?[21] Can a physician be sued for failing to perform amniocentesis or other genetic tests? When does the cost of a procedure exceed its benefit? And who should bear those costs? Can aborted fetal tissues be used to treat Parkinson's disease or juvenile diabetes? How must animals be treated in medical research? Should animals be used to test ballistic effects and new cosmetics?

A related set of concerns centers on the question of whether medicine really is the key to health. Thomas McKeown in the 1970s showed that the rise in life expectancy over the last two centuries in industrial nations owed less to improvements in medicine (the discovery of the pathogens causing measles and whooping cough, for example, or the subsequent development of vaccines) than to improvements in sanitary conditions and diet, including separation of sewage and drinking water. D. P. Burkitt argued that increased consumption of dietary fiber was likely to do far more to eliminate colon cancer than innovations in chemotherapy or surgery; Samuel Epstein postulated that industrial pollution, rather than a failure of medical technique, was the likely cause for the growing incidence of many cancers.[22] Many of these criticisms emerged out of recognition that policy priorities were skewed toward the cure of disease rather than its prevention. Thus critics were not surprised when Richard Nixon's much heralded "war on cancer" (launched in 1971) proved to be a losing one, as cure rates remained little changed while incidence rates actually increased for many cancers. In the Soviet Union, the average life span of adult males actually began to decline in the late 1960s and early seventies, to the point that the Soviet government stopped publishing vital statistics.[23] Critics observing such trends argued that for most people most of the time, human health has less to do with medicine than with diet, exercise, and the cleanliness and safety of the home and workplace. Ivan Illich in 1975 went so far as to suggest that doctors do as much to injure health as to help it.[24]

Another set of concerns has to do with who becomes the object of medical research, and why. Feminists have long protested the medical misreading and misshaping of the female body—from the Galenic conception of women as imperfect men, to nineteenth-century theories that academic studies would "shrivel women's ovaries," to the epidemic of hysterectomies suffered by women in the 1950s and 1960s. Critics point out that diseases of white wealthy males are disproportionately investigated and that women are too often absent from studies of the benefits or risks of pharmaceuticals. Historians have begun to investigate the exclusion of women and minorities from medicine, but why have we had to wait so long for a history of the National Medical Association (representing black American physicians)? How many people even know that, in the 1930s, five thousand black physicians petitioned to join the all-white American Medical Association but were turned down by that august body? (German medical journals reported the rejection and used it to justify the Nazi's exclusion of Jews from the practice of medicine.) Why, as recently as 1988, were women excluded from a major, government-financed study of the effectiveness of aspirin in preventing heart attacks? (In the summer of 1990, the Congress's General Accounting Office condemned the National Institutes of Health for not doing enough to promote studies that would include women as well as men.)[25]

Yet another set of concerns derives from new forms of manipulation of genetic information. It is difficult to name an area more controversial than modern biotechnology. A number of scientists recognized this as early as 1973, when, at the annual Gordon Research Conference in New Hampshire, concerns were voiced that, if unregulated, the new techniques might produce biohazards of a potentially catastrophic nature. Maxine Singer and Dieter Söll led a group of scientists in calling for a moratorium on the most hazardous forms of research until better methods of containment had been developed. Two years later 150 scientists from around the world met at the Asilomar Conference Center in Pacific Grove, California, to develop guidelines restricting the most hazardous forms of recombinant-DNA research involving bacteria that might infect humans. Though subsequent National Institutes of Health (NIH) guidelines regulated only government-funded research (!), the moratorium set an important precedent for scientists questioning the course of scientific innovation before the fact.[26]

Medical defenders of the biotechnology revolution sometimes point to pharmaceuticals as the most likely early fruit of this revolution; here, too, though, critics have pointed out that there are likely to be problems associated with the cost of many of the products. In the mid–1980s, when Merck, Sharp and Dohme launched the world's first genetically engineered vaccine against hepatitis B, critics pointed out that most of those who need it (the poorer peoples of Asia) will never be able to afford it.[27] New techniques (using animal instead of bacterial hosts, for example) will eventually lower costs, but whether those most in need will benefit remains an open question.

The Human Genome Initiative, the largest single undertaking in the history of biology, promises to take these new questions even further. The goal of the project is to map and sequence the entirety of the human genetic material, an undertaking estimated to take about fifteen years and to cost some $3 billion. The project was first proposed in 1984, when scientists at Department of Energy (DOE) nuclear weapons laboratories (Los Alamos and Livermore) suggested that a massive sequencing effort might help identify the kinds of genetic damage suffered by survivors of the Hiroshima bomb. According to one account, the weapons facilities were also looking for ways to turn the huge resources of the labs toward civilian ends should the Reagan-era enthusiasm for military research ever begin to slacken.

Critics have argued that grants for the project are going to be concentrated in the hands of relatively few researchers, drawing scarce funds away from more urgent projects. More troubling ethical questions have been raised concerning how the knowledge gained is likely to be used. What are the implications of being able to diagnose ailments thirty or forty years in advance, as is already possible for Huntington's disease, cystic fibrosis, Tay-Sachs, sickle-cell anemia, and a number of other debilitating diseases? Why has the Department of Energy become a major funder of genome research—and why is it so interested in the so-called radiation repair genes? Will employers try to screen employees to eliminate persons susceptible to specific genetic infirmities? What rights will insurers have to genetic information on their clients, and what rights will persons with "genetic lesions" have to keep that information private? What kinds of fears or stigma will be attached to genetic disease? These are new kinds of questions that courts, hospital review boards, labor leaders, insurance executives, congressional committees, and many other groups are going to have to face.[28]

The Human Genome Project has already drawn flak from groups who fear that the ultimate rationale for the project is a biological determinist one. Project leader James Watson has done little to dispel this concern, defending the initiative as providing us with the ultimate tool for understanding ourselves at the molecular level: "We used to think our fate was in our stars. Now we know, in large measure, our fate is in our genes."

Pointing to the long and seamy tradition of eugenicists exaggerating the role of genes in human behavior, critics argue that it is dangerous to see biology as destiny. Geneticists may identify four or five or six thousand illnesses with a substantial genetic component, but how useful will such knowledge be for providing effective therapies? (The gene for sickle-cell anemia was discovered in the 1960s, and we are still very far from a cure.) There is also a danger that, in the rush to identify genetic components to cancer or heart disease or mental illness, the substantial environmental origins of those afflictions may be slighted. Even where genetic influence is clear, critics worry that widespread availability of genetic testing may generate fears out of proportion to actual risks. In November of 1989, anticipating some of these concerns, the American Society for Human Genetics called for a moratorium on widespread genetic screening for carriers of the cystic fibrosis gene.[29]

Notable in the Human Genome Project, however, is that mechanisms for self-criticism are supposed to have been built into funding for the project—a rather novel phenomenon for both the DOE and the NIH, the two major supporters of the project. Three percent of the multibillion-dollar effort has been earmarked for research on the "social, ethical, legal, and economic implications" of mapping and sequencing the genome, making the NIH's National Center for Human Genome Research the largest single benefactor of bioethics research in the United States. A statement issued by the working group coordinating NIH research in this area warns that "if misinterpreted or misused, these new tools could open doors to psychological anguish, stigmatization, and discrimination for people who carry these genes."[30] Whether the working group has any influence on the directions taken by the project remains to be seen.

Agriculture is another area where research priorities have been subject to scrutiny. Criticisms have been launched against the petrochemical-based, capital-intensive growing methods that emerged in the United States after the Second World War—methods generally requiring heavy doses of herbicides and pesticides. (Chemical inputs were cheap because many chemical companies found themselves with excess capacity after the war; wartime government subsidies aided in the rapid growth of postwar chemical agriculture.)[31] In 1989, the National Academy of Sciences published a report pointing out that the entire system of chemical agriculture was flawed, insofar as pesticides, herbicides, and other products of petrochemical agriculture had produced a negligible return on the investment, not to mention long-standing environmental problems. Others have pointed out that petrochemical agriculture often can't even produce an appetizing product. Jim Hightower, the populist Texas Commissioner of Agriculture, protested as early as 1972 that tomatoes engineered for transport leave much to be desired in terms of taste; the same can be said for many other vegetables.[32] Today, the average food consumed in America travels more than 1,500 miles from where it was produced. Prolonging shelf life has become a major priority of agricultural research, one of the reasons producers in the 1980s shifted to saturated fats (cottonseed, coconut, and palm oils, for example) from the more healthful polyunsaturated fats that do not keep as well.

Biotechnology has been hailed by some as an attractive alternative to petrochemical agriculture. Critics suggest, though, that the "second green revolution" may create as many problems as it solves. Take the example of efforts to produce herbicide-resistant crops. Biotechnology companies have produced herbicide-resistant strains of cotton, tobacco, and soybeans so that fields may be sprayed and have everything but the crops killed. The stakes are high, given that 650,000,000 pounds of herbicides are already applied to U.S. crops annually. Critics point out, however, that the new herbicide-resistant strains will encourage farmers to increase rather than decrease their use of herbicides, further polluting groundwater and exposing farm workers to deadly poisons. The Washington-based Biotechnology Working Group has urged that federal and state governments end their support for developing herbicide-tolerant plants and that encouragement be given instead to nonchemical methods of pest control.[33]

Proposals to release genetically altered microorganisms have met with a different set of concerns—not the least of which has been corporate sloppiness in following (often confusing) procedures for obtaining permits from government agencies. In 1986 the Oakland-based biotech firm Advanced Genetic Sciences (AGS) aroused a storm of protests when, without Environmental Protection Agency (EPA) approval, it injected trees on its rooftop with its "Frost Ban" bacterium designed to improve strawberries' ability to withstand freezing temperatures. The EPA had approved the company's request for permission to spray a test plot outside Monterey the previous year, but citizens' groups (including local farmers) protested the decision, prompting county officials to impose a forty-five-day moratorium on the release. Before the moratorium was over, the unauthorized rooftop experiment in Oakland had become public. The EPA rescinded its permit and fined AGS $20,000 for failing to follow proper EPA procedures. The company eventually satisfied the EPA that the proper pathogenicity tests had been conducted and received its permit to conduct experimental releases; field tests were finally conducted in the spring of 1987 and have continued through the present.[34]

Scientists working in the field of biotechnology have tried to grapple with the possible hazards of environmental release. David Pimentel at the University of Cornell points out that the problem of unforeseeable consequences in environmental release is one that must be met by careful studies of microbial ecology. New variants of caterpillar-killing bacteria may help to control cutworms under the soil, but one has also to be sure that the new bacteria will not kill earthworms or other beneficial fauna. There is a risk that "ice-minus" bacteria could become attached to insect pests, allowing the pests—and not just the crops—to survive freezing temperatures; there is also the question of whether the new strains could have unforeseeable effects on economically important insects such as honeybees. Honeybees pollinate twenty billion dollars' worth of crops every year in the United States, and damage to them could be very costly. Natural ecosystems may be more resilient than many critics are willing to admit, but history is also replete with examples of exotic organisms introduced into new environments with disastrous effects (the gypsy moth in eastern forests, kudzu in California, and so forth). Pimentel thus suggests that manufacturers working in temperate climates confine their engineering to tropical strains that will die out in the winter, assuring that none of the released varieties will gain a permanent foothold in the environment. He also advises that the new technologies not be regulated by agencies whose duties also include the promotion of those technologies; control should therefore be shifted, for example, from the Department of Agriculture to the Environmental Protection Agency and the Occupational Safety and Health Administration (OSHA).[35]

Increased productivity has also been challenged as the single guiding rationale for agricultural research. The widespread use of bovine somatotropin (a growth hormone only recently available in commercial quantities), for example, promises to raise milk production as much as 30 percent (up to twelve or more gallons per cow per day)—but what will the effects of this be on smaller dairies, faced with oversupplies of butter and depressed prices? (The USDA already pays farmers to reduce their herds.) And will cattle remain healthy, given the new strains on their bodies? One economic consequence is likely to be a further reduction in the number of farms—already reduced from about five million in 1950 to just over two million in 1990. Agribusiness supporters argue that the concentration of production is efficient; critics argue that productive diversity is more attractive from the point of view of both the quality of food produced and the preservation of smaller-scale family farms.[36]

There is also the question of impacts on economies abroad. The new agricultural biotechnologies will probably widen further the economic gulf between overfed and underfed peoples. While biosynthetic vanilla or gum arabic or continuous tissue culture (for tobacco, fruits, and other crops) will probably bring jobs and consumer benefits to First World peoples, they will no doubt also deprive some Third World nations of their export markets.[37] Cherry flesh without cherry trees, orange juice without oranges, chili oil without chilies, chocolate without the cocoa bean, and sweeteners without sugar cane, all have become the object of intense biotechnological investigation. First World farmers threatened by such innovations have sometimes managed to delay or halt the process: in Japan, tobacco farmers in the mid-1980s protested the sale of tobacco products made from cell culture, ultimately forcing the manufacturer to stop.[38] In the United States, when the Patent Office approved the patenting of genetically altered animals in 1987, the National Farmers Union representing 250,000 growers in twenty-two states testified at congressional hearings that the move would put many smaller farmers out of business, given that the patents would be controlled by the larger agricultural corporations.[39] Third World farmers are likely to have difficulty mounting challenges to First World innovations in this sphere. Economic relations being what they are, the new genetics is as likely to produce hunger as it is to produce food, insofar as already existing inequalities will simply be amplified.[40]

Capital-intensive agriculture has also aroused criticism with respect to environmental concerns. Pesticides and herbicides have polluted groundwaters, and the great reservoirs that underlie the American Midwest are being pumped out faster than they can be replenished. American agriculture presently uses up 325 gallons of water and 160 pounds of fertilizer per person per day; even if consumption is decreased the day is not far (fifteen years?) when the Ogallala and other great aquifers of the midwest will be too low to pump. Topsoil is being lost more quickly today than in the dust-bowl days of the 1930s—a third of it is already gone.[41]

Wes Jackson, founding director of the Prairie Institute in Kansas, has argued that the roots of modern agricultural troubles go even deeper. He and others have begun to explore how perennials, modeled on prairie grass, might be developed to provide alternate sources of food. Such crops would halt the destruction of topsoil and the water table and eliminate the need for annual plowing.[42] "Alternative agriculture" promises to become a booming business, as governments implement tighter restrictions on pesticides and consumers show themselves willing to pay for organic produce. Low-input agriculture has gained a lot of attention in the research community. Still, critics point out that the USDA program responsible for promoting low-input sustainable agriculture receives only about $4.5 million per year (in 1990), while the Forestry Department continues to fund the development of herbicide-tolerant trees.[43] Research priorities are likely to shift, but only when political support for more ecologically sound agricultural practices tips the scales.

Biological determinism is yet another area where science criticism has flourished. Especially since the 1960s, critics have protested efforts to prove that intelligence, alcoholism, crime, depression, homosexuality, female intuition, and a wide range of other talents or disabilities are the inflexible outcomes of human genes, hormones, neural anatomy, or evolutionary history.⁴⁴ The heat of disputes in this area derives largely from the fact that efforts to trace human capacities to biological predispositions have often emerged as conservative responses to liberation movements. In the 1960s, at the height of civil rights activism, Arthur Jensen proposed that average IQ differences between blacks and whites were due to genetic differences in the two races; in the 1970s, at the height of the feminist movement, sociobiologist E. O. Wilson suggested that women were unlikely ever to achieve equality with men in the spheres of business and science, given their evolutionary adaptation to separate social roles.⁴⁵

The biological response is often a convenient answer to social difficulties. Is terrorism a problem? *Science* magazine reports research suggesting that "most terrorists probably suffer from faulty vestibular functions in the middle ear." Do women become discouraged with college-level math? In 1983 America's leading science publication announced that "Math Genius May Have Hormonal Basis."⁴⁶ Is there bullying in the schools? A popular article in the *New York Times* asserts that the tendency to do physical harm "is intrinsic, fundamental, natural."⁴⁷ Do men cheat on their wives? *Playboy* magazine is able to cite studies claiming that both rape and infidelity are found throughout the animal kingdom.⁴⁸

Critics worry that science in such cases is being used as a proxy for deeply held social values—that women cannot compete, that blacks are inferior, that war or crime or poverty are ineradicable parts of human nature. Critics argue that there is little evidence that terrorism, or sexual preference, or personality traits such as "shyness" or "bullying" are genetically anchored; that it is easy to mistake the intransigence of human cultural qualities (male dominance, for example, or rape) for biological destiny. Critics also argue that determinists often project human qualities onto nature, which they rediscover then in humans. Scientific journals may describe "rape" in mallard ducks or acanthocephalan worms,⁴⁹ but it is more than misleading to compare animal and human behavior in this regard: the danger is that historically specific institutions—such as rape or war or male dominance—will be branded invariant.

The hereditarian argument is not a new one. In the ancient world scholars drew upon the Galenic theory of humors to argue that men were hot and dry (and therefore active) while women were cold and moist (and therefore passive). Polygenists in the seventeenth century argued that blacks and whites were the products of separate creations; David Hume in the eighteenth century postulated that Negroes were a separate species whose intellectual capabilities compared favorably to those of the parrot. In the nineteenth century Hegel compared men with animals and women with plants; craniologists sought to prove that the races of the world could be ranked according to skull size and brain capacity. In the twentieth century the metaphors have changed, but many "results of research" sound much the same. Women and blacks are still passive or incompetent, but now because of deficiencies in androgen or heritable *g* or IQ, or differences in brain structure or "math genes" or sociobiological history.

Critics have argued, by contrast, that hereditarian theories are both technically flawed and politically suspect. Hereditarian theory allows one to argue that social reform is dangerous, futile, or misguided; biological determinism is a "social weapon" used against groups demanding equal access to social opportunities.[50] The hereditarian argument is a political one, insofar as it is used to buttress arguments about the nature or limits of human equality and freedom. Some people are born smart, others stupid. Some are fit to rule, others to follow. Certain groups suffer or succeed because of talents or disabilities stamped indelibly onto the human racial or sexual constitution.

Take, for example, the 1969 essay by Arthur Jensen of Berkeley arguing that most of the average IQ difference between American blacks and whites can be attributed to genetic differences between these two populations.[51] Critics have drawn attention to a host of technical problems in Jensen's argument. Richard Lewontin, for example, pointed out that there is no way to extrapolate from heritability within groups to heritability between groups. The heritability of traits within a population may be taken as an index of heritabilities between those populations only on the assumption that there are no significant differences in the environments of the two populations (diet, education, prejudice, socio-economic status) that might account for differences between the two populations. Yet it is precisely these environmental differences that are in question.

Lewontin and others have also drawn attention to the fallacy in Jensen's presumption that heritability has something to do with changeability.[52] In fact, the proportion of variance of a trait attributable to genetic variance tells us nothing about how that trait might be expressed in a different environment. The fact that a given trait is highly heritable (even 100 percent) in one environment tells us little about whether heritability will be high in a different environment. Heritability is a measure of "unchangeability" only insofar as it can be assumed that all possible trait-relevant environmental variation is present equally and immutably in the environments to be compared. In the example of IQ and educability, heritability says something about the fixity of IQ only if one is confident that "remedial education has been tried, and it has failed." (This was the opening salvo of Jensen's "How Much Can We Boost IQ and Scholastic Achievement?") If, however, one believes that education is not what it could or should be, then heritability says little or nothing about educability with regard to IQ—though it may well say something about the sorry state of education. High heritability of IQ may indicate nothing more than the fact that family background remains more important than schooling in performance on IQ tests.

Many other objections to IQ research have been raised, but the point to keep in mind is that ethical or political objections to hereditarian research stem primarily from concerns about how such research has been used in the past and is likely to be used in the future. Research into the genetics or distribution of intelligence is not prima facie nonsense or reprehensible—any more than research on atomic fission or recombinant DNA or UFOs. But techniques are developed and used in particular contexts, and this is what has led critics to ask: what is the purpose of such research? How will it be used, and who stands to gain or to lose from the results? In the United States, blacks still are underrepresented in science. In 1988, only one American black received a Ph.D. in mathematics, only one American black received a Ph.D. in computer science.[53] In such a climate, where access is already difficult, efforts to blame the victim can only make the already lonely face more lonely.

The danger that hereditarian studies might be used for political purposes is not the product of the critic's imagination. Shortly after the publication of his controversial thesis in February 1969, Jensen's entire 123-page article was read into the *Congressional Record* at the request of a southern congressman. Daniel P. Moynihan briefed the Nixon cabinet on the article, advising that its policy implications should be taken seriously.[54] In Germany, Jensen's views became popular among the far-right nordic supremacists writing for *Neue Anthropologie;* Jensen's work has also figured prominently in British neo-fascist propaganda literature.[55] Jensen has even drawn attention from scholars in socialist countries: in 1980 the East German scientists Volkmar Weiss and Hans-Georg Mehlhorn published an article in the *Biologisches Zentralblatt* arguing that the heritability of intelligence is about 80 percent and that "general intelligence" represents the expression of a single gene.[56]

More research needs to be done on how and why hereditarian views have emerged in socialist countries—whether this has to do with rising rates of crime, or resurgent nationalism, or problems with the integration of ethnic minorities. How can East German endocrinologists argue that homosexuality is a hormonal disorder?[57] What explains the resurgence of sociobiology and Gobineau-style racial theory in what used to be known as the "eastern block"? How does one understand the recent efforts in China to sterilize the mentally retarded?[58] An argument could be made that Soviet efforts to reduce political dissent to psychiatric disorders derives from Stalin's proclamation in the 1930s that class struggle in the Soviet Union had come to an end. Political deviance could then be suppressed as mental illness, as was also the practice in Germany under Hitler.

The ultimate problem with the hereditarian argument is that it is difficult to know when a person or a people has reached their full potential—whatever that might mean. In 1820, who could have predicted Jewish or Japanese success in science a century later? In 600 A.D., at the height of the Gupta empire, who could have predicted India's eclipse and conquest by the British, or Europe's triumph over China as the center of world scientific culture? Who at the height of the Benin empire or the kingdom of Zimbabwe could have predicted the subjugation of African peoples to European rule? The success of a nation is historical; it depends on the configuration of social and economic institutions—not the racial makeup of its citizens.

Not surprisingly, feminists have been among the most persistent critics of biological determinism. The strength of the feminist critique stems from the continued underrepresentation of women in the sciences (women in 1989 earned only 8 percent of all American Ph.D.s in physics), but also from the fact that many scientists still explain the scarcity of female faces by reference to defects in the female body. Skull size and "the wandering womb" are no longer fashionable as predictors of intellectual prowess, but brain lateralization, testosterone, and "math genes" are more in vogue than ever.

There is relatively little controversy today in the view that men and women experience the world differently; Carol Gilligan of Harvard made this a centerpiece of her best-selling book, *In a Different Voice,* and the celebration of "difference" has become a common refrain in many feminist writings. What remains controversial, though, are the origins and consequences of difference. If gender roles or personality traits are rooted in the brain, traceable ultimately to hormones or genes, it is easier to argue that the predominant sexual division of labor is natural and that efforts to alter it are foolhardy. Popular scientific articles on this topic pull no punches: a 1980 article in *Quest Magazine* states that brain lateralization studies "threaten the axioms of militant feminists as well as homosexuals." *Discover* magazine came to similar conclusions in a 1981 article on "The Brain: His and Hers," subtitled "Men and Women Think Differently. Science Is Finding Out Why."[59] On May 18 of that year *Newsweek* devoted a cover story to "The Sexes: How They Differ—And Why," retracing much of this same ground.

The arguments presented in such literature suggest that it is ultimately biology (and specifically hormones) that determines "why women might think intuitively, why men seem better at problem-solving, why boys play rougher than girls."[60] The 1980 *Quest* article describes Jerre Levy's brain lateralization studies at the University of Chicago purportedly showing that whereas "males are good at maps and mazes and math" and are "more narrowly focussed, less distractible," females by contrast have "superior verbal skills," are "sensitive to context, good at picking up information that is incidental to a task that's set them, and distractible." Levy cites the example of a woman at a party who, after talking to a total stranger, can deduce a great deal of information about that person. For Levy, such observations "might mean that she's doing something for which the organization of her brain ideally suits her."[61] *Discover* magazine cited a number of practical implications of such findings:

> If women are indeed at a disadvantage in mastering math, there could be different methods of teaching, or acceptance of the fact that math is not important for certain jobs . . . Eventually, psychiatrists and lawyers may have to assess their male and female clients in a new light. And brain surgeons may have to consider the sex of a patient before operating. For if the two hemispheres of the brain are more intimately connected in women than in men, then women may be able to control a function like speech with either hemisphere. Surgeons could feel confident that a woman would recover the ability to talk, even if her normal speech center were destroyed; they might proceed with an operation that they would hesitate to perform on a man.[62]

Feminists have various reasons to be suspicious of such claims—not the least of which is the extraordinary history of women's sufferings at the hands of physicians misreading or mutilating women's bodies.[63] Critics have countered that if boys are "good at maps" and girls excel in intuition, this is the consequence of socialization rather than of brain structure. Feminists thus reject the oft-cited studies purporting to show that girls born with abnormally high levels of androgens exhibit "distinctively 'tomboyish' behavior," seldom play with dolls, and begin dating at a later age than other girls.[64] Girls born with congenital adrenal hyperplasia (a condition that results in high levels of androgen) do not *look* like normal girls; perhaps this has something to do with how they are treated and hence how they behave. Jerre Levy of Chicago admits as much when she states that "For every experiment in the literature that finds differences, there's another one that doesn't, and vice versa." Even Roger Gorski, an advocate of the hormonal determination theory, concedes that "There's something reductive and scary about a situation in which you *might* be able to ask a mother whether she wants testosterone treatment to avoid having a homosexual son."[65]

Feminists have argued by contrast that masculinity and femininity are socially constructed—that gender roles are in principle independent of sex and that it is foolish therefore to try to trace differences in social behavior (including sexual preferences) to differences in sex—chromosomal, hormonal, or genital. Feminists argue further that whatever role "nature" may play in the question of difference, it is difficult to disentangle this in a political environment where feminist challenges are met with antifeminist rhetoric and policies. Women's success in science is unlikely to match that of men until the social and economic power of women rivals that of men. In Simone de Beauvoir's words, "it is not the inferiority of women that has caused their historical insignificance: it is rather their historical insignificance that has doomed them to inferiority."[66]

Given the world-historical exclusion of women from the sciences, it is not surprising that feminists have become the most steadfast critics also of the ideal of value-neutrality—proclaiming it variously a myth or a mask or the bulwark of liberal ideology. Anne Fausto-Sterling in her study of *Myths of Gender* suggests that "there is no such thing as apolitical science"; science is "a human activity inseparable from the societal atmosphere of its time and place." Ruth Hubbard points out that "the subject of women's biology is profoundly political"; Elizabeth Fee argues that one can only expect "a sexist society to develop a sexist science."[67] Many feminist treatises begin with a caution against the self-serving postures of universalist objectivity; feminists have become wary of such posturings, given that women have so often been saddled with a supposed inability to distance themselves from the objects of scientific study. Objectivity has been considered a distinctively male trait, just as the attributes of science have been portrayed as the attributes of males (cold, hard, impersonal), just as the abstract *man* has tended to stand for males in fact, contrary to its pretence of standing for humans in general.

For most feminists, however, there is a kernel of value in the ideal of objectivity. The goal, as Evelyn Fox Keller has put it, is not so much a science that is "feminist" but rather a gender-free science that is "humanist"—freed from the deformations caused by centuries of exclusion of half of humanity from its ranks. Elizabeth Fee argues that the willingness to question and to have ideas subjected to "the most unfettered critical evaluation" are aspects of scientific objectivity that should be preserved and defended. The point of the feminist critique is to allow us to identify aspects of science that need to be questioned without abandoning the idea that we can come to "an ever more complete understanding of the natural world through a collective and disciplined process of investigation and discovery."[68]

Sandra Harding is another who advocates a selective "colonization" of the ideal of objectivity. In *The Science Question in Feminism*, Harding argues that it is wrong to view objectivity as compromised by taking positions on value questions. Indeed "it has been and should be moral and political beliefs that direct the development of both the intellectual and social structures of science." What is needed is to ensure that it is "participatory" and not "coercive" values informing the practice of science. Coercive values (racism, sexism, classism) undermine objectivity; participatory values (antiracism, antisexism, anticlassism) are preconditions or constituents of objectivity. Paradigm models of objective science are those that challenge oppressive political orders. Harding provocatively speculates that, in this sense, "feminist and similarly estranged inquiries" might be viewed as the true heirs to the legacy of Copernicus, Galileo, and Newton.[69]

* * *

In terms of state and industrial support for science, of course, the "true heirs" of Copernicus, Galileo, and Newton unfortunately lie elsewhere. Indeed if one has to choose one single prominent feature of science in the late twentieth century, it is that it is *militarized*. In the mid-1980s United Nations experts estimated that 20 percent of the world's scientific talent— 500,000 people—are directly involved in military research. In the United States the proportion is even higher: recent estimates are that as many as a third of all American scientists and engineers have some form of military security clearance. From the late 1970s to the late 1980s the fraction of U.S. federal funds for military research and development (R and D) grew from 50 percent to more than 70 percent of all R and D expenditures. In 1990 federal expenditures for military R and D in the Defense Department alone amounted to some $37.5 billion—well over half the total U.S. government R and D outlay of $65 billion.[70]

The Defense Department, however, is not the only conduit for military R and D. The large part of the American nuclear research and development effort is organized by the Department of Energy, which in 1988 employed 90,000 persons at an annual cost of $8 billion to research and produce nuclear weapons. The research part of this effort takes place at four sites: the Los Alamos National Laboratory in the mountains of New Mexico (employing 7,800); the Lawrence Livermore Laboratory near Berkeley (employing 8,000, part of the University of California system); the Sandia National Laboratories in Albuquerque, New Mexico (employing 7,200); and the Nevada Test Site northwest of Las Vegas (employing 8,000).

Nuclear weapons research accounts for a substantial portion of U.S. military R and D, but this is only one among many areas of inquiry. There is also a sizable U.S. research establishment devoted to the production, development, and defense against chemical and biological weapons (CBW). The United States presently has a stockpile of 150,000 tons of chemical weapons, including several million artillery shells and thousands of bomb and missile warheads. The Army depot at Tooele, Utah, alone contains enough chemical weapons to destroy all life on earth. At least ten Pentagon test centers perform research in this area, in cooperation with some 150 university and industrial laboratories.

The Army's Fort Detrick in Maryland was for many years the leading U.S. producer in the area of "biological defense," developing biowarfare agents such as anthrax, Q-fever, brucellosis, yellow fever, Venezuelan equine encephalomyelitis, enterotoxins, and tularemia, along with toxins to destroy wheat, rice, and other crops.[71] Today biowarfare research is pursued at the Army's Aberdeen Proving Ground, the Medical Research Institute of Infectious Diseases, the Walter Reed Army Institute of Research, the Army Medical Research and Development Command, and a number of other institutes and proving grounds. The Navy conducts biowarfare research at the Naval Biosciences Laboratory (until recently part of the University of California, Berkeley), the Naval Medical Research Institute in Bethesda, Maryland, and several other facilities. The Air Force runs similar programs through its own Office of Scientific Research; laboratory work is also done at the United States Air Force Academy and several dozen other labs. The Army has plans to build a Biological Aerosol Testing Facility at Dugway Proving Grounds southwest of Salt Lake City, though opposition from citizen groups and University of Utah scientists has been stiff.[72] In the mid-1980s U.S. research spending on CBW was ten times the amount spent in 1950—this despite the fact that the United States is a signatory to the 1972 Biological Weapons Convention prohibiting the development, production, or stockpiling of biological and toxin weapons.

Research in the area of chemical and biological warfare continues under the pretext that American CBW research is purely for defensive purposes. Critics, however, have pointed out that the scale on which such efforts are undertaken ($90 million for "Biological Defense" in 1986, up from $18 million ten years earlier) effectively blurs the distinction between offensive and defensive research. Vaccines against specific biowarfare agents might be used to defend against an enemy attack; they could also be used in an offensive operation to inoculate friendly forces while spreading germs amongst the enemy. Critics have pointed out that the development of vaccines is destabilizing, insofar as it weakens the primary deterrent to the use of such weapons—namely, the fear that germs distributed amongst one's enemies might spread to one's own population. "Defensive" biowarfare research is destabilizing in the same way that antiballistic missile systems are destabilizing—it makes it easier for a nation to launch an attack without fear of retaliation. Scientists recognizing this have organized a boycott against military biology: in 1987 the Boston-based Committee for Responsible Genetics began circulating a "Pledge Against the Military Use of Biological Research" in response to the rapid growth of the U.S. biodefense budget.[73] By the spring of 1990 more than one thousand scientists had signed the pledge, indicating their commitment not to engage knowingly in research or teaching that would further the development of biological or chemical weapons.

Apart from the bodies already mentioned, there are also within the Department of Defense a number of agencies dedicated to research in specific areas of military interest. The Defense Nuclear Agency (DNA) in northern Virginia receives about $400 million per year to rehearse (or "choreograph") World War III; it also explores "nuclear effects phenomenology." The DNA runs the Armed Forces Radiobiology Research Institute in Bethesda, Maryland, where tests are done (on monkeys) to determine how long troops will be able to fight after exposure to radiation, or how well vehicles or buildings can be expected to stand up under nuclear blasts.[74] The Defense Mapping Agency digitalizes landscape features to prepare routes for cruise and other "smart" missiles; the Defense Communications Agency is responsible for making sure that military information flows smoothly in the event of nuclear war. The Defense Science Board provides expert advice to the Secretary of Defense on a wide range of strategic and scientific issues; the Defense Documentation Center outside Alexandria, Virginia, is the largest clearinghouse for science and technology information in the country—and perhaps in the world. The Defense Advanced Research Projects Agency (DARPA) has overseen the Strategic Computing Initiative and was largely responsible for coordinating research on laser and space weapons until 1983, when the Strategic Defense Initiative Office was formed. With an annual budget of $1.1 billion, DARPA coordinates research in the areas of artificial intelligence, man-machine interfaces, and several other frontier technologies (high-definition TV, for example, and composite plastics). DARPA spends more than $100 million a year on basic research, making it, in the words of one observer, "by far the biggest venture capital fund in the world."[75]

Table 5 Top ten U.S. Department of Defense research, development, test, and evaluation contractors (FY 1989)

Martin Marietta	$1,990 million
McDonnell-Douglas	1,710
General Dynamics	1,180
Boeing	905
General Electric	903
Rockwell International	814
Foundation Health	639
McDonnell-Douglas/General Dynamics Joint Venture	620
Raytheon	614
TRW	578

Source: "Top 100 Department of Defense Research, Development, Test, and Evaluation Contractors: FY 1989," *Defense Daily,* May 4, 1990, p. 202. Note that these figures include only DOD contracts; contractors may also work on military projects for other agencies, such as NASA or DOE.

Most U.S. military research—about three-fourths—is contracted out to industrial laboratories (see Table 5). Military research is also conducted in eleven private "think tanks" (Federally Funded Research and Development Centers—FFRDCs)[76] contracted by the Department of Defense. Though the budgets of such groups may be small compared with more hardware-oriented research, think tanks have played an important role in formulating military policy. The RAND Corporation, established in 1948 in Santa

Table 6 Research and development centers funded by the U.S. Department of Defense, 1990

	Personnel		Annual Budget
	Total	Professional	
MITRE Corporation	5,500	2,800	$500 million
Lincoln Laboratory	2,800	1,300	428
Aerospace Corporation	4,300	2,600	400
RAND Corporation[a]	1,126	603	95
Institute for Defense Analysis	873	383	85
Center for Naval Analysis	550	270	40
Software Engineering Institute	230	150	26
Logistics Management Institute	180	133	21
Institute for Advanced Technology	30	19	3

a. Includes three DOD-funded centers: Project AIR FORCE, the U.S. Army's Arroyo Center; and the National Defense Research Institute. RAND also employs about 400 consultants, adding an equivalent of 82 professional staff.

Monica, California, is the oldest and one of the largest such bodies, with more than 1,100 full-time employees (about half of whom are scientists) involved in research on nuclear strategy, forecasting, counterinsurgency, and other projects for the U.S. Air Force and Army. The corporation administers three of the Department of Defense's eleven FFRDCs: Project AIR FORCE, the U.S. Army's Arroyo Center, and the National Defense Research Institute.

The RAND Corporation is joined by eight other private think tanks (FFRDCs) performing military research for the DOD (see Table 5). The Institute for Defense Analysis in Arlington, Virginia, designs evaluation programs for high-technology military systems. Lincoln Laboratory in Cambridge (part of MIT) does research on "penetration aids" to assist ballistic missiles in reaching their targets. The Software Engineering Institute at Carnegie Mellon University in Pittsburgh is one of the smaller of the Pentagon's FFRDCs, as is the Bethesda-based Logistics Management Institute, with a budget of only $21 million. The largest think tank in the United States is the MITRE Corporation in Bedford, Massachusetts, employing 6,000 workers to conduct research on questions of nuclear strategy, command, control, communications, and intelligence (C^3I), and a host of other projects primarily for the Air Force. (MITRE served as system engineer for the Strategic Air Command's recently completed underground command center.) The newest of the Pentagon's FFRDCs is the Institute for Advanced Technology at the University of Texas at Austin, where research is conducted on "pulsed electromagnetic power" and "electromechanics and hypervelocity physics," areas of interest in the Pentagon's plans to develop directed-energy weapons (particle beams, laser and kinetic weapons, and so forth).

One common criticism of military research is that it is wasteful or inefficient. The press is fond of reporting stories of $800 hammers and $50,000 coffee makers, but the big-ticket items can be even more outlandish. Science for the People, the Cambridge-based radical science group, in 1990 awarded its first annual Twilight Zone Award for "the most ridiculous DOD research program" to the Pentagon's Island Sun program—an $84.6 million project to outfit dozens of lead-lined tractor trailer trucks with electronic equipment to serve as "mini-Pentagons," roaming the public highways in the event of nuclear war.[77] If history is any guide, Island Suns will eventually join a long list of other abandoned military fiascoes, including the early-1950s plans to rearrange cities into linear strings (to make them less vulnerable to atomic blasts) and the vintage civil defense studies of "VD Control in Atom-bombed Areas."[78]

Even Island Suns are small potatoes when compared with the really large defense programs. When future scholars write the history of science in the final decades of the twentieth century, the Strategic Defense Initiative will probably rank as both the most ambitious, and the most foolish, of all research endeavors of our times. It is the largest research effort in history (early estimates of total costs approached the trillion-dollar range), and it is probably the most lampooned. Critics argue that the project is expensive, destabilizing, and ultimately unworkable. Critics have pointed out that even if 99 percent of the incoming missiles could be shot down, more than a hundred would still get through, more than enough to destroy the country. Arguments such as these were put forth even by several of Reagan's political allies, as when British Foreign Secretary Geoffrey Howe likened the program to a "Maginot Line of the 21st century, liable to be outflanked by relatively simpler and demonstrably cheaper countermeasures."[79]

Many—perhaps most—leading scientists in the United States have opposed the project on technical as well as political and moral grounds. By May 1986 more than 6,500 scientists—including fifteen Nobel laureates in physics and chemistry—had signed a petition advocating a boycott of research on SDI and labeling the initiative "ill-conceived and dangerous" as well as technically unfeasible.[80] Never before had so many prominent scientists publicly declared their opposition to a weapons program. Cynics have observed that the technical question of whether Star Wars will "work" or not is not even the central question. The effort has been described as "a public welfare program for major defense contractors" who stand to make substantial profits from the program—whether it works or not. In this sense, as several critics have suggested, the system is "already working."[81]

Apart from the question of waste, critics also commonly argue that the secrecy implicit in most military work leads both to political abuse and ill-treatment of weapons industry workers. Philosophers of science tend to ignore the fact that science has a workplace, and that there are occupational hazards distinctive to that workplace. Some of these have to do with health hazards, others with the hazards of whistle-blowing. In 1988 William J. Broad of the *New York Times* reported that, as a result of political infighting, America's nuclear weapons labs had become "rigid organizations steeped in secrecy, hesitant to pursue the truth if it leads in inconvenient directions and quick to please political masters."[82] For nearly four decades, security-conscious government authorities concealed health files on hundreds of thousands of workers at DOE nuclear plants. Atomic workers at Oak Ridge, Hanford, Portsmouth, and elsewhere had protested health conditions at the plants,[83] though it was not until the late 1980s that the mainstream press picked up on the abuses. In March 1990, after years of prodding by health activists and nuclear workers, DOE Secretary James Watkins agreed to turn over workers' health records to the Department of Health and Human Services.[84]

The most common argument against militarization of science, though, is not that it is wasteful or secretive but rather that it directs resources into areas that could more profitably be used for other, more humane, ends. Is the military threat really so great as to dwarf all other problems for which research funds might be spent? How does one defend the fact that research expenditures for the Department of Defense exceed those for the Environmental Protection Agency by a factor of about 100? How does one justify $40 billion to research means of producing death, and only $400 million to research means of preserving environmental health? J. D. Bernal as early as 1958 suggested that "the militarization of science is one of the main ways in which mankind is now being cheated of wealth and well-being and kept in a state of permanent danger."[85] If this was true in the 1950s, it is certainly no less true in the 1990s.

With the demilitarization of science in the wake of Gorbachev's perestroika and the transformation of eastern Europe, research priorities in the Western alliance have begun to shift. SDI funding has been scaled down (only $20 billion was spent on the program between 1983 and 1990) and some funds have been withdrawn from CBW research. Boeing, McDonnell-Douglas, and most other military producers have begun to lay off professional and technical staff, and most expectations are that U.S. military forces will reduce their personnel. Military research facilities have begun to transfer resources to nondefense projects. Los Alamos National Laboratory in New Mexico, with its extraordinary research and computing capacity (70 Cray equivalents), has begun to expand its genetic research division in concert with the Department of Energy's Human Genome Project. (The laboratory has one of the largest banks of information on human genetics in the world.) Senator John Glenn of Ohio has called for the establishment of an Advanced Civilian Technology Agency parallel to the Defense Advanced Research Projects Agency; and in June of 1990 Senator Sam Nunn of Georgia suggested enlisting military resources as part of a new and ambitious Strategic Environmental Research Program.[86] It remains to be seen whether this will bring about the pacification of the military or the militarization of the environment.

Science criticism in the late twentieth century is distinctive in its organized character. Scientists have organized to promote professional ethics, consciousness-raising, or overt resistance. In the 1960s Scientists and Engineers for Social and Political Action (later Science for the People) publicized the involvement of scientists in the development of antipersonnel weapons for use in Vietnam. Physicians for Social Responsibility and, later, similar groups among educators, computer professionals, lawyers, astronomers, architects, psychologists, social workers, and engineers were formed to educate the public concerning the health effects of atomic testing and the dangers of nuclear war. By 1990 Physicians for Social Responsibility (founded in 1961) had more than 30,000 members; Educators for Social Responsibility (founded in 1981) had 4,000 members; Computer Professionals for Social Responsibility (founded in 1981) had 2,800; Architects, Designers, Planners for Social Responsibility (founded in 1984) had 2,000; and Psychologists for Social Responsibility (founded in 1982) had 1,000 members.

On the environmental front, Ralph Nader–inspired Public Interest Research Groups (PIRGs) in the early 1970s grew out of student and environmentalist movements to research and agitate for environmental improvement and product safety.[87] Today there are twenty-five PIRGs around the country with a total membership of more than one million, making them one of the largest environmentalist organizations in the country. (The National Wildlife Federation claims a membership of five million, Greenpeace three million, and the Sierra Club 550,000.) Local PIRGs have become active in organizing grassroots research and legislation on a broad range of consumer affairs: New York's PIRG, for example, in 1989 organized a successful effort to ban food irradiation in the State of New York; the group also successfully promoted the passage of the 1986 Toxic Victims Access to Justice Law, extending the time people are allowed to sue for exposure to toxics (especially DES and asbestos). PIRGS have helped to enact bottle and recycling bills in a number of states; NYPIRG was instrumental in the passage of the state's 1979 Truth in Testing Law, granting students the right to obtain answers to questions on admissions exams administered by the Educational Testing Service (SATs, GREs,

GMATs, and so forth). NYPIRG successfully opposed plans to build incinerators to dispose of New York City's solid waste; NYPIRG was one of the leading groups responsible for having construction for the Brooklyn Navy Yard's $500 million incinerator delayed indefinitely.

Other groups have organized around publication of alternative and radical science journals: in the 1970s and '80s *Science for the People;* London's *Science for People;* the *Radical Science Journal* and its successor, *Science as Culture;* the Council for Responsible Genetics' *Genewatch;* the *CPSR Newsletter* of the Computer Professionals for Social Responsibility; Italy's *Sapere;* Germany's *Wechselwirkung* and *Kritische Medizin* broke with the older tradition of scientific socialism to argue that what was needed was not simply "more science," nor even the liberation of science from the fetters of custom and tradition, but rather new forms of science—science, as one group urged, "for the people."[88] Groups such as the Union of Concerned Scientists, Mobilization for Survival, The Boston Women's Health Collective, Healthpac, the Union for Radical Political Economics, and many others organized intellectuals around progressive, pacifist, feminist, or other radical or alternative causes. Common to many of these groups has been the notion that the politics of science lies not simply in its "use" or "abuse" but also in its origins and logic.

It would be wrong to exaggerate the unity of these various movements; critics range over the entire political spectrum. Uniting the most substantive critiques, however, are concerns rooted in the practical origins and consequences of particular research agendas. Concerns arise especially in areas where much is at stake: how we grow our food, whether we live at peace or at war, whether we are healthy or sick, treated fairly or unfairly, and so forth. Criticism is directed to the science because in each of these areas the science emerges in response to practical human needs and goals, contributing to their fulfillment or frustration.

Thomas Jefferson, when he stated that the main objects of science were "the freedom and happiness of Mankind," may have failed to distinguish between the real and the ideal. Science can and often does have noble goals, but even if only a fraction of what the critics claim is correct, then there is much that needs to be put right in the structure of our scientific practices and priorities.

Conclusion

Neutrality as Myth, Mask, Shield, and Sword

But the present men of science, recognizing no religion, and so having no grounds on which to pick out, according to their degree of importance, the subjects of study, and to separate them from less important subjects, and finally from that infinite number of subjects which, on account of the limitations of the human mind and the infinitude of these subjects, will always remain unstudied, have formulated for themselves a theory—"science for its own sake"—according to which science does not study what men need, but everything . . . but everything is too large . . . —LEO TOLSTOY, 1905[1]

The ideal of value-neutrality is not a single notion, but has arisen in the course of protracted struggles over the place that science should have in society. The origins of the modern ideal may be traced to four fundamental problems in the relation of science and society: the problem of *utility*, that is, of the relation of theory and practice; the problem of *method*, that is, of securing reliable and objective knowledge; the problem of *value*, that is, of the origins and character of the good and its relation to nature and labor; and the problem of the *security* of knowledge, that is, of the social or institutional conditions necessary for the free and unhampered pursuit of knowledge. Each of these informs the modern ideal of neutrality.

There is, first, the problem of *utility*. For Plato and Aristotle the ideal of *theoria* implied a certain detachment from practical affairs. Science was a luxury of leisure, neither a product of nor servant to utilitarian concerns.

With the rise of the Baconian ideal of science, however, utility becomes a central norm of science. Science was to combine the skills derived from the practical arts with a rational and empirical approach to nature. Science in Descartes's words would make us "masters and possessors of nature." Yet even Bacon, herald of the utilitarian ideal, warned against "mowing the moss, or reaping the green corn"; science would prove of great utility, yet science should also be sought for its own sake, not just for its applications.

There is, second, the problem of *method*. The idea that correct method is the key to the progress of science is fundamentally a modern notion, one that arises with the revolutionary transformation of science in seventeenth-century Europe. The new interest in methods is accompanied by a new subjectivity in science, a recognition that what we see depends upon where we stand and that human understanding is, as Bacon wrote, "no dry light, but receives an infusion from the passions." In the seventeenth century, European natural philosophers devised certain methods to guard against the various "idols" that color or distort human understanding. Central in this development was the distinction of primary and secondary qualities. Philosophers distinguished primary qualities of nature from secondary qualities of the mind in order to separate the additives of human bias from the original of nature. Moral qualities—what is good in a thing—were associated with these secondary and subjective qualities, qualities deriving from subjective human concerns (such as moral knowledge) said to distort and taint the pursuit of natural knowledge. Moral concerns were supposed to be excluded from natural-philosophical discourse so that the true and "primary" qualities of things would be revealed.

There is, third, the problem of *value*. For the ancients, value lay in the structure of the cosmos—a hierarchy of perfection rose from the imperfect earth to the perfect celestial spheres. In the scholastic vision there were final causes or "ends" toward which all things moved. For the moderns, value lies not in the structure of the cosmos or the ends toward which things move but in the products of human agency and design. Value is created not by God or nature but by human labor—there is less value in nature "raw" than in nature transformed by human work. Value is measured against human wants and desires: value in use and value in exchange. Science is neutral because nature itself is neutral; nature in this modern view is "disenchanted" (Weber) or "devalorized" (Koyré)—the causes of things are no longer final but merely efficient; nature is no longer organized according to natural harmonies or hierarchies of rank and perfection. The hierarchical, two-sphere universe of the ancients is replaced by the homogeneous universe of the moderns—everywhere the same and devoid of value.

Finally, in this "prehistory" of neutrality, there is the problem of *security* in the pursuit of knowledge. Euripides argued that knowledge of nature was "safe" knowledge, knowledge that did not tread upon the sensitive ground of politics or ethics. Bacon similarly notes that it was not knowledge of nature but rather "the proud knowledge of good and evil" that gave occasion to the fall of man. Moral knowledge is dangerous knowledge. Despite this caveat, science for Bacon and many of his followers was associated with utopian visions of a New Atlantis that would transform education, society, and the relation of God and man. With the rise of scientific institutions and academies, however, philosophers compromised these reformist goals. Natural philosophers in the Royal Society of London promised not to "meddle in matters of Moralls, Politics, or Rhetorick," in exchange for freedom and funds to pursue science unfettered by state or church controls.

Each of these problems was important in the early modern separation of practical and political concerns from science. The growth of the empirical sciences in the nineteenth century, however, sees the rise of a new and narrower conception of science. The Enlightenment ideal of the cultivated scholar, the *amateur* (lover) or philosopher of nature, is replaced by the ideal and reality of the specialized expert. This changing vision of science is the product of changes in the relation of science and society. Science is transformed from an amateur indulgence into a professional occupation. With the rise of the industrial research laboratory and the exploitation of science for industrial and military purposes, science becomes not just an ideal of life but a way to make a living. With the incorporation of science into industry, scientists cultivate a longing for what they imagine to have been the "purity" of science—science freed from the demands of industrial exploitation.

Social theorists, more so than natural scientists, linked the new ideal of purity to demands for political neutrality. In Germany, the ideal of value-free science emerges early in the twentieth century with the attempt on the part of social philosophers to become scientific. Political economists revolted against what they saw as the political character of social theory, seeking to establish instead a science of society that would no longer concern itself with "what ought to be," but only with "what is." In the new science of sociology, *Wertfreiheit* is defended in a number of different ways. Sombart, Weber, and Tönnies defend the neutrality of science as a methodological canon necessary for securing reliable knowledge. Neutrality is also defended as a consequence of the increasing academic division of labor. Tönnies argues that specialization is a natural and inevitable consequence of the "law of evolution"; it was natural for economics as a practical art to separate itself from economics as a theoretical science. Neutrality was also the sad but inescapable consequence of the disenchantment of the world by science. Weber lamented the fact that one could no longer find, as Swammerdam once found, proof of the glory of God in the anatomy of the louse.

These were some of the reasons provided for neutrality. But reasons are not causes. In Germany, neutrality served to preserve the autonomy of science from its critics—critics from above (government censors) and critics from below (socialists, feminists, and social Darwinians). Neutrality served as a *shield*—helping the young sociologists defend themselves against the charge that "sociology" was only a form of "socialism." The German Society for Sociology required that members renounce the advocacy of "all practical, ethical, religious, political, and aesthetic goals." Sociologists were to be "neither for nor against socialism, neither for nor against the expansion of women's rights, neither for nor against the mixing of the races." Value-neutrality expressed a desire on the part of social philosophers to distance the goals of theory from the demands of social movements. Neutrality was advocated not in the abstract but in response to concrete problems.

Neutrality was not, however, simply a means of protecting the autonomy of science. Neutrality was not just a shield but a sword. Simmel and Tönnies justified the failure of women to participate in science in terms of their inability to maintain an attitude of neutral and detached indifference. Weber rejected scientific socialism, social Darwinism, Ostwald's "energetics," and psychoanalysis as illegitimate confusions of science and *Weltanschauung*, illegitimate intrusions of values into science. Value-neutrality armed sociologists against movements to politicize or moralize the sciences.

Value-neutrality was also put forward, though, as a means of resolving social tensions—between conservatives and socialists, chauvinists and feminists, pacifists and war-mongers. Sombart pleaded that sociologists find some point around which both "believer and disbeliever, pantheist and atheist, Social Democrat and Conservative" could unite their disparate views. It was to science that Sombart looked to rise above these differences—science would arbitrate the great social questions of the day. The neutrality of science, by contrast with the divisiveness of religion or politics, admirably suited it to this task. Value-free science would enable scholars to rise above their differences.

Advocates of value-free science did not deny the importance of values in other spheres of life, however. Weber was primarily worried that professors would use the authority of their position to further their personal values. Weber's argument was a contextual one: so long as there are restrictions on the freedom of teaching (and Wilhelmine Germany was hardly a liberal state), then it was preferable for all values to be silenced than for only some to be heard. Values were to be excluded not out of any disregard for morals or politics—precisely the opposite. Weber did not want to see problems that should be decided by personal conscience or in political struggle reduced to problems of a technical or academic nature. Nor did he want political or personal values foisted onto students in the guise of science. Weberian *Wertfreiheit* was part of a critique of what others have called *scientism, naturalism,* or *historicism*—the view that moral consequences can be derived from laws of nature. Along with Marxists such as Bernstein and conservatives such as Popper and Hayek, Weber was careful to point out that whatever "inevitability" there might be in a social movement, this alone cannot guarantee its conformity to ideals of justice—no more than moral virtue ensures a movement's triumph.

Value-neutrality provided a liberal solution to the problem of the nature and limits of both science and morality. Obverse and reverse of the principle of neutrality was the principle of the subjectivity of values: value for Weber or the neoclassical economist was that about which "men can ultimately only fight." Morality and science were distinguished not just to preserve the autonomy of science but also to preserve a separate sphere of values from the machinations of reason. The complementary principles of neutral science and subjective value together provided the key elements in the liberal conception of the relation of science and society. Science was rendered impervious to moral critique, morals and religion were left unmarked by advances in scientific knowledge.

But the ideal of the political neutrality of science was not confined to liberals. Marxists in the revisionist wing of Germany's Social Democratic party advocated the separation of science from politics on the grounds that confusion of these two spheres was dangerous for both politics and science. Bernstein argued that any attempt to politicize science would degrade the independence of thought on which all science is based. If no one would speak of a "liberal physics" or a "conservative chemistry," why should this be any different for the social sciences? It was dangerous, Bernstein argued, to confuse ethical ideals and scientific principles. If socialist ideals were too closely associated with socialist predictions, then the failure of the latter might serve to discredit the former. Marxism as a science had to be separated from Marxism as a system of ethical ideals. Marxist science would then be open to growth and revision; socialist ideals would be flexible enough to survive the failings of socialists' predictions.

Value-neutrality culminates in the positivist traditions of the mid-twentieth century. Viennese positivists articulated a very narrow view of science, ignoring both the social conditions of its production and the alternate uses to which it might be put. Positivism was radical in its origins, applied to physics, but conservative in its Anglo-American incarnation—especially in the social sciences. Economists used positivist principles to argue that questions of institutional arrangements and the distribution of the social product do not really belong in economic science. Emotivist ethicists applied positivist principles to the analysis of morality, arguing for a value-neutral metaethics that would analyze moral speech but not pass moral judgments.

I have argued that the ideal of value-neutrality must be seen in political context. The neutrality of science is not the consequence of a logical gulf between fact and value, nor the natural outgrowth of the secularization of theory, nor even the outcome of the adoption of physical science methods into the social sciences. It is a reaction to larger political movements, including the changing use of science by government and industry, the professionalization of the separate disciplines, attempts to isolate science from the sensitive questions of the day.

The more general problem of the politics of knowledge has been clouded by failure to agree upon what we mean. When scientists assert that science is apolitical, what is often meant is that viral replication is the same under communism or capitalism, that the boiling point of water or the obliquity of the ecliptic is the product not of particular human passions or institutions but of invariant laws or facts of nature. Parts of science certainly are, or at least can be made, objective in this sense. But science is political in an altogether different sense. Science is political in the sense that science has become a power to be reckoned with in the military and industrial strength of nations. Science is also political in the sense that there are always alternatives to the form and substance of its production and practice. It is important not to confuse science and nature in this regard. Science is political in ways that nature is not, just as astronomy is political in ways the stars are not. Science is also political, though, in the sense that alternatives usually become possible only when they are recognized as alternatives and when people capable of making them come about are moved to make them come about. Most importantly, science is political whenever the objects under investigation are matters of vital human interest— problems of health, security, and the various forms of privilege and exclusion which societies enjoy or from which they suffer. Cancer theory is political not just because both human health and medical profits are at stake but also because the very fact of cancer is largely the product of unregulated industrial life. Physics is political because the greater part of it is (presently) organized to serve the military. Chemistry is political because it is primarily devoted to the development of materials used in industry. So long as the priorities of science are shaped by larger social priorities, the ideal of the "neutrality" of science confronts the reality of the politics of knowledge.

The vision of neutral science persists into recent times. In the spring of 1980, Harvard President Derek Bok stood before a packed audience at Lowell House to defend his appointment of economist Arnold Harberger as Director of the Harvard Institute for International Development (HIID). Harberger had come under attack for his role in training some 150 Ph.D.

economists to help reorganize the Chilean economy along guidelines set by the junta under Augusto Pinochet.² Students questioned Bok about the ethics of appointing such a man to a not insignificant position on the Harvard faculty. In his defense Bok invoked the principle of academic freedom, arguing that political or ethical considerations should have no bearing on academic appointments. (Asked further whether he would appoint as professor of the Harvard faculty "a known member of the Nazi Party," Bok replied that if such a person met scholarly qualifications, then his or her political beliefs or behavior should have no bearing on the issue. Bok conceded that he "would not make affiliation with the National Socialist Party a bar to an academic appointment at this university.") The crucial issue was whether "moral, political and ideological judgments ought to be the criterion in making appointments in the University." Bok said no, decisions on appointments should be based solely on academic excellence because "this University will be better if people feel free to press their judgments hard."³

The philosophical principle invoked by Bok was what he called "institutional neutrality." The principle was described by HIID director Lester Gordon as follows: "We try to get our people to be like the ideal civil servant, who helps the decision-maker identify his problems and outlines alternatives, but leaves him to choose his own objectives."⁴ In his own defense, Harberger argued that economic principles could be used for good or for evil, that moral considerations should have no bearing on the practice of good economic science.⁵ The economist brought to his trade a neutral "tool kit" of principles and relations with which he attempted to solve problems posed to him. His job was scientific, not political.

I began this book noting that the simplest and perhaps the oldest version of the ideal of neutrality is that science may be used for good or for evil. The problem with this view, though, is that it ignores the fact that science has both social origins and social consequences. Who, one can ask, does science serve, and how? Who has gained from "miracle wheat" and who has lost? The power of which countries is augmented by chemical weapons, the power of which others is diminished? Who has been excluded from science, and who has not? Whose economies have benefited from the neoclassical theory of the firm, whose have suffered? Science, in other words, does not always serve the collective *we* or the generic *man* but particular men—often those who control the means of its production and application. Science is not different from other aspects of culture in this sense.

Critics of the critics of science often imply that the only alternative to value-neutrality is blind error or muddled mysticism. Herman Kahn, strategist of thermonuclear war, in his *Thinking the Unthinkable* argued that "critics frequently object to the icy rationality of the Hudson Institute, the RAND Corporation and other such organizations. I'm always tempted to ask in reply, 'Would you prefer a warm, human error? Do you feel better with a nice emotional mistake?'" Kahn preferred an approach he himself called "Byzantine": "no agonizing over cultural and religious values. No moral posturing. It's the kind of thinking that the Hudson Institute and the Institute for Defense Analysis brought into war planning."[6]

Critics of the critics also suggest that it is silly to blame science or technology for the uses to which these may be put. In his 1925 *Callinicus, a Defense of Chemical Warfare*, J. B. S. Haldane described a curious form of honor practiced by the Chevalier Bayard in fifteenth-century France. Bayard extended all comforts and privileges to mounted knights and bowmen captured fighting according to the time-honored customs of traditional warfare. But any man caught using the musket, or any other form of gunpowder, was immediately put to death. It was moral, Bayard believed, to kill by traditional means but barbaric to use the latest methods.[7]

Haldane used the example of *Bayardism* to argue that it is silly to believe that one means of killing is decent, another barbarous. A similar form of Bayardism greeted the first use of the atomic bomb. I. Bernard Cohen, the Harvard historian of science, asked shortly thereafter:

> Who, of all those who cried out at the use of this terrible weapon, previously raised his voice at the bombing of Japanese women and children by detonating or incendiary bombs? It seems almost unthinkable that so-called civilized human beings in this twentieth century, the age of science, object to one form of destruction rather than another. Have we lost all basic sense of the moral idea? The necessities of war may have made ineluctable the large-scale bombings of civilians. But what ethical or moral principle can possibly justify the demolition of a city and its civilian population by the use of 200 "blockbusters," but not by one atomic bomb![8]

Cohen is right of course to condemn the "conventional" bombing of cities. But surely this does not mean we should welcome the development of ever more powerful weapons—weapons which promise new and previously unknown kinds of horrors. People live by their tools, and there are moral consequences to the development of one kind rather than another. The use of tools has consequences—symbolic, practical, political, ecological—and even if every death in war is a tragedy, the means by which such deaths come about are distinguishable. If all means are equally good (or horrible), where does this leave the idea of cruel and unusual punishment? Or the Geneva Convention? Or our revulsion at the idea that the ends justify the means? Where does it leave our revulsion of the military enjoying a research bonanza when more humane priorities go wanting?

The principle of neutral science, together with the doctrine of subjective value, constitutes the fundamental political ideology of modern science.

Science in this view is neutral and public; values are subjective and private. Science is the realm of public reason, values the realm of personal whim. The choice of ends is personal and arbitrary; the means for meeting those ends are public and rational. Science progresses best when governed least: *laissez-innover* in science is the counterpoint to *laissez-faire* in market relations.[9] Science in this vision is a neutral instrument, useless in itself, useful only when applied. The scientist discovers, society applies; values are implicated in the latter but not in the former.

The political significance of this vision has changed, though, with changing social and political circumstances. In the seventeenth century, the neutrality of science toward things political was a liberal, and at this time progressive, solution to the relations of science and society. A truce or compromise was achieved by which the fledgling experimental sciences might carve out their own little corner in the intellectual world, freed from church and state hegemony. By the twentieth century, however, things have changed. Science is no longer a marginal phenomenon fighting for the right to exist. It has, in some sense, become the ethos of an age—a language theory must speak if it is to be heard; grounds on which theory must build if it is to stand. In this context, when science has become a political force in its own right, neutrality serves to camouflage interests, to remove the moral and political from the realm of discourse itself. It is one thing for science to declare itself neutral in the struggle to free itself from feudal fetters. It is quite another when science has become a power in itself, for which neutrality means not a freedom from authority but an escape from commitment—or worse, an instrument to thwart social movement or criticism.

Karl Popper once wrote that all science is cosmology; it is probably as fair to say that all science is politics, or ethics. This is as it should be: science *should* respond to practical problems of human need and suffering. Many will perceive a danger in this dictum: the experience of Galileo before the Inquisition, of Nazi racial science, Lysenko's abuses, or of persistent efforts to subordinate science to industrial and military "needs" makes one realize that the ideal of science "for its own sake" does have its place. But even the ideal of science for its own sake can become a danger if, in the face of human suffering, it is used to escape prudent or critical action. Dante, we should recall, reserved the hottest place in hell for those who remain neutral in a time of crisis.

A century ago it was popular to talk about the "Warfare of Science with Theology." Today, it is social criticism of the origins and effects of science that presents the greatest challenge to science and to the ideal of political or moral neutrality. Scientists may continue to ignore the social context of their research but they do so at their peril, perhaps even at the peril of the rest of the world as well. The subtle dialectic we face is the reconciliation of the values of freedom on the one hand and accountability on the other.

Daniel Greenberg once remarked that science is not like soap, and that politicians wanting to shape science into pre-imagined forms ignore this difference at their peril. Yet neither is science like the Persian God Khoda, whose source is only itself. Science is the product of society and must remain accountable to that society. It is the duty of science, in the face of its public, to understand the conditions of its freedom, but also the responsibilities brought forth by that freedom.

Notes

Introduction. The Dilemma of Science Policy

1. The U.S. government has stockpiled 71,305 pounds of opium for use in the event of nuclear war; civil defense experts in the early 1980s advised increasing this by an additional 58,697 pounds. See the U.S. Civil Defense Agency's *Emergency Management* (Spring 1981).
2. Colin S. Gray and Keith Payne, "Victory Is Possible," *Foreign Policy*, 39 (Summer 1980): 14–27.
3. Michael Klare, *War Without End* (New York, 1972), p. 98; Carol Cohn, "Slick'ems, Glick'ems, Christmas Trees, and Cookie Cutters: Nuclear Language and How We Learned to Pat the Bomb," *Bulletin of the Atomic Scientists*, June 1987, 17–24.
4. "Magnet Design Selected for Giant Atom Device," *New York Times*, September 19, 1985; "Momentum Builds to Map All Genes," *New York Times*, August 25, 1988.
5. Costs to dismantle such devices may ultimately dwarf what it cost to make them: in 1988 the Department of Energy estimated the cost of cleaning up the nation's 28 nuclear bomb plants and testing facilities at between $40 and $110 billion (*New York Times*, July 2, 1988). The General Accounting Office reported shortly thereafter that costs could actually be as high as $175 billion (*New York Times*, July 14, 1988). On MILSTAR see Tim Weiner, *Blank Check: The Pentagon's Black Budget* (New York, 1990), pp. 46–72.
6. The idea that a tool may be used or abused is not of course peculiar to the West. Kuan Yin Tzu in a Taoist text of the ninth century warned of the responsibility that went with the possession of exceedingly sharp swords, cautioning that "only those who have the Tao will be able to perform such actions, and

better still, not perform them, though able to perform them." Cited in Joseph Needham, "History and Human Values; a Chinese Perspective for World Science and Technology," *Centennial Review,* 20 (1976): 23. Compare also the Confucian view that "those who learn but do not think are lost; and those who think but do not learn are very dangerous."

7. See, for example, Michael Polanyi, "The Republic of Science: Its Political and Economic Theory," *Minerva,* 1 (1962): 54-73.
8. Helen E. Longino, *Science as Social Knowledge: Values and Objectivity in Scientific Inquiry* (Princeton, 1989), p. 4; Steven and Hilary Rose, "The Myth of the Neutrality of Science," in *Science and Liberation,* ed. Rita Arditti et al. (Boston, 1980); Ivan Illich, *Medical Nemesis* (Toronto, 1976), p. 71; A. Kukla, "The Logical Incoherence of Value-Free Science," *Journal of Personality and Social Psychology,* 43 (1982): 1014-1017; Anne Fausto-Sterling, "The Myth of Neutrality: Race, Sex and Class in Science," *Radical Teacher,* 19 (1981): 21-25.
9. Abraham Edel, "The Concept of Value and Its Travels," in *Values and Value Theory in 20th-Century America,* ed. Murray G. Murphy and Ivar Berg (Philadelphia, 1988), p. 29; Sandra Harding, "Value-Free Research Is a Delusion," *New York Times,* October 22, 1989.
10. Sandra Harding, *The Science Question in Feminism* (Ithaca, 1986), p. 249.
11. George Stocking, "On the Limits of 'Presentism' and 'Historicism' in the Historiography of the Behavioral Sciences," *Journal of the Behavioral Sciences,* 1 (1965): 211-218; Herbert Butterfield, *The Whig Interpretation of History* (1931; New York, 1964).
12. Kuhn postulates a universal, trans-historical ontogeny of science (normal science, crisis, revolution, etc.) that is the same for all periods in history (ancient, medieval, modern) and all realms of inquiry (biology, physics, etc.). See his *Structure of Scientific Revolutions* (Chicago, 1962).
13. Jean-François Lyotard, *The Postmodern Condition: A Report on Knowledge* (1979; Minneapolis, 1984), p. 9.

1. The Cosmos as Construct

1. Plato, *Philebus,* 55D-E. Citations to Plato are from his *Collected Dialogues,* ed. E. Hamilton and H. Cairns (Princeton, 1961).
2. Plato, *Statesman,* 258D-E; *Philebus,* 59C; *Republic,* 527B.
3. Aristotle, *Nicomachean Ethics,* trans. H. Rackham (Cambridge, Mass., 1934), pp. 333-337; *Metaphysics* in his *Basic Works,* ed. Richard McKeon (New York, 1941), p. 691.
4. Plato, *The Phaedo,* 66C and 82C; also 69B-C.
5. Benjamin Farrington, *Head and Hand in Ancient Greece* (London, 1947); M. I. Finley, "Was Greek Civilization Based on Slave Labour?" in his *Slavery in Classical Antiquity* (Cambridge, Eng., 1960), pp. 57-58.
6. Genevieve Lloyd, *The Man of Reason, "Male" and "Female" in Western Philosophy* (Minneapolis, 1984) pp. 2-3.
7. See A. D. Lindsay's introduction to Plato and Xenophon's *Socratic Discourses* (London, 1910), pp. xii-xiii.

8. Herodotus, in his *Histories* (II, 167), writes in the late fifth century B.C. that "among the Greeks, but also among the Thracians, Scythians, Persians, Lydians, and most other peoples, men who learn trades and their descendants are held in less regard than other citizens, whilst any who need not work with their hands are considered noble, especially if they devote themselves to war" (trans. Harry Carter [London, 1962], p. 159).
9. Cited in Thomas Macaulay, "Francis Bacon," in his *Critical and Historical Essays* (1837; London, 1969), vol. 2, pp. 358–359. Macaulay chides Seneca (author of a book on "anger") for his "labours to clear Democritus from the disgraceful imputation of having made the first arch, and Anacharsis from the charge of having contrived the first potter's wheel." If forced to choose between the philosopher and the shoemaker, says Macaulay, we should choose the shoemaker. For "it may be worse to be angry than to be wet. But shoes have kept millions from being wet; and we doubt whether Seneca ever kept anybody from being angry."
10. *Dinner-table Conversations* [more often referred to as *Table-Talk*], bk. 8, no. 2, cited in Benjamin Farrington, *Science and Politics in the Ancient World* (London, 1939), pp. 29–30.
11. J. B. Bury, *The Idea of Progress* (New York, 1932), p. 115.
12. Francis Bacon, *The New Organon, or True Directions concerning the Interpretation of Nature* (1620), in *The New Organon and Related Writings,* ed. Fulton H. Anderson (New York, 1960), p. xxii; Wolfgang Krohn, "Die 'Neue Wissenschaft' der Renaissance," in Gernot Boehme et al., *Experimentelle Philosophie* (Frankfurt, 1977).
13. Thomas Hobbes, *Leviathan* (1651; New York, 1962), p. 19; Julien de La Mettrie, *L'homme machine* (Leyden, 1748); *Homme plante* (Potsdam, 1748).
14. Edgar Zilsel, "The Sociological Roots of Science," *American Journal of Sociology,* 47 (1942): 245–279.
15. Leonardo Olschski, *Geschichte der neusprachlichen wissenschaftlichen Literatur* (Leipzig, 1919).
16. Edgar Zilsel, "The Origins of the Concept of Scientific Progress," *Journal of the History of Ideas,* 4 (1945): 325–349; Krohn, "Die 'Neue Wissenschaft' der Renaissance," pp. 54–60.
17. Otto Bauer, *Das Weltbild des Kapitalismus* (1916; Frankfurt, 1971), pp. 13–17; also Lewis Mumford's *Technics and Civilization* (New York, 1934).

2. Baconian Caveats, Royalist Compromise

1. Francis Bacon, *The Great Instauration* (1620), in *The New Organon and Related Writings,* p. 7.
2. Bacon to his uncle, Lord Burghley (1592), cited in Catherine Drinker Bowen, *Francis Bacon* (New York, 1963), p. 65; *Great Instauration,* pp. 7 and 15; also his *New Organon,* I. 93, p. 92.
3. Bacon, *Great Instauration,* p. 26; *New Organon,* I. 90, p. 89.
4. Francis Bacon, *The Proficience and Advancement of Learning, Divine and*

Human, in his *Selected Writings*, ed. Hugh G. Dick (New York, 1955), p. 286; also *New Organon*, I. 63, p. 71.
5. Thomas Sprat, *History of the Royal Society of London* (London, 1667), pp. 72, 121-129, 192-194.
6. Ibid., pp. 74 (with three commas removed) and 76; also p. 116.
7. Bacon, *Great Instauration*, pp. 34, 9; *New Organon*, I. 122, p. 112; also I. 61 and 68. Descartes apparently so liked Bacon's parable of the lame runner and the notion of leveling wits through mechanical invention that he copied both from Bacon's original in his *Discourse on Method* (1637; Indianapolis, 1956), p. 3.
8. Sprat, *History of the Royal Society*, pp. 62-64.
9. *Notes and Records of the Royal Society of London* (April, 1946), pp. 39-40; Nicholas Hans, *New Trends in Education in the Eighteenth Century* (London, 1951), pp. 32-33; Michael Hunter, *The Royal Society and Its Fellows, 1600-1700* (Chalfont St. Giles, 1982).
10. Francis Bacon, "On the Plantation of Ireland" (1606), in his *Works* (London, 1838), p. 471; Petty to Southwell, 26 September 1685, *Petty-Southwell Correspondence, 1676-1682* (New York, 1967), p. 168.
11. Descartes, *Discourse on Method*, p. 40.
12. Richard Eden (1574) and George Hakewill (1627) are cited in Robert F. Jones, *The Ancients and the Moderns* (New York, 1936), pp. 12 and 34; Valla is cited in Krohn, "Die 'Neue Wissenschaft' der Renaissance," p. 39. For a history of the "shoulders of giants" metaphor see R. Klibansky, "Notes and Comments," *Isis*, 26 (1936): 147ff. Wolf Lepenies suggests that by the end of the eighteenth century most scientists saw themselves as "giants standing on the shoulders of dwarfs"; see his *Between Literature and Science: The Rise of Sociology* (Cambridge, Eng., 1988), pp. 1-2.
13. Bacon's *New Organon* states that, with proper method, "the discovery of all causes and sciences would be but the work of a few years" (I. 112, p. 104). Perrault is cited in Bury, *Idea of Progress*, pp. 87-90.
14. Bury, *Idea of Progress*, pp. 136-137.
15. See I. Bernard Cohen, "Fear and Distrust of Science in Historical Perspective," in Andrei S. Markovits and Karl Deutsch, eds., *Fear of Science—Trust in Science* (Cambridge, Mass., 1980), pp. 29-58.
16. Sprat, *History of the Royal Society*, pp. 321-322.
17. Samuel Johnson, *The Rambler*, 5 May 1750, p. 80.
18. Benjamin Farrington, *Francis Bacon* (London, 1963), p. 124.
19. Bacon, *Great Instauration*, p. 29.
20. Bacon, *New Organon*, I. 95, p. 93; *Great Instauration*, p. 14.
21. Francis Bacon, *Preparative Toward a Natural and Experimental History* (1622), in *The New Organon and Related Writings*, p. 273; *Great Instauration*, p. 25.
22. Francis Bacon, "Atalanta, or Profit," in *The Wisdom of the Ancients* (1609), in his *Selected Writings*, pp. 416-417; *New Organon*, I. 120, p. 109; *Great Instauration*, pp. 12 and 24. Compare also his claim that "the contemplation of truth is a thing worthier and loftier than all utility and magnitude of works" (*New Organon*, I. 124, p. 113).

23. Sprat, *History of the Royal Society*, p. 68.
24. Bacon, *New Organon*, I. 99, p. 96; I. 80 and 81, pp. 77–78. This is the earliest expression I have found of the ideal of science "for its own sake"; compare also I. 98, p. 95. Bacon, *Advancement of Learning*, pp. 373–374.
25. Bacon, *New Organon*, I. 39–44, pp. 47–49.
26. Ibid., I. 49, p. 52; I. 79, p. 77.
27. Bacon, *Great Instauration*, p. 15.
28. Robert Hooke's 1663 draft statutes for the Royal Society of London are cited in Martha Ornstein, *The Role of Scientific Societies in the Seventeenth Century* (Chicago, 1938), pp. 108–109. Moray is cited in Wolfgang van den Daele, "The Social Construction of Science," in *The Social Production of Scientific Knowledge*, ed. Everett Mendelsohn et al. (Dordrecht, 1977), pp. 27–54. Sprat, *History of the Royal Society*, p. 82.
29. Sprat, *History of the Royal Society*, pp. 33, 62.
30. Accademia del Cimento, *Essayes of Natural Experiments* (1667), trans. Richard Waller (London, 1684), "Preface to the Reader."
31. G. P. Murdock, *Culture and Society*, Twenty-Four Essays (Pittsburgh, 1965), p. 126.
32. Thomas Hobbes, *Elements of Philosophy* (1641), in his *English Works* (London, 1839–1845), vol. 1, pp. 10–11.
33. William Petty, preface to his *Political Arithmetick* (1690) in *The Economic Writings of Sir William Petty*, ed. C. H. Hull (New York, 1963), vol. 1, p. 244; "Political Anatomy of Ireland" (1672), in ibid., p. 129.
34. John Wilkins, *Essay Towards a Real Character and a Philosophical Language* (London, 1668), Epistle Dedicatory; compare also Nigel Fabb's unpublished "Metaphor in the Seventeenth Century" (1980).
35. Bacon, *Great Instauration*, p. 17; Henry Stubbe, *A Censure upon Certain Passages Contained in the History of the Royal Society* (Oxford, 1671), p. 28.
36. Martha Ornstein, *The Role of Scientific Societies in the Seventeenth Century* (1913; Chicago, 1938), p. 45.
37. Andrew Large, *The Artificial Language Movement* (Oxford, 1985), pp. 8–18.
38. Hobbes, *Leviathan*, p. 34; Samuel Parker to Samuel Hartlib (1647), cited in Robert F. Jones et al., *The Seventeenth Century* (Stanford, 1951), p. 117.
39. Wolfgang Lepenies, "Der Wissenschaftler als Autor, Buffons prekärer Nachruhm," in his *Das Ende der Naturgeschichte: Wandel kultureller Selbstverständlichkeiten in den Wissenschaften des 18. und 19. Jahrhunderts* (Frankfurt, 1978), p. 145. Erasmus Darwin, *The Loves of the Plants* (London, 1798); also Londa Schiebinger, "The Private Life of Plants: Sexual Politics in Early Modern Botany," in *Science and Sensibility: Women and Science, 1780–1945*, ed. Marina Benjamin (Oxford, 1990).
40. Londa Schiebinger, *The Mind Has No Sex? Women in the Origins of Modern Science* (Cambridge, Mass., 1989), pp. 151–159.
41. *An Authentic Narrative of the Dissensions and Debates in the Royal Society* (London, 1784), p. 155.
42. J. L. McIntyre, *Giordano Bruno* (London, 1903), p. 21.
43. Robert Merton points out that in 1686, 13 percent of 241 projects undertaken by the Royal Society were concerned with military technology. See his "Inter-

action of Science and Military Technique" (1935), in his *Sociology of Science,* ed. Norman W. Storer (Chicago, 1973), p. 208.

3. The Devalorization of Being

1. Alexandre Koyré, *From the Closed World to the Infinite Universe* (1957; Baltimore, 1969), pp. vii–x, 1–27.
2. Galileo Galilei, *Dialogue Concerning the Two Chief World Systems* (1632), trans. Stillman Drake (Berkeley, 1970), p. 59.
3. Charles Coulton Gillispie, *The Edge of Objectivity* (Princeton, 1960), p. 40.
4. Bacon, *New Organon,* II. 9, p. 129. In his *Advancement of Learning,* Bacon states that the fixation upon final causes in physical inquiry has led to "the great arrest and prejudice of further discovery (p. 259)."
5. Cited in Kenneth J. Franklin's introductory essay to William Harvey, *The Circulation of the Blood and Other Writings* (London, 1963), p. x.
6. Bert Hansen, "Science and Magic," in *Science in the Middle Ages,* ed. David C. Lindberg (Chicago, 1978); compare also Francis Yates, *The Rosicrucian Enlightenment* (Cambridge, Eng., 1974).
7. Keith Thomas, *Religion and the Decline of Magic* (New York, 1971), pp. 190–191, 224; Hansen, "Science and Magic," p. 491.
8. Thomas, *Religion and the Decline of Magic,* p. 645; Descartes, *Discourse on Method,* p. 6.
9. On alchemy in Newton and Galileo see M. L. Righini Bonelli and William R. Shea, eds., *Reason, Experiment, and Mysticism in the Scientific Revolution* (New York, 1975). Robert Merton points out that Vilfredo Pareto found it incredible that Newton should have produced a book on the Apocalypse; Lombroso similarly declared Newton to have been "less than sane when he demeaned his intellect to the interpretation of the Apocalypse." See Merton's "The Puritan Spur to Science," in his *The Sociology of Science* (Chicago, 1973), p. 249.
10. Hobbes, *Leviathan,* pp. 20, 48, 80.
11. Jacques Berriat-Saint-Prix in his "Rapport et Recherches sur le Procès et Jugements relatives aux Animaux" documents 93 cases in which insects, reptiles, rats, mice, pigs or horses were prosecuted and punished from the twelfth to the eighteenth centuries. See the *Mémoires de la Société Royale des Antiquaires de France,* vol. 8 (Paris, 1827), pp. 403–450. Carlo d'Addosio has expanded this to cover 144 cases from the year 824 to 1845; see his *Bestie Delinquenti* (Naples, 1892). For a more recent and subtle analysis of animal trials see J. J. Finkelstein, *The Ox That Gored* (Philadelphia, 1981).
12. Edward Payson Evans, *Criminal Prosecution and Capital Punishment of Animals* (London, 1906), p. 195. Karl von Amira reports that the last trial of an animal before a secular court of law was in 1733; see his *Thierstrafen und Thierprocesse* (Innsbruck, 1891), p. 25.
13. Alexandre Sorel, *Procès contre des animaux et insectes, suivi au moyen age dans la Picardie et le Valois* (Compiègne, 1877), pp. 23–24.
14. Pierre Bayle, sometimes regarded as the first modern atheist, argued in the

1680s and 1690s that comets "are in no sense portents of disaster." See his *Pensées diverses . . . à l'occasion de la comète* (Paris, 1683); also Paul Hazard, *The European Mind, 1680–1715* (1935; New York, 1963), pp. 155–160.
15. Andrew Dickson White, *A History of the Warfare of Science with Theology in Christendom* (1896; New York, 1919), vol. 1, p. 29. White no doubt exaggerates the importance of paleontology in this regard.
16. Ibid., p. 30.
17. William Paley, *Natural Theology: or, Evidences of the Existence and Attributes of the Deity* (1802; London, 1809), pp. 254, 462, 33.
18. Ibid., pp. 467–468. Compare Stephen Jay Gould's discussion of the ichneumon wasp in his "Nonmoral Nature," *Hen's Teeth and Horse's Toes* (New York, 1983), pp. 32–45.
19. W. H. Brock, "The Selection of the Authors of the Bridgewater Treatises," *Notes and Records of the Royal Society*, 21 (1966): 162–179.
20. John Kidd, *On the Adaptation of External Nature to the Physical Condition of Man* (London, 1834), pp. 86–97.
21. Abraham Tucker, *Light of Nature Pursued* (London, 1768–1778), bk. 3, chap. 9, p. 9.
22. Charles Darwin, *Autobiography*, ed. Francis Darwin (New York, 1958), p. 19.
23. Darwin to Lyell, June 17, 1860, in *More Letters*, ed. Francis Darwin and A. C. Seward (London, 1903), vol. 1, p. 154.

4. Secondary Qualities and Subjective Value

1. Bacon, *Advancement of Learning*, p. 295.
2. Robert Merton, "The Puritan Spur to Science," in his *Sociology of Science*; compare also Reijer Hooykaas, *Religion and the Rise of Modern Science* (Grand Rapids, 1972).
3. Galileo Galilei, *The Assayer* (1623), in *Discoveries and Opinions of Galileo*, trans. Stillman Drake (New York, 1973), p. 274. Compare also Burtt, *Metaphysical Foundations*, p. 85.
4. Galileo, *Assayer*, pp. 275–277; see also Robert Boyle's 1666 *Origin of Forms and Qualities* and John Locke's 1690 *Essay Concerning Human Understanding*.
5. Bacon, *Great Instauration*, p. 25.
6. Burtt, *Metaphysical Foundations*, p. 90. Compare Lewis Mumford: "The qualitative was reduced to the subjective: the subjective was dismissed as unreal, and the unseen and unmeasurable nonexistent"; *Technics and Civilization* (1934; New York, 1963), p. 49.
7. Hobbes, *Leviathan*, p. 22.
8. Ibid., p. 48. "The value, or worth of a man, is as of all other things, his price; that is to say, so much as would be given for the use of his power: and therefore is not absolute; but a thing dependent upon the need and judgment of another" (p. 73).
9. John Locke, *An Essay Concerning Human Understanding* (1690), ed. Peter H. Nidditch (Oxford, 1975), pp. 66, 135–138.
10. Ibid., p. 229.

11. Ibid., pp. 549–550; John Hermann Randall, *The Career of Philosophy* (New York, 1962), vol. 1, pp. 718–720, 735. Locke's views are more complex than I've indicated. He considered morality as well as mathematics capable of demonstration, but only upon foundations provided by the idea of a Supreme Being. Given this idea, "measures of right and wrong might be made out" that are as "incontestible as those in mathematics." Locke is not a hedonist: the good is what produces pleasure, but it is also that which is in accordance with God's law. It is ultimately unclear in Locke whether pleasure is the cause or consequence of virtue, whether God's will is primary to or indicative of the moral law.
12. David Hume, *A Treatise of Human Nature* (1739; Oxford, 1978), p. 457.
13. Ibid., pp. 469–470.
14. Alasdair MacIntyre, *After Virtue: A Study in Moral Theory*, 2nd ed. (Notre Dame, 1984), p. 53.
15. Adam Smith, *Theory of Moral Sentiments* (1759; Oxford, 1976), p. 9; compare also James Farr, "Hume, Hermeneutics, and History: A 'Sympathetic Account,' " *History and Theory*, 17 (1978): 289.
16. Stuart Hampshire, "Fallacies in Moral Philosophy," *Mind*, 58 (1949): 466–482; A. C. MacIntyre, "Hume on 'Is' and 'Ought' " (1959), in *Hume: A Collection of Critical Essays*, ed. V. C. Chappell (Garden City, 1966).

5. The German University and the Research Ideal

1. Bernard Mandeville, *Fable of the Bees* (1732; Oxford, 1924), p. 39. Chapter 15 of Machiavelli's *The Prince* (1513) reads: "he who neglects what is being done for what should be done will learn his destruction rather than his preservation"; trans. Mark Musa (New York, 1964), p. 170.
2. The dates I have given here are from the first use as listed in the *Oxford English Dictionary*; several have earlier foreign equivalents.
3. Aristotle, *Eudemian Ethics*, trans. Michael Woods (Oxford, 1982), pp. 6–7; Bacon, *Advancement of Learning*, p. 262; Samuel Johnson, *The Rambler*, May 5, 1750, p. 80.
4. Charles Babbage, *On the Economy of Machinery and Manufactures* (Philadelphia, 1832), pp. 270–271; Justus Liebig, *Die organische Chemie in ihrer Anwendung auf Agricultur und Physiologie* (Brunswick, 1840); Louis Pasteur, "Pourquoi la France n'a pas trouvé d'hommes supérieurs au moment du péril," *Revue scientifique*, 2d ser., 1 (1871): 73–77.
5. See Hajo Halborn, *Germany and Europe* (New York, 1970); Fritz Stern, *The Failure of Illiberalism* (London, 1972); Leonard Krieger, *The German Idea of Freedom* (Boston, 1957).
6. See, for example, Christian von Ferber, "Der Werturteilsstreit 1909/1959," *Kölner Zeitschrift für Soziologie und Sozialpsychologie*, 11 (1959): 21–37; also Paul Honigsheim's "Die Gründung der Deutschen Gesellschaft für Soziologie" in the same volume.
7. Adolf Harnack, *Geschichte der Königlich Preussischen Akademie der Wissenschaften zu Berlin* (Berlin, 1900), vol. 2, p. 556. See also R. S. Turner, "The

Prussian Universities and the Research Imperative, 1806–48," Ph.D. diss., Princeton University, 1973.
8. Wilhelm von Humboldt, "Über die innere und äussere Organisation der höhere wissenschaftlichen Anstalten zu Berlin" (1810), in *Die Idee der deutschen Universität*, ed. Ernst Anrich (Darmstadt, 1956), p. 377.
9. J. G. Herder, *Vom Einfluss der Regierung auf die Wissenschaften und der Wissenschaften auf die Regierung* (Berlin, 1780), p. 7; Humboldt, "Über die innere und äussere Organisation," pp. 255–258, 269.
10. Norbert Elias, *History of Manners* (1939; New York, 1978), pp. 10–14.
11. Helmut Schelsky, *Einsamkeit und Freiheit* (Stuttgart, 1956), pp. 22–25, 39.
12. Wilhelm Kahl, *Geschichtliches und Grundsätzliches aus der Gedankenwelt über Universitätsreform* (Berlin, 1909), p. 9; Friedrich Paulsen, *Die deutschen Universitäten* (Berlin, 1902), p. 57.
13. G. W. F. Hegel, *Phenomenologie des Geistes* (1806; Frankfurt, 1970), pp. 20–22.
14. F. W. J. Schelling, "Vorlesungen über die Methode des akademischen Studiums" (1803), in Anrich, *Die Idee*, pp. 6–18.
15. Friedrich Gren, "Geschichte der Naturwissenschaft," *Annalen der Physik*, 1 (1799): 195.
16. Laetitia Boehm, "Wissenschaft—Wissenschaften—Universitätsreform," *Berichte zur Wissenschaftsgeschichte*, 1 (1978): 19.

6. Empirical Science and Specialized Expertise

1. Gleditsch, *Teutsch-Englisches Lexikon* (Leipzig, 1745). The term *Wissenschaftler* did not arise until the first half of the nineteenth century; scholars in Goethe's time spoke instead of the *Wissenschafter*.
2. Leibniz, letter of July 11, 1700, in Harnack's *Geschichte der Königlich Preussischen Akademie*, pp. 90–94.
3. Boehm, "Universitätsreform," pp. 15–24.
4. Charles McClelland, *State, Society and University in Germany 1700–1914* (Cambridge, Eng. 1980), p. 247.
5. Bacon, *Advancement of Learning*, p. 230; Jean Le Rond d'Alembert, *Discours préliminaire de l'Encyclopédie* (1750; Paris, 1912); McClelland, *State, Society and University*, p. 77.
6. Schelling, "Vorlesungen," pp. 22, 31; Friedrich Lilge, *The Abuse of Learning* (New York, 1948), p. 59; Karl Ernst von Baer, review of Darwin's *Descent of Man*, in *Darwin and His Critics*, ed. David Hull (Cambridge, 1973), p. 421.
7. Lilge, *Abuse of Learning*, p. 71; Friedrich Paulsen, *Die deutschen Universitäten* (Berlin, 1902), p. 76. See also A. W. von Hofmann, "Die Frage der Teilung der philosophischen Fakultäten," Rektoratsrede (Berlin, 1880).
8. McClelland, *State, Society and University*, p. 306.
9. Immanuel Kant, *Critique of Pure Reason* (1781), trans. N. K. Smith (New York, 1929), p. 23.
10. Thomas Willey, *Back to Kant: The Revival of Kantianism in German Social and Historical Thought, 1860–1914* (Detroit, 1978), p. 26.
11. Hermann Lotze, "Streitschriften," cited in Merz, *History of European Thought*,

vol. 4, p. 67; Hans Vaihinger, *Kommentar zu Kants Kritik der reinen Vernunft* (Jena, 1877).
12. Eduard Zeller, "Über die Bedeutung und Aufgabe der Erkenntnistheorie" (1862), in his *Vorträge und Abhandlungen* (Leipzig, 1877), pp. 479-496.
13. Cited in H. Hörz and S. Wollgart's introduction to Helmholtz's *Philosophische Vorträge und Aufsätze* (Berlin, 1971), p. xxvi.
14. Friedrich Allihn and Tuiskon Ziller, "Ankündigung," *Zeitschrift für exacte Philosophie*, 1 (1861): ii–v; Richard Avenarius, "Zur Einführung," *Vierteljahrsschrift für wissenschaftliche Philosophie*, 1 (1877): 2.
15. Mitchel Ash, "Wilhelm Wundt and Oswald Külpe on the Institutional Status of Psychology," in *Wundt Studies/Wundt Studien*, ed. W. G. Bringmann and R. D. Tweeny (Göttingen, 1980).
16. Max Weber, "Wissenschaft als Beruf" (1919), in his *Gesammelte Aufsätze zur Wissenschaftslehre* (Tübingen, 1968), p. 591.
17. Ferdinand Tönnies, "Wege und Ziele der Soziologie," *Verhandlungen des Ersten Deutschen Soziologentages* (Tübingen, 1911), p. 19; Kant, *Critique of Pure Reason*, p. 18.
18. Arnold Hauser, *The Social History of Art* (New York, 1951), vol. 4, pp. 21-22.
19. Auguste Comte, *Cours de philosophie positive* (1830-1842; Paris, 1869), vol. 1, p. 42. Louis Gillet, *Sainte-Beuve et Alfred Vigny, lettres inédites* (Paris, 1929), p. 155. G. F. Nietzsche, ed., *Der werdende Nietzsche* (Munich, 1924), p. 427; Friedrich Nietzsche, *Beyond Good and Evil* (1886), trans. P. J. Hollingdale (New York, 1972), pp. 115 and 118. Charles-Augustin Sainte-Beuve first used the expression *tour d'ivoire* in the poem "A. M. Villemain" in his *Pensées d'Août* (Paris, 1837), p. 152.
20. Jacob Burckhardt, *Weltgeschichtliche Betrachtungen* (1905), cited in James H. Nichols' introduction to the translation of this work as *Force and Freedom* (New York, 1943), p. 53; Burckhardt to Kinkel, cited in Karl Löwith, *Jacob Burckhardt* (Luzern, 1936), pp. 81-82.

7. The *Werturteilsstreit*, or Controversy over Values

1. *Verhandlungen des Ersten Deutschen Soziologentages von 19.-20. Oktober, 1910 in Frankfurt am Main* (Tübingen, 1911; reprinted 1977), p. v. For background on the society see Ferdinand Tönnies, "Die Deutsche Gesellschaft für Soziologie," *Kölner Vierteljahrshefte für Soziologie*, 1 (1921): 42-46.
2. Eugen von Philippovich, "Das Wesen der volkswirtschaftlichen Produktivität und die Möglichkeit ihrer Messung," *Verhandlungen des Vereins für Sozialpolitik in Wien, 1909, Schriften des Vereins für Sozialpolitik*, 132 (1910): 357-358.
3. Ibid., p. 334.
4. *Verhandlungen des Vereins*, p. 559.
5. Joseph Schumpeter, *History of Economic Analysis* (Oxford, 1954), p. 802.
6. *Verhandlungen des Vereins*, pp. 565-566.
7. Ibid., pp. 582-585.

8. Gustav Schmoller, "Volkswirtschaftslehre," *Handwörterbuch der Staatswissenschaften*, ed. J. Conrad et al. (Jena, 1909–1911), vol. 8, pp. 426–501.
9. Ibid., pp. 449, 491–495. Schmoller first attacked Menger in his "Zur Methodologie der Staats- und Sozialwissenschaften," *Jahrbuch für Gesetzgebung, Verwaltung und Volkswirtschaft*, 7 (1883): 239–258; Schmoller's review launched the so-called *Methodenstreit*.
10. The paternalistic *Kathedersozialisten* believed that the state served an important moral function in society; the duty of academics in turn was to serve the state. It was this that earned them the derogatory title "socialists of the chair." The *Kathedersozialisten* themselves (Philippovich and von Wieser, for example) preferred the term *historisch-ethische Schule*.
11. Oscar Engländer, "Die Erkenntnis des Sittlich-Richtigen und die Nationalökonomie," *Jahrbuch für Gesetzgebung und Volkswirtschaft*, 38 (1914): 1799; *Verhandlungen des Vereins*, p. 575.
12. Marianne Weber, *Max Weber: Ein Lebensbild* (Heidelberg, 1950), pp. 420–424.
13. Max Weber, "Äusserungen zur Werturteildiskussion im Ausschuss des Vereins für Sozialpolitik," privately circulated manuscript, 1913 (available from the Archiv der Max Weber-Gesamtausgabe, Munich), p. 84.
14. Tönnies, "Wege und Ziele der Soziologie," pp. 18–19, 20, 23 (emphasis added).
15. Ibid., pp. 26–27 (emphasis added); pp. 37–38.
16. *Verhandlungen des Vereins*, p. 567.
17. Francis Bacon, *Natural and Experimental History for the Foundation of Philosophy* (1622), cited in Fulton Anderson's introduction to *The New Organon*, p. xv; Adam Smith, *Wealth of Nations* (1776; Chicago, 1976), vol. 2, pp. 317–318.
18. Thomas Macaulay, "Francis Bacon" (1837), in his *Critical and Historical Essays* (London, 1967), vol. 2, pp. 375–376.
19. Cited in Karl Bruhns, *Life of Alexander von Humboldt* (London, 1873), vol. 2, p. 130.
20. Lionel Playfair, "Science and Technology as a Source of National Power" (1885), reprinted in *Victorian Science*, ed. George Basalla (New York, 1970), pp. 68, 74–82.
21. L. F. Haber, *The Chemical Industry during the Nineteenth Century* (Oxford, 1958), pp. 120–128.
22. John J. Beer, "Coal Tar Dye Manufacture and the Origins of the Modern Research Laboratory," *Isis*, 49 (1958): 129–133.
23. "International Science," *Smithsonian Report for 1906*, no. 1770 (Washington, D.C., 1907), p. 495.
24. A. W. von Hofmann, "Allgemeine Sitzung," *Verhandlungen der Gesellschaft Deutscher Naturforscher und Ärzte* (Leipzig, 1890), pp. ix–x.
25. Cohen, "Fear and Distrust of Science," p. 51; McClelland, *State, Society and University*, pp. 307–312; Haber, *Chemical Industry*, p. 72; Craig, *Germany*, p. 200.
26. "Vorschläge zur Minderung der wissenschaftlichen 'Sprachverwirrung,'" *Naturwissenschaftliche Wochenschrift*, new series, 1 (1902): 239.
27. Alfred Weber, *Der Anfang der Volkswirtschaftslehre als Wissenschaft* (Tübingen, 1909), pp. 75–76.

28. John Theodore Merz, *A History of European Thought in the Nineteenth Century* (New York, 1899-1912), vol. 3, pp. 62-63.

8. The Social Context of German Social Science

1. The term *scientist* is first used in William Whewell, "On the Connexion of the Physical Sciences, by Mrs. Somerville," *Quarterly Review*, 51 (March 1834): 65.
2. William Cunningham, "A Plea for Pure Theory," *Economic Review* (January 1892); F. Y. Edgeworth, "Pure Theory of Taxation," *Economic Journal* (June 1897). Tönnies's *Gemeinschaft und Gesellschaft* in the first two editions after 1887 bore the subtitle *Abhandlung des Kommunismus und des Sozialismus als empirischer Kulturformen;* in later editions (the 3rd revised edition of 1920, for example) he changed this to *Grundbegriffe der reinen Soziologie.*
3. Charles P. Snow, *The Two Cultures and the Scientific Revolution* (New York, 1959), p. 34.
4. Ferdinand Tönnies, "Comtes Begriff der Soziologie," *Monatsschrift für Soziologie,* 1 (1909): 50. The term *sociologie* was first used by Auguste Comte in 1839 in Book 4 of his *Cours de philosophie positive.* See Victor V. Branford, "On the Origin and Use of the Word Sociology," *Sociological Papers* (London, 1905), vol. 1, pp. 3-24. The *Oxford English Dictionary* lists the first English use of the term *sociology* as 1843, in an article in *Blackwell's Magazine* where the new science is equated with "Social Ethics." In Germany, the earliest use of the term appears in reference to Ludwig Gumplowicz's gropings for a racial theory of society; Tönnies reports that Gumplowicz's 1883 *Grundriss der Soziologie* represented "the first noteworthy book in the German language to use the word *Soziologie* in its title" (cited in Gumplowicz's *Geschichte der Staatstheorien* [Innsbruck, 1892], pp. i-ix). Comte's *Cours de philosophie positive* was translated into German in 1873; Spencer's *Introduction to the Study of Sociology* was translated in 1875 and his *Principles* in 1877.
5. Lukács, *Zerstörung der Vernunft,* vol. 3, p. 42.
6. Karl Erich Born, "Structural Changes in German Social and Economic Development at the End of the Nineteenth Century," in *Imperial Germany,* ed. James Sheehan (New York, 1976), p. 17.
7. David Ricardo, *The Principles of Political Economy and Taxation* (1817; London, 1911), p. 81.
8. Born, "Structural Changes," pp. 23-28.
9. Carl Schorske, *German Social Democracy: The Origins of the Great Schism* (Cambridge, Mass., 1955), pp. 7-31.
10. Friedrich Engels, "Einleitung zu Karl Marx: Die Klassenkämpfe in Frankreich 1848 bis 1850," in Marx and Engels, *Ausgewählte Schriften* (Berlin, 1953), vol. 1, pp. 113-114.
11. Wilhelm Lexis, *Die deutschen Universitäten* (Berlin, 1893), p. 141; Carl Schorske, "Higher Education in Germany in the Nineteenth Century," *Journal of Contemporary History,* 2 (1967): 136-137.
12. Fritz Ringer, *The Decline of the German Mandarins* (Cambridge, Mass., 1969); compare Craig, *Germany,* p. 205.

13. Weber, *Max Weber*, pp. 420-424; Christian von Ferber, "Der Werturteilsstreit 1909/1959," *Kölner Zeitschrift für Soziologie und Sozialpsychologie*, 11 (1959): 21-37; and Else Frohnhäuser, *Das Werturteil in der Volkswirtschaftslehre* (Giessen, 1929).
14. *Verhandlungen des Vereins*, p. 1.
15. Max Weber, "Die 'Objektivität' der soziologischen und ökonomischen Wissenschaften" (1904), in his *Gesammelte Aufsätze zur Wissenschaftslehre* (Tübingen, 1968), p. 167.
16. Tomás Masaryk, *Versuch einer konkreten Logik* (Vienna, 1887), pp. 138ff.
17. Ludwig Gumplowicz, *Die sociologische Staatsidee* (1892), 2d ed. (Innsbruck, 1902), p. 38; also his *Sociologie und Politik* (Leipzig, 1892), pp. 14-15.
18. Ringer, *Decline of the German Mandarins*, pp. 142-143; Craig, *Germany*, p. 202.
19. Craig, *Germany*, p. 201; Ringer, *Decline of the German Mandarins*, pp. 142-143.
20. As late as 1929 there was only one chair of sociology in Prussia. This contrasts with the situation in America, where sociology was taught in universities as early as 1889 (at the University of Kansas) and 1893 (at the University of Chicago). In 1901, 198 instructors at 169 institutions of higher learning in America were teaching sociology. See Anthony Oberschall, ed., *The Establishment of Empirical Sociology* (Paris, 1965), p. 221; also Bert Hardin, *The Professionalization of Sociology, a Comparative Study: Germany-USA* (Frankfurt, 1977).
21. C. H. Becker, *Gedanken zur Hochschulreform* (Leipzig, 1920), pp. 4-9.
22. Georg von Below, "Soziologie als Lehrfach," reprinted separately (Munich, 1920), pp. 28-30; also his "Soziologie und Hochschulreform," *Weltwirtschaftliches Archiv*, 16 (1920): 512-527.
23. Abroteles Eleutheropulos, "Was ist das Objekt bzw. die Aufgabe der Soziologie?" *Monatsschrift für Soziologie*, 1 (1909): 319.
24. Herbert Spencer, *Principles of Sociology* (London, 1896), p. 244; *Introduction to the Study of Sociology* (London, 1873), p. 22; also his *Progress: Its Law and Cause* (London, 1857), p. 56.
25. René Worms, *La science et l'art en économie politique* (Paris, 1896), p. 54.
26. Hermann Siemens, "Die Proletarisation unseres Nachwuchses, eine Gefahr unrassenhygienische Bevölkerungspolitik," *Archiv für Rassen- und Gesellschaftsbiologie*, 12 (1916/18): 43-55.
27. Henri Potoniés, "Über die Entstehung der Denkformen," *Die Naturwissenschaftliche Wochenschrift*, 6 (1891): 146-147; compare also Richard Avenarius, *Philosophie als Denken der Welt* (Leipzig, 1876); Ernst Mach, "On the Part Played by Accident in Invention and Discovery," *Monist*, 6 (1896): 161-175; Ludwig Boltzmann, *Populäre Schriften* (Leipzig, 1905), pp. 338-344.
28. "Über die gegenwärtige Lage des biologischen Unterrichts an höheren Schulen," *Die Naturwissenschaftliche Wochenschrift*, 18 (1902): 205.
29. H. E. Ziegler, "Über die Beziehungen der Zoologie zur Soziologie," *Verhandlungen der Deutschen Zoologischen Gesellschaft*, 3 (1893): 51-55. Leading works in the German organic sociology movement include Albert Schäffle, *Bau und Leben des socialen Körpers*, 4 vols. (Tübingen, 1875-1878); Paul von

Lilienfeld, *Zur Vertheidigung der organischen Methode in der Soziologie* (Berlin, 1898).
30. Paul Weindling, *Health, Race and German Politics between National Unification and Nazism, 1870–1945* (Cambridge, Eng. 1989), p. 140. Ferdinand Tönnies, "Eugenik," *Schmollers Jahrbuch*, 29 (1905): 273–290; also his "Studies in National Eugenics," *Sociological Papers* (London, 1906), vol. 2, pp. 40–42.
31. *Verhandlungen des Zweiten Deutschen Soziologentages 20–22 Oct., 1912 in Berlin* (Tübingen, 1913), p. 74. Weber's contribution to the first (1910) meeting of the Deutsche Gesellschaft für Soziologie has been published in translation in *Social Research*, 38 (1971): 30–41. See also Moritz Manasse, "Max Weber on Race," *Social Research*, 14 (1947): 191–221.
32. See my "From *Anthropologie* to *Rassenkunde*: Concepts of Race in German Physical Anthropology," in *Bones, Bodies, Behavior: Essays on Biological Anthropology*, ed. George Stocking (Madison, 1988), pp. 138–179.
33. Fritz Lenz, *Die Rasse als Wertprinzip* (1917; Munich, 1933); also my *Racial Hygiene*, pp. 59–63.
34. Craig, *Germany*, p. 212; Richard Evans, "Bourgeois Feminists and Women Socialists in Germany, 1894–1914," *Women's Studies International Quarterly*, 3 (1980): 363–365.
35. Schiebinger, *The Mind Has No Sex?* pp. 245–277.
36. Jean-Jacques Rousseau, *Emile, or On Education* (1762), trans. Allan Bloom (New York, 1979), p. 386.
37. G. W. F. Hegel, *Philosophy of Right*, trans. T. M. Knox (Oxford, 1967), pp. 263–264.
38. Arthur Kirchhoff, ed., *Die akademische Frau, hervorragender Universitätsprofessoren, Frauenlehrer und Schriftsteller über die Befähigung der Frau zum wissenschaftlichen Studium und Berufe* (Berlin, 1897).
39. Ibid., pp. 256–257.
40. Ibid., pp. 270–273.
41. Wundt nevertheless argued that neither the military nor politics nor certain fields of medicine (surgery and gynecology) were appropriate for women, insofar as these endeavors are based upon "political strength and moral will" (ibid., pp. 179–181).
42. Ibid., pp. 241, 260–261.
43. Ibid., p. 177.
44. Ibid., pp. 33, 85–86, 198.
45. Max Funke, *Sind Weiber Menschen?* (Halle, 1910), p. 16.
46. See, for example, Kirchhoff, *Die akademische Frau*, p. 207.
47. Ferdinand Tönnies, *Community and Society*, trans. Charles P. Loomis (New York, 1963), pp. 151–154.
48. Ibid., pp. 154–182.
49. Georg Simmel, *Philosophische Kultur* (Leipzig, 1911), pp. 67–68.
50. Ibid., p. 69.
51. A failure to achieve objectivity was used to explain why other groups failed at science. Max Lenz in his *Römische Glaube und freie Wissenschaft* (Berlin, 1902) declared that Roman Catholics were incapable of doing objective science because they had failed to cultivate the attitude of *Vorurteilslosigkeit*.

52. Lujo Brentano argued that women should have access to university education and that women had made substantial contributions to political economy (he mentions Beatrice Webb). See *Die akademische Frau,* pp. 193, 209.
53. Marianne Weber, *Ehefrau und Mutter in der Rechtsentwicklung* (Tübingen, 1907).
54. Weber, *Max Weber,* pp. 233-237, 263, 408-412, 473-475.
55. Ibid., pp. 216-218.
56. Kurt Singer, "Die Krisis der Soziologie," *Weltwirtschaftliches Archiv,* 16 (1920-21): 247.
57. Fritz Stern, *The Failure of Illiberalism* (London, 1972), p. xxiii.

9. Neutral Marxism

1. See, for example, F. A. Hayek, *The Counter-Revolution of Science* (New York, 1955).
2. Mortin J. Plotnik, *Werner Sombart and His Type of Economics* (New York, 1937), p. 34.
3. Werner Sombart, *Die römische Campagna* (Leipzig, 1888).
4. Friedrich Engels, "Nachtrag," in Karl Marx, *Das Kapital,* vol. 3 (1894; Berlin, 1953), p. 28.
5. Werner Sombart, *Sozialismus und soziale Bewegung im 19. Jahrhundert* (Jena, 1896), pp. 11-12, 121.
6. Ludwig von Mises, "Antimarxismus," *Weltwirtschaftliches Archiv,* 21 (1925): 284. Plotnik, *Werner Sombart,* p. 28.
7. Werner Sombart, review of Julius Wolf's *Sozialismus und kapitalistische Gesellschaftsordnung* in the *Archiv für soziale Gesetzgebung und Statistik,* 5 (1892): 487-489.
8. Sombart, *Sozialismus und soziale Bewegung,* pp. 73-77.
9. This correction to Marx is found in the 1897 English edition of Sombart's *Socialism and Social Movement* (pp. 10-11) but not in the 1896 German edition. Sombart revised every edition of this work, often with fairly substantial changes, over some thirty years.
10. Werner Sombart, "Studien zur Entwicklungsgeschichte des italienischen Proletariats," *Archiv für soziale Gesetzgebung und Statistik,* 6 (1893): 178.
11. *Schriften der Deutschen Gesellschaft für Soziologie,* 6 (1929): 25.
12. Plotnik, *Werner Sombart,* p. 35. Werner Sombart, *Händler und Helden* (Munich, 1915); *Deutscher Sozialismus* (Leipzig, 1934).
13. Sombart, *Sozialismus und soziale Bewegung,* pp. 1, 118.
14. Werner Sombart, *Dennoch! Aus Theorie und Geschichte der gewerkschaftlichen Arbeiterbewegung* (Jena, 1900); Werner Krause, *Werner Sombarts Weg vom Kathedersozialismus zum Faschismus* (Berlin, 1962), pp. 40ff.
15. Werner Sombart, *Die Juden und das Wirtschaftsleben* (Leipzig, 1911), p. 293; *Volk und Raum* (Hamburg, 1928).
16. Franz Boese, *Geschichte des Vereins für Sozialpolitik 1872-1932* (Berlin, 1939), pp. 274-275.

17. Werner Sombart, *Weltanschauung, Wissenschaft und Wirtschaft* (Berlin, 1938), pp. 21–23.
18. Eduard Bernstein, "Wie ist wissenschaftlicher Sozialismus möglich?" (1901), in *Ein revisionistisches Sozialismusbild,* ed. Helmut Hirsch (Hanover, 1966), pp. 13–17.
19. Ibid., p. 22. Ludwig Gumplowicz devoted an entire chapter of his *Sociologie und Politik* to "Politik als angewandte Sociologie" (pp. 101–134).
20. Bernstein, "Wie ist wissenschaftlicher Sozialismus möglich?" p. 32.
21. Ibid., pp. 35, 38.
22. Ibid., pp. 32–35.
23. Eduard Bernstein, *Evolutionary Socialism* (1899; New York, 1962), p. xxvii.
24. Ibid., pp. 2–3 and 5–7.
25. Max Adler, "Marx und die Dialektik," *Der Kampf,* 1 (1907–08): 256. Compare Werner Sombart's rejection of dialectics as "etwas altfränkisch," in his *Sozialismus und soziale Bewegung,* pp. 71–72.
26. Max Adler, *Marxistische Probleme* (Stuttgart, 1913), pp. 105–106.
27. Karl Kautsky, *Ethik und materialistische Geschichtsauffassung* (Stuttgart, 1906), pp. 78–79.
28. Adler, *Marxistische Probleme,* pp. 109–115, 211. Compare also Otto Bauer, "Marxismus und Ethik," *Neue Zeit,* 24 (1906, part 2): 485–499; Sombart, *Sozialismus und soziale Bewegung,* pp. 105–107.
29. Bernstein, "Wie ist wissenschaftlicher Sozialismus möglich?" p. 50.
30. Bogdanov distinguished bourgeois and proletarian science as part of his effort to establish a distinctively "proletarian culture"—including workers' schools, theaters, and libraries. In 1926, when the Austrian socialist Paul Kammerer committed suicide after having been accused of falsifying his data to prove the inheritance of acquired characters, SPD leaders managed to avoid embarrassment by pleading neutrality on questions of a purely technical nature. Karl Kautsky thus in 1927 stated that "the materialist conception of history does not require that one choose between the Darwinian or Lamarckian conceptions of nature. Such a decision could only be established through scientific research and scientific arguments." See his *Die materialistische Geschichtsauffassung* (Berlin, 1927), p. 199; also my *Racial Hygiene,* pp. 30–38, 258–261.

10. Max Weber and *Wertfreie Wissenschaft*

1. The literature on Weber is enormous; exemplary works include Wolfgang J. Mommsen, *Max Weber und die deutsche Politik 1890–1920* (Tübingen, 1959); Karl Loewenstein, *Max Weber's Political Ideas in the Perspective of Our Time* (Cambridge, Mass., 1966); the introductory essay by Hans Gerth and C. Wright Mills in their *From Max Weber* (New York, 1946); J. P. Mayer, *Max Weber and German Politics* (1944; London, 1955); and Stephen P. Turner and Regis A. Factor, *Max Weber and the Dispute over Reason and Value* (London, 1984). For Weber's bibliography see Dirk Käsler, "Max-Weber-Bibliographie," *Kölner Zeitschrift für Soziologie und Sozialpsychologie,* 27 (1975): 703–730.
2. Max Weber, "Die 'Objektivität' sozialwissenschaftlicher und sozialpolitischer

Erkenntnis" (1904), in his *Gesammelte Aufsätze zur Wissenschaftslehre*, 3d ed. (Tübingen, 1968), p. 148; compare p. 185. References here will be to the German original; the article has been translated as "Objectivity in Social Science and Social Policy," in *The Methodology of the Social Sciences*, ed. Edward A. Shils and Henry A. Finch (Glencoe, Illinois, 1949).
3. Georg Lukács, *History and Class Consciousness* (1922; Cambridge, Mass., 1971); Karl Popper, *The Poverty of Historicism* (1957; New York, 1964); F. A. Hayek, *The Counter-Revolution of Science* (New York, 1955).
4. Weber, "Objektivität," p. 157. Compare Weber's claim for the need to open eyes "blinded for a thousand years" by the "grandiose moral fervor of Christian ethics." Cited in Alan Sica, *Weber, Irrationality, and Social Order* (Berkeley, 1988), p. 125.
5. Weber, "Objektivität," p. 158.
6. Max Weber, "Der Sinn der 'Wertfreiheit' der soziologischen und ökonomischen Wissenschaften" (1917), in his *Gesammelte Aufsätze zur Wissenschaftslehre*, pp. 510–511. This essay has been translated as "The Meaning of 'Ethical Neutrality' in Sociology and Economics" in *The Methodology of the Social Sciences*; references here will be to the German original.
7. Anthony Oberschall, *Empirical Social Research in Germany, 1848–1914* (Paris, 1965), pp. 17–18, 20.
8. Max Weber, "Die ländliche Arbeitsverfassung" (1893), in his *Gesammelte Aufsätze zur Sozial- und Wirtschaftsgeschichte*, pp. 455–457.
9. Max Weber, "Der Nationalstaat und die Volkswirtschaftspolitik" (1894), in his *Gesammelte politische Schriften* (Munich, 1921), p. 17.
10. Marianne Weber, *Max Weber*, p. 253.
11. Ibid., p. 258; also Max Weber, *Jugendbriefe*, ed. Marianne Weber (Tübingen, 1936), p. 298.
12. Weber, "Der Nationalstaat," p. 19.
13. Max Weber, "Parlamentarisierung und Föderalismus" (1917), in his *Gesammelte politische Schriften*, pp. 258–259.
14. Max Weber, *Economy and Society*, ed. Guenther Roth and Claus Wittich (New York, 1968), p. 56; "Politik als Beruf" (1917), in his *Gesammelte politische Schriften*, pp. 397, 445.
15. Weber, "Der Nationalstaat," pp. 19–20.
16. Cited in J. E. T. Eldridge, "Weber's Approach to the Sociological Study of Industrial Workers," in *Max Weber and Modern Sociology*, ed. Arun Shay (London, 1971), pp. 99–100, translation slightly modified.
17. Weber, "Der Nationalstaat," pp. 24–29.
18. Weber to Tönnies, October 15, 1914, in Weber's *Gesammelte politische Schriften*, p. 458.
19. Marianne Weber, *Max Weber*, pp. 572, 582.
20. Ibid., p. 136.
21. Ibid., pp. 630, 702–703.
22. Max Weber, "Der Sozialismus," in his *Gesammelte Aufsätze zur Soziologie und Sozialpolitik* (Tübingen, 1924), p. 508.
23. Georg Lukács, "Max Weber and German Sociology" (1955), *Economy and Society*, 1 (1972): 389–394.

24. Schorske, *German Social Democracy*, pp. 1-18.
25. Ibid., pp. 93-94; Mayer, *Max Weber*, p. 65.
26. Marianne Weber, *Max Weber*, p. 49; Weber to Baumgarten, July 5, 1887, in his *Jugendbriefe*, p. 261.
27. Max Weber, "'Energetische' Kulturtheorien" (1909), in his *Gesammelte Aufsätze zur Wissenschaftslehre*, p. 401. Weber's review is of Ostwald's *Die energetische Grundlage der Kulturwissenschaft* (Leipzig, 1909).
28. Cited in Marianne Weber, *Max Weber*, pp. 376-377.
29. Max Weber, "Die Grenznutzenlehre und das 'psychophysische Grundgesetz'" (1908), in his *Gesammelte Aufsätze zur Wissenschaftslehre*.
30. Max Weber, "Die logische Probleme der historischen Nationalökonomie" (1903-06), in his *Gesammelte Aufsätze zur Wissenschaftslehre*, pp. 100-101.
31. Weber, *Economy and Society*, p. 15.
32. Heinrich Rickert, *Kulturwissenschaft und Naturwissenschaft* (Freiburg, 1899), pp. 20, 28.
33. Ibid., pp. 28-30, 118, 318.
34. Weber, "Die 'Objektivität,'" p. 184.
35. Weber, "Wissenschaft als Beruf," pp. 594-598.
36. Weber, "Die 'Objektivität,'" p. 213.
37. Weber, "Wissenschaft als Beruf," pp. 599-601.
38. Weber, "Der Sinn der 'Wertfreiheit,'" pp. 496-497; compare also his "Die sogenannte 'Lehrfreiheit' an den deutschen Universitäten," *Frankfurter Zeitung*, September 20, 1908, p. 3, morning ed.
39. Weber, "Wissenschaft als Beruf," pp. 600-602.
40. *Verhandlungen des Vereins*, p. 584.
41. Weber, "Wissenschaft als Beruf," p. 598.
42. Weber, "Wissenschaft als Beruf," pp. 603-604; "Der Sinn der 'Wertfreiheit,'" p. 493.
43. Stephen Turner and Regis Factor in their insightful *Max Weber and the Dispute over Reason and Value* describe Weber as a kind of "establishment existentialist" (p. 99).
44. Georg Lukács in 1922 argued that the neo-Kantian confusion of the empirically real with the historically necessary linked both conservative and Marxist forms of positivism. See his *History and Class Consciousness*, especially the first four chapters.
45. Leo Tolstoy, preface to Carpenter's article, "Modern Science," in Leo N. Tolstoy, *Works*, trans. Leo Wiener (Boston, 1905), vol. 6, p. 107.
46. Marianne Weber, *Max Weber*, pp. 468-469.
47. See, for example, Otto Neurath, "Zur Theorie der Sozialwissenschaften," *Jahrbuch für Gesetzgebung, Verwaltung und Volkswirtschaft*, 34 (1910): 56-62.
48. Paul Honigsheim, "Max Weber als Soziologe," *Kölner Vierteljahrshefte für Soziologie*, 1 (1921): 35.
49. Weber, "Die 'Objektivität,'" p. 156.
50. "Soziologische Fragen," *Neue Zeit*, 29 (1911): 852.
51. *Verhandlungen des Vereins*, pp. 568-569.
52. Ibid., p. 572; Weber, "Der Sinn der 'Wertfreiheit,'" p. 501.

11. Catholicism without Christianity

1. Edmund Husserl, *The Crisis of the European Sciences* (1936; Evanston, 1970); Siegfried Kracauer, "Die Wissenschaftkrisis," in his *Das Ornament der Masse* (1923; Frankfurt, 1963), p. 197.
2. Erich von Kahler, *Der Beruf der Wissenschaft* (Berlin, 1920), pp. 20–29, 37, 49, 82; Arthur Salz, *Für die Wissenschaft* (Munich, 1921). See also Stephen P. Turner and Regis A. Factor, *Max Weber and the Dispute over Reason and Value* (London, 1984), pp. 90–120; Dirk Käsler, *Die frühe deutsche Soziologie 1909–1933* (Opladen, 1984).
3. M. Rainer Lepsius, ed., *Soziologie in Deutschland und Oesterreich 1918–1945* (Opladen, 1981), p. 17; *Kölner Zeitschrift für Soziologie,* 36 (1984): 421.
4. Franz Boese, *Geschichte des Vereins für Sozialpolitik 1872–1932* (Berlin, 1939), pp. 282–286. On German sociology under the Nazis see Otthein Rammstedt, "Theorie und Empirie des Volksfeinds—zur Entwicklung einer 'deutschen Soziologie,' " in *Wissenschaft im Dritten Reich,* ed. Peter Lundgreen (Frankfurt, 1985), pp. 253–313.
5. Leszek Kolakowski, *The Alienation of Reason: A History of Positivist Thought* (New York, 1968), pp. 1–6.
6. Kant, *Critique of Pure Reason,* p. 9.
7. Auguste Comte, *Auguste Comte and Positivism: The Essential Writings,* ed. Gertrud Lenzer (New York, 1975), p. 46.
8. Ibid., pp. 73, 94, 465–474, and 84. Comte's Latin American followers were important in early nationalist movements in several countries; today the Brazilian flag echoes Comte's motto with the words *Ordem e progresso.*
9. J. S. Mill, *On Liberty* (1859; Oxford, 1948), p. 12; Thomas Huxley, *The Physical Basis of Life* (London, 1883), p. 140; Jürgen Habermas, *Knowledge and Human Interests,* trans. Jeremy Shapiro (Boston, 1971), p. 4.

12. Logical Positivism

1. Ernst Mach, *The Science of Mechanics* (1883; La Salle, 1960), pp. 278–284.
2. Herbert Feigl, "The Wiener Kreis in America," in Donald Fleming and Bernard Bailyn, *The Intellectual Migration* (Cambridge, Mass., 1969), p. 631.
3. Gerald Holton, "Mach, Einstein, and the Search for Reality," *Daedalus* (Spring 1968): 636–673.
4. Joseph Petzoldt, "Metaphysikfreie Wissenschaft," *Naturwissenschaftliche Wochenschrift,* n.s. 1 (1902): 364; "Positivistische Philosophie," *Zeitschrift für positivistische Philosophie,* 1 (1913): 3–4.
5. The recent revival of interest in human origins, paleontology, macroevolution, extinction theory, and cosmology derives partly from recognition of the value of historical insight in physical and biological science. Stephen Jay Gould has been one of the leaders in this revival; see his *Ontogeny and Phylogeny* (Cambridge, Mass., 1977).
6. Otto Neurath, "Empirical Sociology," in *Empiricism and Sociology,* ed. Marie Neurath and Robert S. Cohen (Dordrecht, 1973), p. 410.

7. Henri Poincaré, "The Value of Science" (1905), in *The Foundations of Science,* trans. G. B. Halsted (New York, 1913), p. 206.
8. Henri Poincaré, "La morale et la science," in his *Dernière Pensées* (Paris, 1913), pp. 223-225.
9. For a history of the founding of the Vienna Circle, see Neurath, *Empiricism and Sociology,* pp. 1-83.
10. Herbert Feigl and Albert E. Blumberg coined the term *logical positivism* in 1931 (Feigl, "Wiener Kreis," p. 630). Feigl notes that although Moritz Schlick preferred the term *konsequenter Empirismus,* most members of the group after 1936 preferred the term *logical empiricism* or *scientific empiricism* (ibid., pp. 653, 630). Others spoke in terms of *neo-positivism* or *wissenschaftliche Philosophie.*
11. Otto Neurath, Rudolf Carnap, and Hans Hahn, "Die Wissenschaftliche Weltauffassung: Der Wiener Kreis" (1929), in Neurath, *Wissenschaftliche Weltauffassung, Sozialismus und logischer Empirismus,* ed. Rainer Hegselmann (Frankfurt, 1979); also Feigl, "Wiener Kreis," p. 632. The manifesto appears in English translation in Neurath, *Empiricism and Sociology.*
12. Neurath et al., "Wissenschaftliche Weltauffassung," p. 87.
13. Carnap, "On the Character of Philosophic Problems," *Philosophy of Science,* 1 (1934): 14.
14. The Neuraths' authorized translation of Galton's *Hereditary Genius* appeared as *Genie und Vererbung* (Leipzig, 1910).
15. Otto Neurath, "Einführung in die Kriegswirtschaftslehre" (1914), in his *Durch die Kriegswirtschaft zur Naturalwirtschaft* (Munich, 1919), pp. 43, 25.
16. Marx W. Wartofsky, "Positivism and Politics: The Vienna Circle as a Social Movement," *Grazer Philosophische Studien,* 16/17 (1982): 100.
17. Peter Galison, "Aufbau/Bauhaus: Logical Positivism and Architectural Modernism," *Critical Inquiry,* 16 (1990): 741.
18. Hans Reichenbach, "In eigener Sache," *Erkenntnis,* 4 (1934): 76.
19. See, however, Hans Sluga's forthcoming *The Sense of Order;* also his "Metadiscourse: German Philosophy and National Socialism," *Social Research,* 56 (1989): 795-818; and Wolfgang F. Haug, ed., *Deutsche Philosophen 1933* (Berlin, 1989). On Heidegger, see Hugo Ott's *Martin Heidegger, Unterwegs zu seiner Biographie* (Frankfurt, 1988).
20. Helmut Kuhn, "German Philosophy and National Socialism," *Encyclopedia of Philosophy,* 8 vols. (New York, 1967), vol. 3, p. 309.
21. John Dewey, *German Philosophy and Politics* (1915; New York, 1942), pp. 20, 35. Dewey included a long introductory chapter on Nazi philosophy in the 1942 edition of his book, suggesting that intellectual "compartmentalization" had allowed Hitler to reduce all forms of science to "sheer tools of Nazi policy" (p. 46).
22. Hermann Glockner, *Vom Wesen der deutschen Philosophie* (Stuttgart, 1941), p. 2. Other systematic treatises on contemporary philosophy published during the Nazi period include Gerhard Lehmann's *Die deutsche Philosophie der Gegenwart* (Stuttgart, 1943) and Walter Del-Negro's *Die Philosophie der Gegenwart in Deutschland* (Leipzig, 1942).

23. "Allgemeiner Bericht," *Mitteilungen der Gesellschaft Deutscher Naturforscher und Ärzte,* 10 (1934): 19.
24. Fritz Lenz, *Menschliche Auslese und Rassenhygiene,* 3d ed. (Munich, 1931), p. 417.
25. Del-Negro, *Philosophie der Gegenwart,* p. 1.
26. Max Wundt, *Die Würzeln der deutschen Philosophie in Stamm und Rasse* (Berlin, 1944).
27. Ibid., pp. 11–22.
28. Hermann Glockner, "Deutsche Philosophie," *Zeitschrift für deutsche Kulturphilosophie,* 1 (1934): 30–36.
29. Del-Negro, *Philosophie der Gegenwart,* pp. 1–24.
30. Lehmann, *Deutsche Philosophie,* pp. 150–152, 161.
31. Del-Negro, *Philosophie der Gegenwart,* pp. 5–7, 13.
32. V. J. McGill, "Notes on Philosophy in Nazi Germany," *Science and Society,* 4 (1940): 12. Some influential Nazi philosophers appealed to positivist doctrines before it became taboo to do so. Alfred Bäumler, named professor of philosophy and political pedagogy at the University of Berlin in the spring of 1933 (at which time he also joined the Nazi party), had cited Reichenbach, Carnap, and Schlick approvingly in a 1931 speech, reprinted in his *Männerbund und Wissenschaft* (Berlin, 1934). Bäumler was later sharply reprimanded for having supported what one of his opponents called the "Jewish-Marxist" members of "Einstein's spiritual body-guard" (Berlin Document Center, Bäumler file).
33. See, for example, Edgar Zilsel, "Moritz Schlick," *Die Naturwissenschaften,* 25 (1937): 161–167. Zilsel was living in Austria at the time.
34. Viktor Kraft, *Der Wiener Kreis* (Vienna, 1968), p. 7.
35. See A. J. Ayer, "Editor's Introduction," in his *Logical Positivism* (New York, 1959), p. 7.
36. Edgar Zilsel, *Die sozialen Ursprünge der neuzeitlichen Wissenschaft,* ed. Wolfgang Krohn (Frankfurt, 1976), p. 44.
37. This is an oversimplification, given that positivist contacts with the United States began even before 1933. Moritz Schlick lectured at Stanford in 1929; Herbert Feigl emigrated to the United States in the fall of 1930 and received an appointment at Iowa in 1931 and then at Minnesota in 1940. Albert Blumberg and Feigl introduced American philosophers to the movement in their "Logical Positivism: A New Movement in European Philosophy," *Journal of Philosophy,* 28 (1931): 281–296. Clark Hull adopted the term *logical empiricism* after attending the Third International Congress for the Unity of Science in Paris, in 1937; he compared the work of Yale's Institute for Human Relations to positivism shortly thereafter. See his "Logical Positivism as a Constructive Methodology in the Social Sciences," *Einheitswissenschaft,* 6 (1938): 35–38. B. F. Skinner was a charter subscriber to *Erkenntnis* beginning in 1930, by which time he was already committed to a Mach-style empiricism. See Laurence D. Smith, *Behaviorism and Logical Positivism* (Stanford, 1986), pp. 188–195, 279–280.
38. Fritz Lenz, *Die Rasse als Wertprinzip, Zur Erneuerung der Ethik* (Munich, 1933).
39. See the essays by George Geiger, Herbert Feigl, C. E. Ayres, and others in the

Journal of Social Issues, 6 (1950); Ray Lepley, ed., *Value: A Co-operative Inquiry* (New York, 1949); Otto Stammer, ed., *Max Weber and Sociology Today* (New York, 1971), pp. 27–82; Richard Rudner, "No Science Can Be Value-Free," *Philosophy of Science,* 20 (1953): 1–6.
40. Benno Müller-Hill, "Genetics after Auschwitz," *Holocaust and Genocide Studies,* 2 (1987): 3–20.
41. "Which Way for American Science?" *Business Week,* September 14, 1946; reprinted in the *Bulletin of the Atomic Scientists,* 2 (Sept.–Oct. 1946): 1.
42. E. Rabinowitch, "International Cooperation of Scientists," *Bulletin of the Atomic Scientists,* 2 (May-June 1946): 1. Rabinowitch warned that the power released by the atom had caused "large groups of American scientists to forget their past aloofness in politics and to organize for a fight to prevent science from becoming an executioner of mankind."
43. Norbert Wiener, "A Scientist Rebels," *Bulletin of the Atomic Scientists,* 3 (January 1947): 31; Philip Morrison, "The Laboratory Demobilizes," *Bulletin of the Atomic Scientists,* 2 (November 1946): 5–6.
44. Merle Miller, "The Atomic Scientists in Politics," *Bulletin of the Atomic Scientists,* 3 (September 1947): 242.
45. George Lundberg, *Can Science Save Us?* (New York, 1947); J. A. Passmore, "Can the Social Sciences Be Value-Free?" *Proceedings of the Tenth International Congress of Philosophy* (Amsterdam, 1949); Joseph Schumpeter, "Science and Ideology," *American Economic Review,* 39 (1949): 345–359. Compare also Ernest Nagel, *The Structure of Science* (New York, 1961), pp. 485–502.
46. Robert Bierstedt, "Social Science and Social Policy," *Bulletin of the American Association of University Professors,* 34 (1948): 312–316.
47. Schumpeter, "Science and Ideology," pp. 346–348.
48. John Dewey, "Challenge to Liberal Thought," *Fortune,* 30 (1944): 188; Felix Kaufmann, "The Issue of Ethical Neutrality in Political Science," *Social Research,* 16 (1949): 344–352.
49. George Geiger, "Values and Social Science," *Journal of Social Issues,* 6 (1950): 8; Bierstedt, "Social Science and Social Policy," p. 317.
50. Leo Strauss, *Natural Right and History* (Chicago, 1953), p. 47; also his *What Is Political Philosophy? and Other Studies* (New York, 1959), pp. 18–26.
51. Strauss, *Natural Right and History,* pp. 35–80; *What Is Political Philosophy?* pp. 18–26. Lundberg had suggested that "the services of *real* social scientists would be as indispensable to Fascists as to Communists and Democrats, just as are the services of physicists and physicians." It made no more sense to ask about the politics of social scientific research than to ask "whether the law of gravity is Catholic, Protestant, or Pagan." See his *Can Science Save Us?* pp. 48, 21.
52. Max Horkheimer, "Value-Freedom and Objectivity," in *Max Weber,* ed. Stammer, pp. 51–53.
53. Jürgen Habermas, "Value-Freedom and Objectivity," in *Max Weber,* ed. Stammer, pp. 59–66.
54. Alvin W. Gouldner, "Anti-Minotaur: The Myth of a Value-Free Sociology" (1961), in *The Relevance of Sociology,* ed. Jack D. Douglas (New York, 1970), p. 66.

55. Ibid., pp. 69-84.
56. Hans Gerth and C. Wright Mills, eds., *From Max Weber* (New York, 1946); Edward A. Shils and Henry A. Finch, eds., *The Methodology of the Social Sciences* (Glencoe, 1949). The first major works of Weber to appear in English were his *General Economic History,* translated by Frank Knight (London, 1927), and his *Protestant Ethic and the Spirit of Capitalism,* translated by Talcott Parsons (London, 1930).

13. Positive Economics

1. John Neville Keynes, *The Scope and Method of Political Economy* (London, 1891), pp. 34-35; Milton Friedman, *Essays in Positive Economics* (Chicago, 1953), p. 4. T. W. Hutchison notes that Nassau Senior and J. S. Mill were the first to distinguish normative and positive economics; see his *'Positive' Economics and Policy Objectives* (London, 1964), p. 18.
2. For the history of the marginal revolution, see the 1972 special issue of *History of Political Economy;* John Maloney, *Marshall, Orthodoxy and the Professionalisation of Economics* (Cambridge, Eng., 1985); and the essays collected in R. D. Collison Black et al., eds., *The Marginal Revolution in Economics* (Durham, 1973).
3. Paul Samuelson notes that "If one were looking for a single criterion by which to distinguish modern economic theory from its classical precursors, he would probably decide that this is to be found in the introduction of the so-called subjective theory of value into economic theory." See his *Foundations of Economic Analysis* (Cambridge, Mass., 1947), p. 90. Ronald Meek suggests that *subjectivism* might have been a better word than marginalism to describe neoclassical orthodoxy; see his "Marginalism and Marxism" in Black et al., *Marginal Revolution,* p. 234.
4. Richard Whately, *Introductory Lectures on Political Economy* (London, 1855), p. 167.
5. W. S. Jevons, *Theory of Political Economy* (1871; Harmondsworth, 1971), p. 77.
6. The first substantial effort to relegate the Marxian system to the classical economic past was Eugen von Böhm-Bawerk's *Karl Marx and the Close of His System,* published in German in 1896.
7. John Stuart Mill, *Principles of Political Economy* (Boston, 1848), vol. 2, pp. 355-356. Mill used this principle to defend a modestly "graduated" (or, as we would say, "progressive") income tax. Following Bentham, Mill argued that income should be taxed only when it exceeds a certain minimum level (say, 50 pounds per annum)—after which it should be taxed a standard percentage (say, 10 percent). Mill argued against the use of taxation to redistribute income in a more dramatic fashion and called proposals to do so "a mild form of robbery," imposing as it does a penalty on people "for having worked harder and saved more than their neighbors" (p. 357).
8. Jevons, *Theory of Political Economy,* p. 85.
9. Ibid., pp. 102, 105. Hermann Heinrich Gossen as early as 1854 argued that

"nothing in the external world possesses absolute value"; the value of a thing depended solely upon the desire of an individual to obtain that object. See his *Entwicklung der Gesetze des menschlichen Verkehrs* (1854; Berlin, 1889), pp. 46–47.
10. Jevons, *Theory of Political Economy*, p. 130.
11. V. K. Dmietriev, *Economic Essays on Value, Competition and Utility* (1898–1902), trans. D. Fry, ed. D. M. Nuti (Cambridge, Eng., 1974), p. 207; Jevons is cited in Wesley C. Mitchell, *Types of Economic Theory* (New York, 1966), vol. 2, p. 30.
12. Cited in Maloney, *Marshall*, p. 207.
13. John Bates Clark, "The Law of Wages and Interest," *Annals of the American Academy of Political and Social Science*, 1 (July 1890): 43–65.
14. John Bates Clark, *The Distribution of Wealth* (New York, 1899), p. 7.
15. Ibid., pp. 1–8.
16. Ricardo, *Principles of Political Economy*, p. 52; Mark Blaug, *Economic Theory in Retrospect* (1962), 3d ed. (London, 1978).
17. Maloney, *Marshall*, pp. 208–210. Outside the field of professional economics there are occasional examples of socialists using marginalist methods. The Fabian George Bernard Shaw utilized neoclassical principles; Joseph Schumpeter once suggested that Marx, had he lived fifty years later, would have employed the new tools.
18. Maloney, *Marshall*, pp. 75, 192–193. Cannan later became much more conservative; by 1931 he was arguing that economic theory could serve as a political "peacemaker" only insofar as it helped "to get rid of that stupid cry for 'rights' and 'justice' which causes and exacerbates industrial and commercial quarrels."
19. Maloney, *Marshall*, p. 207.
20. Lionel Robbins, *An Essay on the Nature and Significance of Economic Science* (1932; London, 1935), p. 91.
21. Ibid., p. vii, 16, 24–28.
22. Ibid., pp. 137–139.
23. Ibid., pp. 134–135, 140–143. Interestingly, Robbins utilizes a similar argument to show that one cannot derive normative conclusions from the operations of the free market, either. Equilibrium theory enables us to understand market stability but it does not "by itself provide any ethical sanctions." Our focus on equilibrium does not mean that we necessarily approve of the market system: "Equilibrium is just equilibrium" (p. 143).
24. A. C. Pigou, *Wealth and Welfare* (London, 1912), p. 24. Pigou distinguished "ethically superior" from "ethically inferior" tastes (literature and art versus gambling and opium eating, for example), but he rejected the argument that a redistribution of wealth to the poor would result in the dissipation of funds in "worthless forms of exciting waste" (pp. 28–29).
25. J. R. Hicks, "The Foundations of Welfare Economics," *Journal of Economics*, 49 (1939): 696–697, 711. Compare also G. J. Stigler, "The New Welfare Economics," *American Economic Review*, 33 (1943): 355. For a critical view, see Maurice Dobb, *Welfare Economics and the Economics of Socialism* (Cambridge, Eng., 1969); also Robert Cooter and Peter Rappoport, "Were the Ordinalists

Wrong about Welfare Economics?" *Journal of Economic Literature*, 22 (1984): 507-530.
26. Hicks, "Foundations," p. 701. Hicks and R. G. D. Allen argued for an ordinal analysis of utility in their "A Reconsideration of the Theory of Value," *Economica*, n.s. 1 (1934): 51-76. Developing this view further, Samuelson identified economic welfare with Vilfredo Pareto's principle that market competition produces a *maximum d'utilité collective* "regardless of the distribution of income." See his *Foundations*, pp. 203-253.
27. Hicks, "Foundations," pp. 704, 711-712.
28. J. R. Hicks, *Value and Capital* (1939), 2d ed. (Oxford, 1946), p. 17. Compare also Paul Samuelson's statement that the concept of utility has been undergoing "throughout its entire history a purging out of objectionable, and sometimes unnecessary, connotations . . . One clearly delineated drift in the literature has been a steady tendency towards the rejection of utilitarian, ethical, and welfare connotations of the Bentham, Sidgwick, Edgeworth variety" (*Foundations*, p. 90). Samuelson himself promised to remove from consumer choice theory "any vestigial traces" of the notion of utility. See his "A Note on the Pure Theory of Consumer's Behavior," *Economica*, 5 (1938): 61-71.
29. Maloney, *Marshall*, pp. 203-204, 77, 193.
30. Joan Robinson, *Economic Philosophy* (London, 1962), p. 52. P. H. Wicksteed, review of H. J. Davenport's *The Economics of Enterprise*, in the *Economic Journal*, 24 (1914): 422. Gunnar Myrdal, *The Political Element in the Development of Economic Theory* (1930; New York, 1954), pp. vi, 154, v.
31. Wesley Mitchell originally described Veblen's work as "evolutionary" and later "genetic economics"; only later (around 1917-1918) did he settle on the name "institutional economics" (*Types of Economic Theory*, vol. 2, p. 610).
32. Veblen's contrast between making goods and making money should be understood in light of the fact that he came from a small and isolated farming community of Norwegian Lutheran immigrants in Wisconsin (and later Minnesota); he did not learn English well until he was in his teens. The principles expressed in his 1904 *Theory of Business Enterprise* and his subsequent *Engineers and the Price System* can be understood as a reaction against the commercialization that must have seemed so strange to him when he emerged from a community of subsistence farmers in the northern Midwest. See Mitchell, *Types of Economic Theory*, vol. 2, pp. 618-619.
33. Mitchell, *Types of Economic Theory*, vol. 2, p. 764.
34. C. E. Ayres, "Moral Confusion in Economics," *Ethics*, 45 (1935): 170-199; F. H. Knight, "Intellectual Confusion on Morals and Economics," *Ethics*, 45 (1935): 200-220.
35. Knight to Ayres, January 22, 1935, in Warren J. Samuels, ed., "Knight-Ayres Correspondence" (unpub. ms.), p. 20.
36. C. E. Ayres, *Theory of Economic Progress* (Chapel Hill, 1944). In his last letter to Knight, on July 18, 1969, Ayres reported his dissatisfaction with Thomas Kuhn's *Structure of Scientific Revolutions:* "Kuhn I thought is nonsense. It isn't by paradigms that science proceeds; it's by instruments. Some instruments open vistas, and this gives the illusion of paradigmatic progression" (in Warren J. Samuels, ed., "Knight-Ayres Correspondence" [unpub. ms.], p. 39).

37. Samuelson, *Foundations*, pp. 90–95.
38. Robbins, *Nature and Significance*, pp. 89–91, 148. Ludwig von Mises was another who appealed to Weber to defend value-neutrality; see his *Grundprobleme der Nationalökonomie* (Jena, 1933), pp. 34–49. Mises and Philip Wicksteed are the only two figures credited by Robbins as having helped inspire his *Nature and Significance*.
39. Terence Hutchison, *The Significance and Basic Postulates of Economic Theory* (1938; New York, 1960), p. 10.
40. Ibid., pp. 11–13.
41. Ludwig von Mises, the prolific libertarian Viennese economist, linked the ideal of *wertfreie Wissenschaft* to the ideal of science in a liberal political order. See his *Kritik des Interventionismus* (Jena, 1929); also his *Nationalökonomie* (Geneva, 1940).
42. Samuelson's *Foundations of Economic Analysis* was submitted to a prize committee at Harvard University in 1941 with the subtitle "The Operational Significance of Economic Theory," reflecting the influence of P. W. Bridgman's operationalism.
43. Maloney, *Marshall*, pp. 187, 203.
44. See Schumpeter's *History of Economic Analysis*, p. 37, where he argues that the facts with which one works may be ideologically charged, even if one refrains from passing value judgments.
45. Robbins, *Nature and Significance*, pp. 24–28.
46. Maloney, *Marshall*, p. 65.
47. Ibid., pp. 216–217.

14. Emotivist Ethics

1. Much of modern positivist writing has been concerned with ethical problems; the bibliography at the end of A. J. Ayer's *Logical Positivism* contains references to some fifty works on ethical theory.
2. I. A. Richards and C. K. Ogden, *The Meaning of Meaning* (1923), 2d ed. (New York, 1946), p. 125.
3. A. J. Ayer, *Language, Truth and Logic* (1936; New York, 1952), p. 32. Ayer also expressed a debt to Rudolf Carnap, Gilbert Ryle, and Isaiah Berlin. Ayer's first published article was a 1933 essay in *Oxford Outlook* entitled "The Case for Bolshevism."
4. Ayer, *Language, Truth and Logic*, pp. 32, 103, 133, 153.
5. Ibid., pp. 20, 103–104.
6. Ibid., p. 108.
7. Graham MacDonald and Crispin Wright, eds., *Fact, Science and Morality: Essays on A. J. Ayer's Language, Truth and Logic* (New York, 1986), pp. 1–2; A. J. Ayer, *Part of My Life* (London, 1977), p. 298.
8. George Edward Moore, *Principia Ethica* (Cambridge, Eng. 1903), p. ix. Moore's is the first of several twentieth-century works purporting to follow, at least by virtue of their titles, the model of Newton's *Principia Mathematica*. Compare Bertrand Russell and A. N. Whitehead's *Principia Mathematica*

(Cambridge, Eng., 1910–1913); also Heilbronn and Kosswig's "Principia Genetica," *Erkenntnis,* 8 (1939): 228–255.
9. Stuart Hampshire pointed out in 1949 that "Hume never denied that our moral judgments are based on arguments about matters of fact; he only showed that these arguments are not logically conclusive or deductive arguments." See his "Fallacies in Moral Philosophy," *Mind,* 58 (1949): 466. Interestingly, Moore does not cite Hume as the discoverer of the naturalistic fallacy; Hume is not even mentioned in the *Principia Ethica.* See also W. K. Frankena, "The Naturalistic Fallacy," *Mind,* 48 (1939): 464–477.
10. Abraham Edel, "The Logical Structure of G. E. Moore's Ethical Theory," in *The Philosophy of G. E. Moore,* ed. P. A. Schilpp (Evanston, 1942), p. 137.
11. Cited in Paul Levy, *Moore: G. E. Moore and the Cambridge Apostles* (New York, 1979), p. 234.
12. Moritz Schlick, *Fragen der Ethik* (Vienna, 1930); Rudolf Carnap, *Philosophy and Logical Syntax* (London, 1936), pp. 24–25. Carnap in his "Die Überwindung der Metaphysik durch logische Analyse," *Erkenntnis,* 2 (1932): 227, states that "Einen Satz, der ein Werturteil ausspricht, kann mann überhaupt nicht bilden." In 1963 Carnap reasserted that "If a statement on values or valuations is interpreted neither as factual nor as analytic (or contradictory), then it is noncognitive; that is to say, it is devoid of cognitive meaning, and therefore the distinction between truth and falsity is not applicable to it." See his "Replies and Systematic Expositions," in *The Philosophy of Rudolf Carnap,* ed. Paul A. Schilpp (LaSalle, 1963), p. 999.
13. C. L. Stevenson, "The Emotive Meaning of Ethical Terms," *Mind,* 46 (1937): 14–31; *Ethics and Language* (London, 1944).
14. G. J. Warnock, "Gilbert Ryle's Editorship," *Mind,* 85 (1976): 53; C. Lewy, " 'Mind' under G. E. Moore," *Mind,* 85 (1976): 45.
15. MacDonald and Wright, *Fact, Science and Morality,* p. 6.
16. E. W. F. Tomlin, "Logical Negativism," *Scrutiny,* 5 (1936): 200–218; Ayer, *Part of My Life,* p. 155.
17. W. D. Hudson, *Modern Moral Philosophy* (1970), 2d ed. (New York, 1983), pp. 66–126; also J. O. Urmson, *The Emotive Theory of Ethics* (New York, 1968).
18. Hudson, *Modern Moral Philosophy,* p. 1. Efforts to make a value-neutral "science of ethics" date back at least to Henry Sidgwick's *Methods of Ethics* (London, 1874), where "moralizing" is explicitly abandoned in favor of a rigorous and logical analysis of human beliefs. Otto Neurath later conceived of sociology as a replacement for ethics: "In place of 'ethics', we have systematic sociology, in place of 'ethical' behavior, behavior within a group, determined socially" (*Empiricism and Sociology,* p. 328).
19. R. M. Hare, *The Language of Morals* (Oxford, 1952), p. v.
20. Hampshire, "Fallacies in Moral Philosophy," pp. 466–482; compare also MacIntyre, *After Virtue,* pp. 11–14, 16–35.
21. Cited in Kai Nielsen, "Ethics, History of," *Encyclopedia of Philosophy* (New York, 1968), vol. 3, p. 110.
22. John Dewey, *Theory of Valuation* (Chicago, 1939), pp. 33–34.

23. For just one example of this broader interest, see Philip Hallie, *The Paradox of Cruelty* (Middletown, Conn., 1969).
24. John Rawls, *Theory of Justice* (Cambridge, Mass., 1971).
25. Alan Gewirth, "The Is-Ought Problem Resolved," *Proceedings and Addresses of the American Philosophical Association*, 47 (1974): 34–61.

15. Social Theory of Science

1. John Passmore, "Positivism," *Encyclopedia of Philosophy* (New York, 1968), vol. 5, p. 56.
2. C. Lewy, " 'Mind' under G. E. Moore, 1921–1947," *Mind*, 85 (1976): 46.
3. Hans Reichenbach, "Kausalität und Wahrscheinlichkeit," *Erkenntnis*, 1 (1930): 186.
4. Karl Popper, 1959 preface to his *Logic of Scientific Discovery* (1934; New York, 1968), p. 21.
5. Popper, *Logic of Scientific Discovery*, pp. 30–31; *Conjectures and Refutations* (1962; New York, 1968), pp. 22–45; *Objective Knowledge* (Oxford, 1972), pp. 341–342. Rudolf Carnap distinguished contexts of discovery and justification in his 1928 *Logical Structure of the World*, trans. Rolf A. George (Berkeley, 1969), p. xvii.
6. See Popper's chapter, "Who Killed Logical Positivism?" in his *Unended Quest* (1974; La Salle, Ill., 1976), pp. 87–90; also the papers from the 1961 Tübingen meeting of the Deutsche Gesellschaft für Soziologie, reprinted in Theodor W. Adorno et al., *The Positivist Dispute in German Sociology* (New York, 1976).
7. Not that Popper denies these origins. It is simply that one cannot easily investigate them; nor do they have any fundamental bearing on the validity of scientific principles. Popper distinguishes sharply between the logic of knowledge and the psychology or sociology of knowledge. Discovery may very well be influenced by sociological and psychological variables, but these are not amenable to systematic investigation. See his *Conjectures and Refutations*, pp. 30–31; also *The Poverty of Historicism* (New York, 1964), pp. 35–55.
8. Thomas Kuhn, *The Structure of Scientific Revolutions* (1962; Chicago, 1970), p. 116.
9. Ibid., pp. 26, 77. Kuhn's conception of the growth of science contrasts with that of Henri Poincaré, for whom "the advance of science is not comparable to the changes of a city, where old edifices are pitilessly torn down to give rise to new, but rather to the continuous evolution of zoologic types, which develop ceaselessly and end by becoming unrecognizable to the average sight." See his "La moral et la science," in his *Dernière Pensées* (Paris, 1913), p. 14.
10. Kuhn, *Structure of Scientific Revolutions*, p. x.
11. Karl Popper, *The Open Society and Its Enemies* (1945), 5th rev. ed. (New York, 1966), vol. 2, p. 217; compare also Michael Polanyi's "The Republic of Science."
12. Popper, *Conjectures and Refutations*, p. 221.
13. Karl Popper, "The Bucket and the Searchlight" (1949), in his *Objective Knowledge* (Oxford, 1972), pp. 341–361; also his *Conjectures and Refutations*, p. 46;

Friedman, *Essays in Positive Economics,* p. 34; Comte, *Essential Writings,* p. 73; Bacon, *Great Instauration,* pp. 21–22; also his *New Organon,* I. 62–64. Compare also Gunnar Myrdal's claim that "facts do not organize themselves into concepts and theories just by being looked at; indeed, except within the framework of concepts and theories, there are no scientific facts but only chaos" (*The Political Element,* p. vii). Kuhn was not the first historian of science to build upon the Gestalt analogy; see, for example, Norwood Hanson's *Patterns of Discovery* (1958; Cambridge, Eng., 1972), pp. 4–30.
14. Hobbes, *Leviathan,* p. 34.
15. Georg Lukács, *History and Class Consciousness* (1922; Cambridge, Mass., 1971); Otto Bauer, "Das Weltbild des Kapitalismus" (1916), in *Der lebendige Marxismus,* ed. Otto Jennsen (Jena, 1924); Franz Borkenau, *Der Übergang vom feudalen zum bürgerlichen Weltbild* (Paris, 1934); Thorstein Veblen, *The Place of Science in Modern Civilization and Other Essays* (New York, 1919); E. A. Burtt, *The Metaphysical Foundations of Modern Science* (New York, 1924); Nikolai Bukharin, *Historical Materialism* (New York, 1925); Edgar Zilsel, *Die Entstehung des Geniebegriffs* (Tübingen, 1926); Max Scheler, *Die Wissensformen und die Gesellschaft* (Leipzig, 1926); Karl Mannheim, *Ideology and Utopia* (1929; New York, 1936); Lewis Mumford, *Technics and Civilization* (New York, 1934); Ludwik Fleck, *Genesis and Development of a Scientific Fact* (1935; Chicago, 1979); J. D. Bernal, *The Social Function of Science* (New York, 1939).
16. Max Scheler, "Die positivistische Geschichtsphilosophie des Wissens und die Aufgaben einer Soziologie der Erkenntnis," *Kölner Vierteljahrshefte für Sozialwissenschaften,* 1 (1921): 22–31.
17. N. I. Bukharin, *Historical Materialism: A System of Sociology* (1921; New York, 1925), pp. 162–163.
18. Bacon, *Advancement of Learning,* p. 287.
19. Boris Hessen, "The Social and Economic Roots of Newton's Principia," in N. I. Bukharin et al., *Science at the Cross Roads* (1931; London, 1971), p. 154.
20. Ibid., p. 172.
21. N. I. Bukharin, "Theory and Practice from the Standpoint of Dialectical Materialism," in Bukharin et al., *Science at the Cross Roads,* p. 32.
22. See also J. B. S. Haldane, *Daedalus, or Science and the Future* (New York, 1923); Bertrand Russell, *Icarus, or the Future of Science* (New York, 1924).
23. Edward A. Shils, "A Critique of Planning—The Society for Freedom in Science," *Bulletin of the Atomic Scientists,* 3 (March 1947): 80–82; Congress for Cultural Freedom, *Science and Freedom* (London, 1955). For a history of this movement see Geoffrey Price, "The 'Freedom or Planning in Science' Debate in Britain, 1931–50," in the SISCON pamphlet, *Politics of Planning and the Problems of Science Policy,* Unit 1 (Leeds, 1976).
24. Richard Lewontin and Richard Levins, "The Problem of Lysenkoism" (1976), in their *The Dialectical Biologist* (Cambridge, Mass., 1985), pp. 163–196. Lysenko's report was translated into English as *The Situation in Biological Science: Proceedings of the Lenin Academy of Agricultural Science of the USSR: July 31st-August 7th, 1948* (Moscow, 1949).
25. Loren Graham, *Science, Philosophy and Human Behavior in the Soviet Union*

(New York, 1987); Greta Jones, "British Scientists, Lysenko and the Cold War," *Economy and Society*, 8 (1979): 26-58.
26. Loren Graham, "The Socio-political Roots of Boris Hessen," *Social Studies of Science*, 15 (1985): 705-722.
27. There is no mention of Hessen in the Moscow State University commemoration of the tricentennial of Newton's birth, nor in S. I. Vavilov's biography of Newton published in 1943 and revised in 1945. See S. I. Vavilov, ed., *Isaac Newton, 1643-1727* (in Russian, Moscow, 1943); also S. I. Vavilov, *Isaac Newton*, 2d ed. (Moscow, 1945). Even more curious is the fact that the materialist analysis so striking in the Russian delegation papers of the 1931 London Congress is absent from both of these works. Vavilov's 230-page biography is a rather orthodox, internalist chronicle of Newton's life and science. The Moscow University volume, with contributions by A. N. Krylov, S. Y. Luria, A. M. Deborin, and others (edited by Vavilov, president of the Soviet Academy of Sciences and brother of the persecuted geneticist, N. I. Vavilov), discusses Newton's optics, calculus, tidal theory, and so forth, with only a single unimpressive article by Deborin on the broader context of the physicist's work.
28. See, for example, A. Rupert Hall, "Merton Revisited," *History of Science*, 2 (1963): 1-16.
29. Max Weber, "Die protestantische Ethik und der Geist des Kapitalismus" (1904-1905), in his *Gesammelte Aufsätze zur Religionssoziologie* (Tübingen, 1920), p. 188.
30. Robert Merton, *Science, Technology and Society in Seventeenth-Century England* (1938; New York, 1970). Alphonse de Candolle in his *Histoire des sciences et des savants depuis deux siècles* (Geneva, 1885) had noted the disproportionate number of Puritans among early English scientists; compare also Dorothy Stimson, "Puritanism and the New Philosophy in 17th-Century England," *Bulletin of the Institute of the History of Medicine*, 3 (1935): 321-334.
31. Merton, *Science, Technology and Society*, 1970 Preface, pp. xx-xxi.
32. B. J. Shapiro, "Latitudinarianism and Science in Seventeenth Century England," *Past and Present*, 40 (1968): 16-41; T. K. Rabb, "Puritanism and Experimental Science in England," *Journal of World History*, 7 (1962): 46-67.
33. Merton, *Science, Technology and Society*, p. vii.
34. Robert Merton, "Note on Science and Democracy," *Journal of Legal and Political Sociology*, 1 (1942): 115-126; reprinted as "The Normative Structure of Science" in *The Sociology of Science*, ed. Norman W. Storer (Chicago, 1973), pp. 270-278. In his 1952 *Science and the Social Order*, Bernard Barber suggested that Merton's norm of *communism* be renamed *communalism* in step with political sensitivities of the age (p. 268n7).
35. Robert Merton, "Science and the Social Order," *Philosophy of Science*, 5 (1938): 321-337, reprinted in his *Sociology of Science*, pp. 260-261.
36. Ibid., pp. 261-263.
37. See David A. Hollinger, "The Defense of Democracy and Robert K. Merton's Formulation of the Scientific Ethos," in Robert A. Jones and Henrika Kuklick, eds., *Knowledge and Society* (Greenwich, 1983), vol. 4, pp. 1-15.
38. Merton, "The Normative Structure of Science," pp. 267-270.
39. Merton, "Science and the Social Order," p. 261.

40. Norman W. Storer, Prefatory Note in Merton's *Sociology of Science*, p. 223.
41. See my *Racial Hygiene*, esp. pp. 289–297.

16. Realism versus Moralism

1. John Durham Peters, "Revising the 18th-Century Script," *Gannett Center Journal*, 3 (1989): 163.
2. Steven Shapin and Simon Schaffer, *Leviathan and the Air-Pump: Hobbes, Boyle, and the Experimental Life* (Princeton, 1985), p. 5.
3. Joseph Ben-David, *The Scientist's Role in Society: A Comparative Study* (Englewood Cliffs, N.J., 1971), pp. 7–13.
4. Steven Shapin, "History of Science and Its Sociological Reconstructions," *History of Science*, 20 (1982): 195.
5. I choose the term *realist* (by analogy with political realism) to designate that school of thought which denies the value of an evaluative approach to science theory, championing instead a "reflexive" and value-neutral sociology of scientific knowledge. Realism in a philosophical sense is sometimes contrasted with instrumentalism, but this contrast is of no concern to me here. Indeed many sociological "realists" consider themselves instrumentalists. The contrast that interests me here is between neutral sociological realism and activist political moralism. The sociological realism I shall be discussing is represented most perfectly in the "strong program" of the Science Studies Unit at Edinburgh, but it can also be found in most other research centers for the sociology of science. Apart from the texts discussed above, representative realist texts would include David Bloor's *Knowledge and Social Imagery* (London, 1976); Bruno Latour and Steve Woolgar's *Laboratory Life: The Construction of Scientific Facts* (1979; Princeton, 1986); and K. D. Knorr-Cetina's *The Manufacture of Knowledge: An Essay on the Constructivist and Contextual Nature of Science* (Oxford, 1981).
6. Shapin, "History of Science," pp. 196–198.
7. There are of course such studies, but they are too few and far between. See, though, the sources cited in Chapter 17.
8. Bruno Latour states, for example, that "agnosticism in matters of science is the only way to start without being trapped on one side of the many wars being fought by the guardians of science's borders." See *The Pasteurization of France* (Cambridge, Mass., 1988), p. 6.
9. Barry Barnes, *Scientific Knowledge and Sociological Theory* (London, 1974), pp. viii–xi.
10. Ibid., pp. 3, 128–129.
11. See Melville J. Herskovits, *Cultural Relativism: Perspectives in Cultural Pluralism*, ed. Frances Herskovits (New York, 1972).
12. Barnes, *Scientific Knowledge*, p. 154.
13. Ibid., pp. 5–8. Barnes does not make clear what beliefs might be other than "natural." The naturalistic approach is presumably opposed to the view which sees beliefs as autonomous. At one point he contrasts his "naturalistic" method with the "philosophical," suggesting that whereas the latter is evaluative, the former is not (p. 8).

14. Ibid., pp. 125–127, emphasis in original.
15. Ibid., p. 145.
16. See Thorstein Veblen, *The Place of Science in Modern Civilization* (New York, 1919).
17. Compare also Jean-François Lyotard's thesis that there is no "discourse free of domination," no "noise-free communication" (*The Postmodern Condition: A Report on Knowledge* [Minneapolis, 1984]). As a self-styled herald of the "postmodern condition" Lyotard argues that the modernist "narrative archetypes" of *totality* (e.g., the Hegelian dialectic) and of *emancipation* (e.g., Marxism and the Enlightenment revolutionary spirit) have been transcended. Texts such as Lyotard's make one wary of deriving philosophical significance from architectural fashions.
18. Social constructivism generally shares with subjectivist value theories a conception of politics as a problem of allegiance to some particular class or group or (in the case of neoclassical economics) individual interests—as if consciousness of interests were sufficient to guarantee right reasoning and action about them. Part of this reduction can be attributed to the notion that it is the duty of theory to inform but not to command, to teach but not to preach; that given the facts, each of us has the right to make up our own mind. The dilemma of ethics within a paradigm of class allegiance is well formulated in Otto Bauer, "Marxismus und Ethik," *Neue Zeit*, 24 (1906, part 2): 485–499. We need a systematic and critical history of social constructivism.

17. Critiques of Science

1. Plato, *Lesser Hippias*, 366D; *Republic*, I 333E.
2. J. R. Ravetz, "Criticisms of Science," in *Science, Technology and Society*, ed. Ina Spiegel-Rösing and Derek de Solla Price (London, 1977), pp. 71–73.
3. Robert S. Lopez, "Wisdom, Science, and Mechanics: The Three Tiers of Medieval Knowledge and the Forbidden Fourth," in *Fear of Science*, ed. Markovits and Deutsch, p. 16; Augustine, *Confessions*, trans. R. S. Pine-Coffin (Harmondsworth, 1961), pp. 241–242.
4. Cited in Wolfgang Krohn, "Die 'Neue Wissenschaft' der Renaissance," in Böhme et al., *Experimentelle Philosophie*, p. 46.
5. Jonathan Swift, *Gulliver's Travels* (1726), ed. Louis A. Landa (Cambridge, Mass., 1960), pp. 145–148.
6. Locke is cited in Marshall McLuhan, *The Mechanical Bride* (1951; New York, 1967), p. 34. Thomas Carlyle, *The Nigger Question* (1849), ed. Eugene R. August (New York, 1971); *Sartor Resartus* (1831), ed. K. McSweeney and P. Sabor (Oxford, 1987), pp. 3, 53. Goethe's *Faust* (1790–1832), part 2, act 1, scene 2, reads:

> Daran erkenn' ich den gelehrten Herrn!
> Was ihr nicht *tastet,* steht euch meilenfern,
> Was ihr nicht *fasst* fehlt euch ganz und gar,
> Was ihr nicht *rechnet,* glaubt ihr, sei nicht wahr,
> Was ihr nicht *wägt,* hat für euch kein Gewicht.
> Was ihr nicht *münzt,* das, meint ihr, gelte nicht.

7. White, *Warfare of Science*, vol. 1, pp. 126–128.
8. Cohen, "Fear and Distrust of Science."
9. Plato, *The Phaedrus*, 275A. On the crossbow see Lopez, "Wisdom, Science, and Mechanics," p. 16. On gunpowder see Joseph Needham et al., *Hand and Brain in China and Other Essays* (London, 1971), pp. 2–3.
10. Eric Hobsbawm, "The Machine Breakers," *Past and Present*, 1 (1953): 57–70; also Malcolm I. Thomis, *The Luddites* (New York, 1970).
11. *The Grand Concern of England* (London, 1673), pp. 25–41.
12. Haber, *Chemical Industry*, p. 204; Francis Moore Lappé and Joseph Collins, *Food First* (New York, 1980), pp. 119–131.
13. Passano and Cushing are cited in Peter J. Kuznick, *Beyond the Laboratory: Scientists as Political Activists in 1930s America* (Chicago, 1987), pp. 58–59. Sir Josiah Stamp, chairman of the London Midland and Scottish Railway, proposed that the advancement of science be "retarded" to a rate that could be absorbed without economic disruption. See his "Must Science Ruin Economic Progress?" *Nature*, 132 (1933): 429. Compare also Carroll W. Pursell, " 'A Savage Struck by Light': The Idea of a Research Moratorium, 1927–37," *Lex et Scientia*, 10 (1974): 147–158.
14. Theodore Koppanyi, "The Case against a National Science Foundation," *Bulletin of the American Association of University Professors*, 34 (1948): 307.
15. Arthur J. Bachrach, "Ethics of Tachistoscopy," *Bulletin of the Atomic Scientists*, 15 (1959): 212–215.
16. John May, *The Greenpeace Book of the Nuclear Age: The Hidden History, the Human Cost* (New York, 1989), pp. 82–83.
17. John Marks, *The Search for the Manchurian Candidate* (New York, 1980); James H. Jones, *Bad Blood: The Scandalous Story of the Tuskegee Experiment* (New York, 1981); S. R. Bernard, "Maximum Permissible Amounts of Natural Uranium in the Body, Air and Drinking Water Based on Human Experimental Data," *Health Physics*, 1 (1958): 288–305.
18. Leonard Cole, *Clouds of Secrecy: The Army's Germ Warfare Tests Over Populated Areas* (Totowa, N.J., 1988).
19. George J. Annas, *The Rights of Patients*, 2d ed. (Carbondale, 1989), pp. 145–146; Stanley Milgram, "Behavioral Study of Obedience," *Journal of Abnormal Social Psychology*, 67 (1963): 371–378.
20. The concept of *antiscience* dates back at least to early-twentieth-century efforts to combat the suppression of Darwinian evolution. See C. Rowell, "The Cancer of Ignorance; The Spread of Anti-Science in an American Commonwealth," *Survey*, 55 (1925): 159–161; also B. Mausner and Judith Mausner, "A Study of the Anti-scientific Attitude," *Scientific American*, 192 (1955): 35–39. Compare also Edward Shils, "Anti-science: Observations on the Recent 'Crisis' of Science," in The Ciba Foundation, *Civilization and Science in Conflict or Collaboration?* (London, 1972).
21. See Ralph W. Moss, *The Cancer Industry* (New York, 1989).
22. Thomas McKeown, *The Role of Medicine: Dream, Mirage, or Nemesis?* (1976; Princeton, 1979); D. P. Burkitt and H. C. Trowell, eds., *Refined Carbohydrate Foods and Disease: Some Implications of Dietary Fibre* (London, 1975); Samuel Epstein, *The Politics of Cancer* (New York, 1979).

23. Richard Cooper, "Rising Death Rates in the Soviet Union," *New England Journal of Medicine*, 304 (1981): 1259–1265.
24. Ivan Illich, *Medical Nemesis* (New York, 1975).
25. "Keine Negerärzte in der amerikanischen Standesorganisation," *Archiv für Rassen- und Gesellschaftsbiologie*, 33 (1939–1940): 276; also 33 (1939–1940): 96. See also Kenneth Manning's forthcoming *A Social History of Blacks in American Medicine*. Joan O'C. Hamilton, "When Medical Research Is for Men Only," *Business Week*, July 16, 1990, p. 33; Andrew Purvis, "A Perilous Gap," *Time*, Special Issue, Fall 1990, pp. 66–67.
26. Marc Lappé, *Broken Code: The Exploitation of DNA* (San Francisco, 1984), p. 22. The moratorium on scientific research was not entirely unprecedented. In 1676 Isaac Newton wrote to Henry Oldenburg, Secretary of the Royal Society, asking for a moratorium on Robert Boyle's research into "The Incalescence of Quicksilver with Gold." Newton was concerned that Boyle's research might have dangerous alchemical implications and was "not to be communicated without immense dammage to ye world if there should be any verity in the Hermetick writers." See H. W. Turnbull, ed., *The Correspondence of Isaac Newton* (Cambridge, Eng., 1960), pp. 1–2.
27. Stephanie Tanchinski, "Boom and Bust in the Bio Business," *New Scientist*, January 22, 1987, p. 47.
28. Natalie Angier, "Vast, 15-Year Effort to Decipher Genes Stirs Opposition," *New York Times*, June 5, 1990; Nancy S. Wexler, "The Oracle of DNA," in *Molecular Genetics in Diseases of Brain, Nerve, and Muscle*, ed. L. P. Rowland et al. (Oxford, 1989); Dorothy Nelkin and Laurence Tancredi, *Dangerous Diagnostics: The Social Power of Biological Information* (New York, 1989).
29. Leslie Roberts, "To Test or Not to Test?" *Science*, 247 (1990): 17–19. Watson is cited in Leon Jaroff, "The Gene Hunt," *Time*, March 20, 1989, p. 67.
30. "Draft Statement of the First Workshop of the Joint Working Group on the Ethical, Legal and Social Issues Related to Mapping and Sequencing the Human Genome," Williamsburg, Virginia, February 5–6, 1990.
31. Richard Levins and Richard Lewontin, "The Political Economy of Agricultural Research," in their *Dialectical Biologist*, pp. 209–224.
32. National Research Council, *Alternative Agriculture* (Washington, D.C., 1989); Jim Hightower, *Hard Tomatoes, Hard Times: The Failure of the Land Grant College Complex* (Washington, D.C., 1972).
33. Rebecca Goldburg et al., *Biotechnology's Bitter Harvest: Herbicide-Tolerant Crops and the Threat to Sustainable Agriculture* (Washington, D.C., 1990).
34. Sheldon Krimsky et al., "Environmental Release of Genetically Engineered Organisms," *Genewatch*, 5 (March–June 1988): 1–5.
35. David Pimentel, "Down on the Farm: Genetic Engineering Meets Ecology," *Technology Review*, January 1987, pp. 24–30.
36. See the essays in Catherine Lerza and Michael Jacobson, eds., *Food for People Not for Profit* (New York, 1975).
37. Calestous Juma, "The Last Harvest? Biotechnology and Third World Agriculture," *Genewatch*, 6 (March–April 1990): 1–11.
38. Andrew Pollack, "It May Taste Like Vanilla, But Is It Vanilla?" *New York Times*, June 24, 1987.

39. See Keith Schneider, "Agency and Congress Face Clash over Patenting of Animals," *New York Times,* July 23, 1987.
40. On this general point see Richard Levins, "Genetics and Hunger," *Genetics,* 78 (1974): 67-76; also Richard Levins and Richard Lewontin, *The Dialectical Biologist* (Cambridge, Mass., 1985), pp. 209-224.
41. John Opie, *Ogallala: Water for a Dry Land* (Lincoln, 1991).
42. Evan Eisenberg, "Back to Eden," *Atlantic,* 264 (November 1989): 57-89.
43. Goldburg et al., *Biotechnology's Bitter Harvest,* p. 24.
44. R. C. Lewontin, Steven Rose, and Leon J. Kamin, *Not in Our Genes: Biology, Ideology, and Human Nature* (New York, 1983).
45. Arthur Jensen, "How Much Can We Boost IQ and Scholastic Achievement?" *Harvard Educational Review,* 38 (1969): 1-123; Edward O. Wilson, "Human Decency Is Animal," *New York Times Magazine,* October 12, 1975, p. 50.
46. Constance Holden, "Study of Terrorism Emerging as an International Endeavor," *Science,* 203 (1979): 33-35. "Math Genius May Have Hormonal Basis," *Science,* 222 (1983): 1312.
47. Melvin Konner, "The Aggressors," *New York Times Magazine,* August 14, 1988, pp. 33-34.
48. Scot Morris, "Do Men *Need* to Cheat on their Wives? A New Science Says YES: Darwin and the Double Standard," *Playboy* (May 1978): 109ff.
49. David Barash, "Sociobiology of Rape in Mallards," *Science,* 197 (1977): 788-789; L. G. Abele and S. Gilchrist, "Homosexual Rape and Sexual Selection in Acanthocephalan Worms," *Science,* 197 (1977): 81-83.
50. See the Ann Arbor Science for the People Editorial Collective's *Biology as a Social Weapon* (Minneapolis, 1977).
51. Arthur Jensen, "How Much Can We Boost IQ and Scholastic Achievement?" *Harvard Educational Review,* 38 (1969): 1-123.
52. Richard Lewontin, "Race and Intelligence" (1970), in *The IQ Controversy,* ed. N. J. Block and Gerald Dworkin (New York, 1976), pp. 78-92.
53. Robert Pool, "Who Will Do Science in the 1990s?" *Science,* 248 (1990): 435.
54. Gar Allen, "A History of Eugenics," *Biology as Destiny: Scientific Fact or Social Bias?* ed. Science for the People Sociobiology Study Group (Cambridge, Mass., 1984), pp. 13-19.
55. Jon Beckwith, "The Political Uses of Sociobiology in the United States and Europe," *Philosophical Forum,* 13 (1981-82): 311-321; also my "Eugenics among the Social Sciences: Hereditarian Thought in Germany and the United States," in *The Estate of Social Knowledge,* ed. JoAnne Brown and David K. van Keuren (Baltimore, 1991), pp. 194-197.
56. Volkmar Weiss and Hans-Georg Mehlhorn, "Der Hauptgenlocus der Allgemeinen Intelligenz," *Biologisches Zentralblatt,* 99 (1980): 297-310.
57. G. Dörner et al., "A Neuroendocrine Predisposition for Homosexuality in Men," *Archives of Sexual Behavior,* 4 (1975): 1-8.
58. Nicholas D. Kristof, "Chinese Region Uses New Law to Sterilize Mentally Retarded," *New York Times,* November 21, 1989; Loren R. Graham, *Science, Philosophy and Human Behavior in the Soviet Union* (New York, 1987), pp. 220-265.

59. Jo Durden-Smith, "Male and Female—Why?" *Quest Magazine*, October 1980, p. 15; Pamela Weintraub, "The Brain: His and Hers," *Discover*, April 1981, pp. 15–20.
60. David Gelman et al., "Just How the Sexes Differ," *Newsweek*, May 18, 1981, p. 72.
61. Jo Durden-Smith, "Male and Female—Why?" pp. 17, 93.
62. Pamela Weintraub, "The Brain: His and Hers," p. 20.
63. Barbara Ehrenreich and Deirdre English, *For Her Own Good: 150 Years of the Experts' Advice to Women* (Garden City, 1978).
64. David Gelman et al., "Just How the Sexes Differ," p. 78. The studies referred to include John Money and Anke Ehrhardt, *Man and Woman, Boy and Girl* (New York, 1972); also Anke Ehrhardt and S. W. Baker, "Fetal Androgens, Human Central Nervous System Differentiation, and Behavior Sex Differences," in *Sex Differences in Behavior*, ed. R. C. Friedman et al. (New York, 1974). For the feminist critique see Ruth Bleier, *Science and Gender: A Critique of Biology and Its Theories on Women* (New York, 1984); also Anne Fausto-Sterling, *Myths of Gender: Biological Theories About Women and Men* (New York, 1985), esp. pp. 90–156.
65. Jo Durden-Smith, "Male and Female—Why?" pp. 18, 96.
66. Simone de Beauvoir, *The Second Sex* (1949), trans. H. M. Parshley (New York, 1953), p. 148.
67. Fausto-Sterling, *Myths of Gender*, pp. 1–12 and 205–213; also her "Myth of Neutrality." Elizabeth Fee, "Women's Nature and Scientific Objectivity," in *Woman's Nature: Rationalizations of Inequality*, ed. Marian Lowe and Ruth Hubbard (New York, 1983), p. 22; Ruth Hubbard, *The Politics of Women's Biology* (New Brunswick, 1990).
68. Evelyn Fox Keller, *Reflections on Gender and Science* (New Haven, 1985), pp. 3–13 and 177–179; Fee, "Women's Nature and Scientific Objectivity," pp. 16–17.
69. Harding, *The Science Question*, pp. 248–250.
70. "R & D Budget: Civilian Gains Outpace Defense," *Science News*, 137 (February 3, 1990): 71.
71. Robert Harris and Jeremy Paxman, *A Higher Form of Killing: The Secret Story of Chemical and Biological Warfare* (New York, 1982).
72. J. B. Neilands, "Navy Alters Course at Berkeley," *Science for the People*, November–December 1988, pp. 10–12, 24; Jonathan King, "The Threat and Fallacy of a Biological Arms Race," *Genewatch*, May–August 1985; pp. 13–20; Nachama L. Wilker, "Army Fails to Address Dugway Risks," *Genewatch*, January–February 1988; pp. 1–7.
73. Nachama L. Wilker and Nancy Connell, "Update on the Military and Biotechnology," *Genewatch*, July–October 1987, pp. 1–3; Jonathan King, "Resisting the Militarization of Biomedical Research," *Genewatch*, July–October 1987, pp. 3–5; Charles Piller and Keith R. Yamamoto, *Gene Wars: Military Control over the New Genetic Technologies* (New York, 1988); Susan Wright, ed., *Preventing a Biological Arms Race* (Cambridge, Mass., 1990).
74. Rick Atkinson, "Rehearsing for Nuclear War," *Washington Post National Weekly Edition*, June 18, 1984.

75. Thomas G. Donlan, "Redoubtable DARPA: It Shapes the Future of U.S. Technology," *Barron's,* April 3, 1989, pp. 14–20.
76. There are currently 40 FFRDCs in the U.S.: 11 working for the DOD, 22 for the DOE, 4 for NSF, 1 for NASA, 1 for the Nuclear Regulatory Commission, and 1 for Health and Human Services. FFRDCs altogether account for about 10 percent of total federal R & D.
77. Paul Sullivan, "Pentagon Gets 'Twilight Zone' Award," *Boston Herald,* February 8, 1990; also MIT's *Thistle,* February 26, 1990, pp. 8–9.
78. See C. W. Clarke, "VD Control in Atom-bombed Areas," *Journal of Social Hygiene,* 37 (January 1951): 3–7.
79. "Britain Questions 'Star Wars' Plan," *New York Times,* March 16, 1985; Sanford Lakoff and Herbert F. York, *A Shield in Space? Technology, Politics, and the Strategic Defense Initiative* (Berkeley, 1989).
80. "6,500 Scientists Vow to Boycott Studies Aided by 'Star Wars'," *Chronicle of Higher Education,* May 21, 1986. A group calling itself the "Science and Engineering Committee for a Secure World," headed by Frederick Seitz of Rockefeller University, organized a pro–Star Wars recruitment drive in response to the boycott; by June of 1986 the group had managed to recruit only 80 scientists and engineers (ibid.).
81. Coalition Against the Space Arms Race, *Star Wars: First Strike against the World* (San Francisco, 1987), p. 15; Vincent Mosco, "Star Wars Is Already Working," *Science as Culture,* Pilot Issue (1987): 12–34.
82. William J. Broad, "Beyond the Bomb, Turmoil in the Labs," *New York Magazine,* October 9, 1988, p. 93.
83. Geoffrey Bernstein, "Cascade: Worker Resistance at the Portsmouth Uranium Plant," Senior Thesis, Harvard University, 1981.
84. *Bulletin of the Atomic Scientists,* May 1990, p. 4.
85. J. D. Bernal, *World without War* (London, 1958), p. 163.
86. Gary Chapman, "Computer Science and the Bush Administration," *The CPRS Newsletter,* 8 (Winter–Spring, 1990): 10; Philip Shabecoff, "Senator Urges Military Resources Be Turned to Environmental Battle," *New York Times,* June 29, 1990.
87. Ralph Nader and Donald Ross, *Action for a Change* (New York, 1971); Kelley Griffin, *Ralph Nader Presents More Action for a Change* (New York, 1987).
88. Bill Zimmermann et al., "Science for the People," pamphlet published by the Science for the People Collective (Jamaica Plain, Mass., 1971).

Conclusion. Neutrality as Myth, Mask, Shield, and Sword

1. Leo Tolstoy, "Preface to Carpenter's Article, 'Modern Science,' " in Tolstoy's *Works,* vol. 23 (Boston, 1905), p. 110.
2. Harberger boasted that "there are nearly 100 of my former students in Chile now—you just try to form a government without one of my students" (*Harvard Crimson,* May 6, 1981, p. 3).
3. *Harvard Crimson,* February 22, 1980, pp. 1 and 4. Bok told students at this meeting he "would have no qualms about appointing a Nazi to an academic

position at Harvard if the appointee were qualified." Bok defended his decision to hire Harberger with the hypothetical scenario that if the United States were to take "a sharp turn to the right," then he would have to resist pressures against appointing left-leaning faculty.
4. Cella Dugger, "Experts Debate Harberger's Appointment," *Harvard Crimson*, January 31, 1980.
5. *Harvard Crimson*, February 1, 1980, p. 4.
6. Herman Kahn, *Thinking the Unthinkable* (New York, 1962), cited in the anonymous *Report from Iron Mountain* (New York, 1967), p. xx.
7. J. B. S. Haldane, *Callinicus, a Defense of Chemical Warfare* (New York, 1925), pp. 28–29; also Cohen, *Science, Servant of Man,* pp. 285–293.
8. Cohen, *Science, Servant of Man,* p. 286.
9. David A. Hollinger, "Free Enterprise and Free Inquiry: The Emergence of Laissez-Faire Communitarianism in the Ideology of Science in the United States," *New Literary History,* 21 (Autumn, 1990): 897–919.

Bibliography

Adler, Max. *Marxistische Probleme.* Stuttgart, 1913.
Adorno, Theodor, ed. *The Positivist Dispute in German Sociology* (1969). New York, 1976.
Alexander, Robert E. "Metaethics and Value Neutrality in Science." *Philosophical Studies,* 25 (1974): 391-401.
Anrich, Ernst, ed. *Die Idee der deutschen Universität und die Reform der deutschen Universitäten.* Darmstadt, 1956.
Arditti, Rita, Pat Brennan, and Steve Cavrak, eds. *Science and Liberation.* Boston, 1980.
Aristotle. *Basic Works,* ed. Richard McKeon. New York, 1941.
Avenarius, Richard. "Zur Einführung." *Vierteljahrsschrift für wissenschaftliche Philosophie,* 1 (1877): 1-14.
Ayer, Alfred Jules. *Language, Truth and Logic* (1936). New York, 1952.
——— ed. *Logical Positivism.* New York, 1959.
——— *Part of My Life.* London, 1977.
Bacon, Francis. *The New Organon, or True Directions concerning the Interpretation of Nature* (1620), published together with *The Great Instauration* and *Preparative Toward a Natural and Experimental History,* ed. Fulton H. Anderson. New York, 1960.
——— *The Proficience and Advancement of Learning, Divine and Human,* in *Selected Writings of Francis Bacon,* ed. Hugh G. Dick. New York, 1955.
——— *Selected Writings of Francis Bacon,* ed. Hugh G. Dick. New York, 1955.
Barber, Bernard. *Science and the Social Order.* New York, 1952.
Barnes, Barry. *Scientific Knowledge and Sociological Theory.* London, 1974.
Bauer, Otto. "Marxismus und Ethik." *Neue Zeit,* 24 (1906, part 2): 485-499.
Becker, Carl H. *Gedanken zur Hochschulreform.* Leipzig, 1920.

Beer, John J. "Coal Tar Dye Manufacture and the Origins of the Modern Research Laboratory." *Isis,* 49 (1958): 123-131.
Bendix, Reinhard, and Guenther Roth, eds. *Scholarship and Partisanship: Essays on Max Weber.* Berkeley, 1971.
Bernal, John Desmond. *The Social Function of Science.* London, 1939.
Bernstein, Eduard. *Evolutionary Socialism* (1899). New York, 1962.
——— "Wie ist wissenschaftlicher Socialismus möglich?" (1901). In *Ein revisionistisches Sozialismusbild,* ed. Helmut Hirsch. Hanover, 1966.
Bierstedt, Robert. "Social Science and Social Policy." *Bulletin of the American Association of University Professors,* 34 (1948): 310-319.
Black, R. D. Collison, et al., eds. *The Marginal Revolution in Economics.* Durham, 1973.
Blume, Stuart S. *Toward a Political Sociology of Science.* New York, 1974.
Boehm, Laetitia. "Wissenschaft—Wissenschaften—Universitätsreform." *Berichte zur Wissenschaftsgeschichte,* 1 (1978): 7-36.
Boese, Franz. *Geschichte des Vereins für Sozialpolitik 1872-1932.* Berlin, 1939.
Böhme, Gernot. *Alternativen in der Wissenschaft.* Frankfurt, 1980.
Böhme, Gernot, Wolfgang van den Daele, and Wolfgang Krohn. *Experimentelle Philosophie.* Frankfurt, 1977.
Born, Karl Erich. "Structural Changes in German Social and Economic Development at the End of the Nineteenth Century." In *Imperial Germany,* ed. James Sheehan. New York, 1976.
Bukharin, Nikolai I., et al. *Science at the Cross Roads* (1931). London, 1971.
Burtt, E. A. *The Metaphysical Foundations of Modern Science* (1924). New York, 1954.
Bury, John Bagnell. *The Idea of Progress* (1920). New York, 1932.
Butterfield, Herbert. *The Whig Interpretation of History* (1931). New York, 1964.
Carnap, Rudolf. *Der logische Aufbau der Welt.* Berlin, 1928.
Chappell, V. C., ed. *Hume: A Collection of Critical Essays.* Garden City, 1966.
Clark, G. N. *Science and Social Welfare in the Age of Newton.* Oxford, 1937.
Clegg, Arthur. "Craftsmen and the Origin of Science." *Science and Society,* 43 (1979): 186-201.
Cohen, I. Bernard. "Fear and Distrust of Science in Historical Perspective." In *Fear of Science—Trust in Science,* ed. Andrei S. Markovits and Karl Deutsch. Cambridge, Mass., 1980.
——— *Science, Servant of Man.* Boston, 1948.
Cole, Stephen. "Continuity and Institutionalization in Science: A Case Study in Failure." In *The Establishment of Empirical Sociology,* ed. Anthony Oberschall. New York, 1972.
Comte, Auguste. *Auguste Comte and Positivism: The Essential Writings,* ed. Gertrud Lenzer. New York, 1975.
Congress for Cultural Freedom. *Science and Freedom.* London, 1955.
Craig, Gordon. *Germany 1849-1945.* Oxford, 1978.
Daniels, George H. "The Pure Science Ideal and Democratic Culture." *Science,* 156 (1967): 1699-1705.
Del-Negro, Walter. *Die Philosophie der Gegenwart in Deutschland.* Leipzig, 1942.
Descartes, René. *Discourse on Method* (1637). Indianapolis, 1956.

Dickson, David. *The New Politics of Science*. New York, 1984.
Easlea, Brian. *Liberation and the Aims of Science*. London, 1974.
Ehrenreich, Barbara, and Deirdre English. *For Her Own Good: 150 Years of the Experts' Advice to Women*. Garden City, 1978.
Eisenberg, Evan. "Back to Eden." *Atlantic*, 264 (November 1989): 57–89.
Evans, Edward Payson. *Criminal Prosecution and Capital Punishment of Animals*. London, 1906.
Fabb, Nigel. "Metaphor in the Seventeenth Century." Unpublished manuscript. Cambridge, England, 1980.
Farr, James. "Hume, Hermeneutics, and History: A 'Sympathetic Account.'" *History and Theory*, 17 (1978): 285–310.
Farrington, Benjamin. *Head and Hand in Ancient Greece*. London, 1947.
——— *Science and Politics in the Ancient World*. London, 1939.
Fausto-Sterling, Anne. *Myths of Gender: Biological Theories about Women and Men*. New York, 1985.
——— "The Myth of Neutrality: Race, Sex and Class in Science." *Radical Teacher*, 19 (1981): 21–25.
Fee, Elizabeth. "Women's Nature and Scientific Objectivity." In *Woman's Nature: Rationalizations of Inequality*, ed. Marian Lowe and Ruth Hubbard. New York, 1983.
Feigl, Herbert. "The Wiener Kreis in America." In *The Intellectual Migration*, ed. Donald Fleming and Bernard Bailyn. Cambridge, Mass., 1969.
Ferber, Christian von. "Der Werturteilsstreit 1909/1959. Versuch einer wissenschaftsgeschichtlichen Interpretation." *Kölner Zeitschrift für Soziologie und Sozialpsychologie*, 11 (1959): 21–37.
Feyerabend, Paul. *Science in a Free Society*. New York, 1982.
Finkelstein, J. J. *The Ox That Gored*. Philadelphia, 1981.
Finley, M. I. "Was Greek Civilization Based on Slave Labour?" In his *Slavery in Classical Antiquity*. Cambridge, England, 1960.
Friedman, Milton. *Essays in Positive Economics*. Chicago, 1953.
Fuller, Steve. *Social Epistemology*. Bloomington, Ind., 1988.
Furner, Mary O. *Advocacy and Objectivity*. Lexington, Ky., 1975.
Galileo Galilei. *The Assayer* (1623). In *Discoveries and Opinions of Galileo*, trans. Stillman Drake. New York, 1973.
Galison, Peter. "History, Philosophy, and the Central Metaphor." *Science in Context*, 2 (1988): 197–212.
——— "Aufbau/Bauhaus: Logical Positivism and Architectural Modernism." *Critical Inquiry*, 16 (1990): 709–752.
Gerth, Hans, and C. Wright Mills eds. *From Max Weber*. New York, 1946.
Gillispie, Charles Coulton. *The Edge of Objectivity*. Princeton, 1960.
Goldburg, Rebecca, et al. *Biotechnology's Bitter Harvest: Herbicide-Tolerant Crops and the Threat to Sustainable Agriculture*. Washington, D.C., 1990.
Gould, Stephen Jay. "Nonmoral Nature." In his *Hen's Teeth and Horse's Toes*. New York, 1983.
Gouldner, Alvin W. "Anti-Minotaur: The Myth of a Value-Free Sociology" (1961). In *The Relevance of Sociology*, ed. J. D. Douglas. New York, 1970.

Graham, Loren R. *Science, Philosophy and Human Behavior in the Soviet Union* (1966). 2d rev. ed. New York, 1987.

────── "The Socio-political Roots of Boris Hessen." *Social Studies of Science*, 15 (1985): 705–722.

Greenberg, Daniel S. *The Politics of Pure Science.* New York, 1967.

Gren, Friedrich Carl. "Geschichte der Naturwissenschaft." *Annalen der Physik,* 1 (1799): 167–204.

Gumplowicz, Ludwig. *Grundrisse der Soziologie* (1885). Innsbruck, 1926.

────── *Die sociologische Staatsidee* (1892). 2d ed. Innsbruck, 1902.

────── *Sociologie und Politik.* Leipzig, 1892.

────── *Der Rassenkampf, Sociologische Untersuchungen.* Innsbruck, 1883.

Haber, L. F. *The Chemical Industry during the Nineteenth Century.* Oxford, 1958.

Habermas, Jürgen. *Knowledge and Human Interests* (1968), trans. Jeremy Shapiro. Boston, 1971.

Hansen, Bert. "Science and Magic." In *Science in the Middle Ages,* ed. David C. Lindberg. Chicago, 1978.

Hardin, Burt L. *The Professionalization of Sociology.* Frankfurt, 1977.

Harding, Sandra. *The Science Question in Feminism.* Ithaca, 1986.

────── and Jean O'Barr, eds. *Sex and Scientific Inquiry.* Chicago, 1987.

Harnack, Adolf. *Geschichte der Königlich Preussischen Akademie der Wissenschaften zu Berlin.* 3 vols. Berlin, 1900.

Haskell, Thomas L. *The Emergence of Professional Social Science: The American Social Science Association and the Nineteenth-Century Crisis of Authority.* Urbana, Ill., 1977.

Hauser, Arnold. *The Social History of Art.* 4 vols. New York, 1951.

Hayek, Friedrich A. *The Counter-Revolution of Science.* New York, 1955.

Hegel, G. W. F. *Phenomenologie des Geistes* (1806). Frankfurt, 1970.

Helmholtz, Ludwig Hermann von. *Philosophische Vorträge und Aufsätze* (1872). Berlin, 1971.

Herder, J. G. *Vom Einfluss der Regierung auf die Wissenschaften und der Wissenschaften auf die Regierung.* Berlin, 1780.

Hessen, Boris. "The Social and Economic Roots of Newton's 'Principia.'" In N. I. Bukharin et al., *Science at the Cross Roads* (1931). London, 1971.

Hicks, J. R. "The Foundations of Welfare Economics." *Journal of Economics,* 49 (1939): 696–697.

Hobbes, Thomas. *Leviathan, or the Matter, Forme and Power of a Commonwealth Ecclesiasticall and Civil* (1651). New York, 1962.

Honigsheim, Paul. "Die Gründung der Deutschen Gesellschaft für Soziologie in ihren geistesgeschichtlichen Zusammenhängen." *Kölner Zeitschrift für Soziologie und Sozialpsychologie,* 11 (1959): 3–10.

Hook, Sidney. *Reason, Social Myths and Democracy* (1940). New York, 1966.

Hubbard, Ruth. *The Politics of Women's Biology.* New Brunswick, 1990.

Hudson, William Donald. *Modern Moral Philosophy.* 2d ed. New York, 1983.

────── ed. *The Is-Ought Question: A Collection of Papers on the Central Problems in Moral Philosophy.* London, 1969.

Humboldt, Wilhelm von. "Über die innere und äussere Organisation der höhere

wissenschaftlichen Anstalten zu Berlin" (1810). In *Die Idee der deutschen Universität*, ed. Ernst Anrich. Darmstadt, 1956.
Hume, David. *A Treatise of Human Nature* (1739). Oxford, 1978.
Hutchison, Terence W. *The Significance and Basic Postulates of Economic Theory*. London, 1938.
——— *'Positive' Economics and Policy Objectives*. London, 1964.
Janik, Allan, and Stephen Toulmin. *Wittgenstein's Vienna*. New York, 1973.
Jevons, William Stanley. *The Theory of Political Economy* (1871), ed. R. D. Collison Black. Harmondsworth, 1970.
Jones, Robert F. *Ancients and Moderns*. New York, 1936.
Kahl, Wilhelm. *Geschichtliches und Grundsätzliches aus der Gedankenwelt über Universitätsreform*. Berlin, 1909.
Kahler, Erich von. *Der Beruf der Wissenschaft*. Berlin, 1920.
Kant, Immanuel. *Critique of Pure Reason* (1781), trans. Norman Kemp Smith. New York, 1929.
Käsler, Dirk. *Die frühe deutsche Soziologie 1909–1933*. Opladen, 1984.
Kautsky, Karl. *Ethik und materialistische Geschichtsauffassung*. Stuttgart, 1906.
Keller, Evelyn Fox. *Reflections on Gender and Science*. New Haven, 1985.
Kirchhoff, Arthur, ed. *Die akademische Frau. Gutachten hervorragender Universitätsprofessoren, Frauenlehrer und Schriftsteller über die Befähigung der Frau zum Wissenschaftlichen*. Berlin, 1897.
Kolakowski, Leszek. *The Alienation of Reason: A History of Positivist Thought*. New York, 1968.
Koyré, Alexandre. *From the Closed World to the Infinite Universe* (1957). Baltimore, 1968.
Krause, Werner. *Werner Sombarts Weg vom Kathedersozialismus zum Faschismus*. Berlin, 1962.
Kuhn, Thomas. *The Structure of Scientific Revolutions* (1962). Chicago, 1970.
Kuznick, Peter J. *Beyond the Laboratory: Scientists as Political Activists in 1930s America*. Chicago, 1987.
Lehmann, Gerhard. *Die deutsche Philosophie der Gegenwart*. Stuttgart, 1943.
Lenz, Fritz. *Die Rasse als Wertprinzip, Zur Erneuerung der Ethik* (1917). Munich, 1933.
Lepenies, Wolf. *Das Ende der Naturgeschichte: Wandel kultureller Selbstverständlichkeiten in den Wissenschaften des 18. und 19. Jahrhunderts* (1976). Frankfurt, 1978.
——— *Between Literature and Science: The Rise of Sociology* (1985), trans. R. J. Hollingdale. Cambridge, England, 1988.
Levins, Richard. "Genetics and Hunger." *Genetics*, 78 (1974): 67–76.
Levins, Richard, and Richard Lewontin. *The Dialectical Biologist*. Cambridge, Mass., 1985.
Lewis, John. *Max Weber and Value-Free Sociology*. London, 1975.
Lilge, Friedrich. *The Abuse of Learning: The Failure of the German University*. New York, 1948.
Lobkowitz, Nicolas. *Theory and Practice*. New York, 1968.
Locke, John. *An Essay Concerning Human Understanding* (1690), ed. Peter H. Nidditch. Oxford, 1975.

Loewenstein, Karl. *Max Weber's Political Ideas in the Perspective of Our Time.* Cambridge, Mass., 1966.
Longino, Helen E. *Science as Social Knowledge: Values and Objectivity in Scientific Inquiry.* Princeton, 1990.
Lopez, Robert S. "Wisdom, Science, and Mechanics: The Three Tiers of Medieval Knowledge and the Forbidden Fourth." In *Fear of Science—Trust in Science,* ed. Andrei S. Markovits and Karl W. Deutsch. Cambridge, Mass., 1980.
Lukács, Georg. *History and Class Consciousness* (1922). Cambridge, Mass., 1971.
——— "Max Weber and German Sociology" (1955). *Economy and Society,* 1 (1972): 386–398.
——— *Die Zerstörung der Vernunft,* vol. 3, *Irrationalismus und Soziologie* (1954). Darmstadt, 1974.
Lundberg, George A. *Can Science Save Us?* New York, 1947.
Lyotard, Jean-François. *The Postmodern Condition: A Report on Knowledge* (1979). Minneapolis, 1984.
Macaulay, Thomas. *Critical and Historical Essays* (1837). 2 vols. London, 1969.
McClelland, Charles. *State, Society and University in Germany, 1700–1914.* Cambridge, England, 1980.
MacDonald, Graham, and Crispin Wright, eds. *Fact, Science and Morality: Essays on A. J. Ayer's Language, Truth, and Logic.* New York, 1986.
MacIntyre, Alasdair. *After Virtue: A Study in Moral Theory.* Notre Dame, 1984.
McLuhan, Marshall. *The Mechanical Bride* (1951). New York, 1967.
Maloney, John. *Marshall, Orthodoxy and the Professionalisation of Economics.* Cambridge, England, 1985.
Marcuse, Herbert. *One-Dimensional Man: Studies in the Ideology of Advanced Industrial Society.* Boston, 1964.
Mayer, J. P. *Max Weber and German Politics* (1944). London, 1955.
Mendelsohn, Everett. "The Emergence of Science as a Profession in Nineteenth-Century Europe." In *The Management of Scientists,* ed. Karl B. Hill. Boston, 1964.
Mendelsohn, Everett, Peter Weingart, and Richard Whitley, eds. *The Social Production of Scientific Knowledge.* Dordrecht, Holland, 1977.
Merchant, Carolyn. *The Death of Nature* (1980). New York, 1983.
Merton, Robert. "Science and the Social Order" (1938). In his *The Sociology of Science,* ed. Norman W. Storer. Chicago, 1973.
——— "Note on Science and Democracy" (1942). Reprinted as "The Normative Structure of Science" in his *The Sociology of Science,* ed. Norman W. Storer. Chicago, 1973.
——— *Science, Technology and Society in Seventeenth-Century England* (1938). New York, 1970.
——— *The Sociology of Science,* ed. Norman W. Storer. Chicago, 1973.
Merz, Theodore. *A History of European Thought in the Nineteenth Century.* 4 vols. New York, 1899–1912.
Mitchell, Wesley C. *Types of Economic Theory* [from class notes, 1926–1935], ed. Joseph Dorfman. 2 vols. New York, 1967–1969.
Mommsen, Wolfgang J. *Max Weber und die deutsche Politik 1890–1920.* Tübingen, 1959.

Moore, George Edward. *Principia Ethica* (1903). Cambridge, England, 1951.
Mühlmann, W. E. "Die Hitler-Bewegung, Bemerkungen zur Krise der bürgerlichen Kultur." *Sociologus,* 9 (1933): 129–140.
Mumford, Lewis. "Authoritarian and Democratic Technics." *Technology and Culture,* 5 (1964): 1–8.
——— *Technics and Civilization* (1934). New York, 1963.
Myrdal, Gunnar. *The Political Element in the Development of Economic Theory* (1930). New York, 1954.
Nagel, Ernest. *The Structure of Science.* New York, 1961.
Needham, Joseph. "History and Human Values; A Chinese Perspective for World Science and Technology." *Centennial Review,* 20 (1976): 1–35.
Neurath, Otto. *Empiricism and Sociology* (1931), ed. Marie Neurath and Robert S. Cohen. Dordrecht, Holland, 1973.
Neurath, Otto, Rudolf Carnap, and Hans Hahn. "Die Wissenschaftliche Weltauffassung: Der Wiener Kreis." In Otto Neurath, *Wissenschaftliche Weltauffassung, Sozialismus und logischer Empirismus,* ed. Rainer Hegselmann. Frankfurt, 1979.
Nietzsche, Friedrich. *Beyond Good and Evil* (1886), trans. R. J. Hollingdale. New York, 1972.
Novick, Peter. *That Noble Dream: The "Objectivity Question" and the American Historical Profession.* Cambridge, England, 1988.
Oberschall, Anthony. *Empirical Social Research in Germany, 1848–1914.* Paris, 1965.
Olschki, Leonardo. *Geschichte der neusprachlichen wissenschaftlichen Literatur.* Leipzig, 1919.
Paley, William. *Natural Theology; or, Evidences of the Existence and Attributes of the Deity* (1802). 12th ed. London, 1809.
Passmore, John. *Science and Its Critics.* New Brunswick, 1978.
Paulsen, Friedrich. *Die deutschen Universitäten.* Berlin, 1902.
Petty, William. *Petty-Southwell Correspondence, 1676–1682.* New York, 1967.
Petzoldt, Joseph. "Positivistische Philosophie." *Zeitschrift für positivistische Philosophie,* 1 (1913): 1–16.
Plato. *Collected Dialogues,* ed. Edith Hamilton and Huntington Cairns. Princeton, 1963.
Playfair, Lionel. "Science and Technology as a Source of National Power" (1885). In *Victorian Science,* ed. George Basalla. New York, 1970.
Plotnik, Martin J. *Werner Sombart and His Type of Economics.* New York, 1937.
Poincaré, Henri. "La Morale et la science." In his *Dernière Pensées.* Paris, 1913.
Polanyi, Michael. *The Logic of Liberty.* Chicago, 1951.
——— "The Republic of Science: Its Political and Economic Theory." *Minerva,* 1 (1962): 54–73.
Popper, Karl. *The Logic of Scientific Discovery* (1934). New York, 1968.
——— *Objective Knowledge* (1972). Oxford, 1975.
——— *The Open Society and Its Enemies* (1945). New York, 1967.
——— *The Poverty of Historicism* (1957). New York, 1964.
——— *Unended Quest* (1974). La Salle, Ill., 1976.

Proctor, Robert N. *Racial Hygiene: Medicine under the Nazis.* Cambridge, Mass., 1988.
Ravetz, Jerome R. *Scientific Knowledge and Its Social Problems.* Oxford, 1971.
——— "Criticisms of Science." In *Science, Technology and Society,* ed. Ina Spiegel-Rösing and Derek de Solla Price. London, 1977.
Ricardo, David. *The Principles of Political Economy and Taxation* (1817). London, 1911.
Rickert, Heinrich. *Kulturwissenschaft und Naturwissenschaft.* Freiburg, 1899.
Ringer, Fritz. *The Decline of the German Mandarins.* Cambridge, Mass., 1969.
Robbins, Lionel. *An Essay on the Nature and Significance of Economic Science* (1932). London, 1935.
Rorty, Richard. *Philosophy and the Mirror of Nature.* Princeton, 1979.
Rossi, Paolo. *Philosophy, Technology, and the Arts in the Early Modern Era,* trans. Salvator Attanasia. New York, 1970.
Rouse, Joseph. *Knowledge and Power: Toward a Political Philosophy of Science.* Ithaca, 1987.
Russell, Bertrand. *Icarus, or the Future of Science.* New York, 1924.
Salz, Arthur. *Für die Wissenschaft: Gegen die Gebildeten unter ihren Verächtern.* Munich, 1921.
Samuelson, Paul A. *Foundations of Economic Analysis.* Cambridge, Mass., 1947.
Scheler, Max. "Die positivistische Geschichtsphilosophie des Wissens und die Aufgaben einer Soziologie der Erkenntnis." *Kölner Vierteljahrshefte für Sozialwissenschaften,* 1 (1921): 22–31.
Schelling, F. W. J. "Vorlesungen über die Methode des akademischen Studiums" (1803). In *Die Idee der deutschen Universität,* ed. Ernst Anrich. Darmstadt, 1956.
Schelsky, Helmut. *Einsamkeit und Freiheit.* Stuttgart, 1956.
Schiebinger, Londa. *The Mind Has No Sex? Women in the Origins of Modern Science.* Cambridge, Mass., 1989.
Schmoller, Gustav. "Volkswirtschaftslehre." In *Handwörterbuch der Staatswissenschaften,* ed. Joseph Conrad. 3d ed. Jena, 1909–1911.
Schorske, Carl. *German Social Democracy, 1905–1917: The Development of the Great Schism.* Cambridge, Mass., 1955.
Schumpeter, Joseph A. *History of Economic Analysis.* Oxford, 1954.
——— "Science and Ideology." *American Economic Review,* 39 (1949): 345–359.
Shapin, Steven. "History of Science and Its Sociological Reconstructions." *History of Science,* 20 (1982): 157–211.
Sombart, Werner. *Sozialismus und Soziale Bewegung.* Jena, 1896.
Sprat, Thomas. *History of the Royal Society of London.* London, 1667.
Stammer, Otto, ed. *Max Weber and Sociology Today* (1965), trans. Kathleen Morris. New York, 1971.
Steding, Christoph. *Politik und Wissenschaft bei Max Weber.* Breslau, 1932.
Strauss, Leo. *Natural Right and History.* Chicago, 1953.
——— *What Is Political Philosophy? and Other Studies.* New York, 1959.
Taylor, Charles. "Neutrality in Political Science" (1967). In his *Philosophy and the Human Sciences.* Cambridge, England, 1985.
Thomas, Keith. *Religion and the Decline of Magic.* New York, 1971.

Tolstoy, Leo. "Preface to Carpenter's Article 'Modern Science.' " In *Tolstoy's Works*, trans. Leo Wiener, vol. 23. Boston, 1905.
Tönnies, Ferdinand. *Community and Society* (1883), trans. Charles P. Loomis. New York, 1963.
────── "Comtes Begriff der Soziologie." *Monatsschrift für Soziologie*, 1 (1909): 42–50.
────── "Eugenik." *Schmollers Jahrbuch*, 29 (1905): 273–290.
────── "Wege und Ziele der Soziologie." *Verhandlungen des Ersten Deutschen Soziologentages*. Tübingen, 1911.
Turner, Stephen P., and Regis A. Factor. *Max Weber and the Dispute over Reason and Value*. London, 1984.
Unger, Roberto Mangabeira. *Knowledge and Politics*. New York, 1975.
Veatch, Robert M. *Value-Freedom in Science and Technology*. Missoula, Montana, 1976.
Verhandlungen des Ersten Deutschen Soziologentages vom 19.–20. Oktober, 1910 in Frankfurt am Main. Tübingen, 1911.
Verhandlungen des Vereins für Sozialpolitik in Wien 1909, Schriften des Vereins für Sozialpolitik, 132 (1910): 329–425.
Weber, Marianne. *Max Weber: Ein Lebensbild*. Tübingen, 1926.
Weber, Max. "Äusserungen zur Werturteildiskussion im Ausschuss des Vereins für Sozialpolitik." Privately circulated manuscript; now available at the Archiv der Max Weber-Gesamtausgabe, Munich. Munich, 1913.
────── *Gesammelte Aufsätze zur Sozial- und Wirtschaftsgeschichte*. Tübingen, 1924.
────── *Gesammelte Aufsätze zur Soziologie und Sozialpolitik*, ed. Marianne Weber. Tübingen, 1924.
────── *Gesammelte Aufsätze zur Wissenschaftslehre*. 3d ed. Tübingen, 1968.
────── *Gesammelte politische Schriften*. Munich, 1921.
────── *Jugendbriefe*, ed. Marianne Weber. Tübingen, 1936.
────── "Der Nationalstaat und die Volkswirtschaftspolitik" (1894). In his *Gesammelte politische Schriften*. Munich, 1921.
────── "Die 'Objektivität' sozialwissenschaftlicher und sozialpolitischer Erkenntnis" (1904). In his *Gesammelte Aufsätze zur Wissenschaftslehre*. 3d ed. Tübingen, 1968. Translated as " 'Objectivity' in Social Science and Social Policy," in *The Methodology of the Social Sciences*, ed. Edward A. Shils and Henry A. Finch. Glencoe, Ill., 1949.
────── "Der Sinn der 'Wertfreiheit' der soziologischen und ökonomischen Wissenschaften" (1917). In his *Gesammelte Aufsätze zur Wissenschaftslehre*. 3d ed. Tübingen, 1968. Translated as "The Meaning of 'Ethical Neutrality' in Sociology and Economics," in *The Methodology of the Social Sciences*, ed. Edward A. Shils and Henry A. Finch. Glencoe, Ill., 1949.
────── "Wissenschaft als Beruf" (1919). In his *Gesammelte Aufsätze zur Wissenschaftslehre*. 3d ed. Tübingen, 1968. Translated as "Science as a Vocation" in *From Max Weber*, ed. Hans Gerth and C. Wright Mills, New York, 1958.
────── *Wirtschaft und Gesellschaft* (1921). 2d ed. Tübingen, 1925. Translated as *Economy and Society*. New York, 1968.

Weindling, Paul. *Health, Race and German Politics between National Unification and Nazism, 1870-1945*. Cambridge, England, 1989.

White, Andrew Dickson. *A History of the Warfare of Science with Theology in Christendom* (1896). 2 vols. New York, 1919.

Wiese, Leopold von. "Die Deutsche Gesellschaft für Soziologie. Persönliche Eindrücke in den ersten Fünfzig Jahren (1909-1959)." *Kölner Zeitschrift für Soziologie und Sozialpsychologie*, 11 (1959): 11-20.

Winner, Langdon. "Do Artifacts Have Politics?" (1980). In his *The Whale and the Reactor: A Search for Limits in an Age of High Technology*. Chicago, 1986.

Zeller, Eduard. "Über die Bedeutung und Aufgabe der Erkenntnistheorie" (1862). In his *Vorträge und Abhandlungen*. Leipzig, 1877.

Zilsel, Edgar. *Die Entstehung des Geniebegriffs, ein Beitrag zur Ideengeschichte der Antike und des Frühkapitalismus*. Tübingen, 1926.

——— "The Genesis of the Concept of Physical Law." *Philosophical Review*, 51 (1942): 245-279.

——— "The Origins of the Concept of Scientific Progress." *Journal of the History of Ideas*, 4 (1945): 325-349.

——— "The Sociological Roots of Science." *American Journal of Sociology*, 47 (1942): 245-279.

Index

Absolute space and time, 164
Academic freedom, 5, 149, 182
Académie Royale des Sciences, 34
Accademia del Cimento, 34
Adler, Max, 121, 131–132
Aeschylus, 19
Agriculture, 94–95, 237, 240, 244–247
Akademie der Wissenschaften, Berlin, 37, 68, 75–76
Alberti, Leon Battista, 22
Alchemy, 43, 236
Alembert, Jean Le Rond d', 77
American Medical Association, 242
American Society for Human Genetics, 244
Ancients vs. moderns, 21, 28–30, 55–56
Antiquarianism, 12
Antiscience, 240, 305n20
Anti-Semitic party, 138
Anti-Socialist Laws, 102–103, 141
Archimedes, 20
Aristophanes, 233
Aristotle, 18, 21, 41, 43, 44, 69, 206, 215, 224, 232
Arons, Leo, 106
Art (*techne*), 17, 22
Art for its own sake, 82–83
Artist-engineers, 23, 53
Ash, Mitchel, 282n15

Astrology, 43
Atom bomb, 4, 175–176, 178, 238, 258, 269, 294n42. *See also* Nuclear war
Augustine of Hippo, 46–47, 233
Avenarius, Richard, 81, 109
Ayer, Alfred J., 174, 201–208, 298n1, 298n3
Ayres, Clarence Edwin, 195–196, 297n36

Babbage, Charles, 69
Bachrach, Arthur, 238
Bacon, Francis, 1, 15, 25–35, 41, 47, 93; dangers of moral knowledge, 32; death of, 29; experiments, 55; "faith of eyes," 54; idols of the tribe, cave, marketplace, and theater, 6, 32, 54; knowledge is power, 31, 62; on Irish, 27; on origins of arts, 215; on perfectibility of science, 28, 276n13; on pure and mixed maths, 69; on science for its own sake, 5, 31, 263, 277n24; optimism of, 93
Badische Anilin- und Soda-Fabrik (BASF), 95
Baker, John R., 218
Barber, Bernard, 221, 302n34
Barnes, Barry, 201, 226–230, 303n13
Baronius, Cardinal Caesar, 45
Bauch, Bruno, 174–175
Bauer, Otto, 24, 168, 213, 304n18
Bauhaus, 169

322   Index

Bäumler, Alfred, 293n32
Bavarian Soviet Republic, 168–169
Bayardism, 269
Bayer, Friedrich & Co., 95
Bayle, Pierre, 37, 278n14
Beauvoir, Simone de, 252
Bebel, August, 120, 121
Becker, C. H., 107
Ben-David, Joseph, 225
Bentham, Jeremy, 177, 186, 295n7
Bernal, John Desmond, 215, 217, 259
"Bernalism," 218
Bernstein, Eduard, 121, 127–131, 266
Biblical criticism, 78
Bierstedt, Robert, 176–178
Big Science, 4–5, 243, 254–260
Biological determinism: brain lateralization studies, 251; early views, 248; feminist criticisms, 251–252; hormones and homosexuality, 252; in Eastern Europe, 250; Jensen's thesis, 249–250; Lewontin's criticism, 249; rationale for Human Genome Project, 243
Biological warfare. *See* Chemical and biological warfare
Biotechnology, 240–241, 245–247
Bishop of Ripon, 238
Blacks in mathematics, 250
Blaug, Mark, 188
Blumberg, Albert E., 292n10, 293n37
Bogdanov, Alexander, 133, 288n30
Böhm-Bawerk, Eugen von, 89, 187, 295n6
Bok, Derek, 267–268, 309n3
Book of Daniel, 21, 25
Bovine growth hormone, 240, 246
Boyle, Robert, 35, 46, 236, 306n26
Brentano, Lujo, 81, 88, 99, 106, 144, 287n52
Bridgewater Treatises, 49–50
British Association for the Advancement of Science, 93–94
Broad, William J., 259
Bruno, Giordano, 42, 235
Buckland, William, 49
Buffon, Georges Louis Leclerc, comte de, 36–37
Bukharin, Nikolai I., 132, 214–215, 217
Bulletin of the Atomic Scientists, 175–176, 238, 294n41–44
Bund Deutscher Frauenvereine, 111, 118
Burckhardt, Jacob, 83–84
Burkitt, D. P., 241

Burtt, E. A., 56
Butterfield, Herbert, 12, 224

C^3I, 257
Cancer, 241, 267
Candolle, Alphonse de, 302n30
Cannan, Edwin, 189, 198–199, 296n18
Carlyle, Thomas, 190, 194, 235
Carnap, Rudolf, 166, 167, 169, 172, 174, 205, 210, 293n32, 299n12, 300n5
Carson, Rachel, 240
Cassel, Gustav, 194, 196
Catholic Church and theology, 45, 60, 214, 235
Causes, 41–42, 278n4
Central Intelligence Agency, 239–240
Chemical and biological warfare (CBW), 239–240, 254–255, 259, 269
Chemical industry, 94–95, 244
Chicago School of economics, 183, 195, 267–268
Clark, John B., 187–188, 195
Cohen, Hermann, 172
Cohen, I. Bernard, 236, 269
Committee for Freedom in Science (Congress for Cultural Freedom), 218
Committee for Responsible Genetics, 255, 261
Communism and communist parties, 168, 218
Comte, Auguste: coins term *sociologie*, 284n4; on guidance of theory, 213; on hierarchy of sciences, 147, 161; on specialization, 83; positive philosophy of, 66, 80, 160–162; rejects theology and metaphysics, 160; Scheler on, 214; "social physics," 100; studied engineering, 99
Copernicus, Nicolaus, 22, 42, 211, 235
Cosmology: ancient vs. modern, 21–22, 24, 39–40; Chinese, 211; Copernican, 21–22, 235; 3°K black body radiation, 4
Crisis of science, 157–158
Culpepper, Nicholas, 36
Curiosity, 233
Cyrano de Bergerac, 233

Darwin, Charles, 45, 50–51, 109
Darwin, Erasmus, 37
Defense Advanced Research Projects Agency (DARPA), 256, 260
Defense Mapping Agency, 256

Defense Nuclear Agency (DNA), 255–266
Delbrück, Hans, 106–107
Del-Negro, Walter, 172–173
de Man, Paul, 170
Department of Energy (DOE), 239, 243, 254, 259, 273n5
Department of Defense (DOD), 254–259
Descartes, René: copies Bacon, 276n7; Latin obsolete, 35–36; methods to level men's wits, 26; need for practical philosophy, 27–28; on alchemy, 43; on God, 46; perfectibility of science, 28; pineal gland seat of soul, 22
Deutsche Gesellschaft für Soziologie (German Society for Sociology): excludes value judgments, 85–86; founding, 86, 90, 125; opposes socialism, 103–104; Weber's disappointment with, 151–152
Devalorization of being, 7, 39–53, 263
Dewey, John, 165, 170, 177, 207, 292n21
Dilthey, Wilhelm, 100–101, 145, 170, 172
Dingler, Hugo, 169
Du Bois-Reymond, Emil, 78
Dühring, Eugen, 80
Dürer, Albrecht, 23
Durkheim, Emile, 99, 227

Ecole Polytechnique, 68, 73, 99
Economic theory. *See* Institutional economics; Marginalism and the marginal revolution; Welfare economics
Edel, Abraham, 9, 204
Edgeworth, F. Y., 187
Edinburgh School, 228–231. *See also* Barnes, Barry
Educability, 249
Einstein, Albert, 164, 218–219, 293n32
Eisner, Kurt, 168
Eleutheropulos, Abroteles, 107–108
Emotivism, 181, 201–208
Engels, Friedrich, 103, 121, 122–123, 128–131, 184
Enlightenment ideals, 77, 101, 264
Environmental Protection Agency (EPA), 245–246, 259
Epistemology, 10, 13, 80
Epstein, Samuel, 241
Erkenntnis, 169–170, 173
Erxleben, Dorothea, 112
Ethics: absent from Marxism, 124; Aristotelian, 206; emotivist, 181, 201–208;

moral philosophy vs. moralism, 206; naturalistic, 204; science-free, 149, 201; socialist, 128–132; utilitarian, 185–192; value-free, 201, 203, 206, 299n18; vs. metaethics, 207–208
Eugenics, 101, 110, 229. *See also* Racial hygiene
Euripides, 19, 263–264
Evil, problem of, 46–49, 61
Evolution, 50–52
Experiment, 55–56

Fact-value dualism, 150, 178, 205–209, 226. *See also* "Ought" and "is"
Factor, Regis, 290n43
Fall of man, 35, 46–47, 60
Farr, James, 280n15
Fausto-Sterling, Anne, 253
Fechner, Gustav, 81, 144
Federally Funded Research and Development Centers (FFRDCs), 256–257, 309n76
Fee, Elizabeth, 253
Feigl, Herbert, 164, 166, 292n10, 293n37
Feminism, 70, 111, 117–118, 242, 251–253
Ferri, Enrico, 106
Fludd, Robert, 43
Fontenelle, Bernard Le Bovier de, 29, 93
Forster, E. M., 218
Frankfurt School, 139, 158, 179
Franklin, Benjamin, 15, 236
French Revolution, 37, 72, 108, 161
Friedman, Milton, 183, 213
Funke, Max, 115

Galilei, Galileo: and alchemy, 43; and positivism, 167; cosmology and fear of death, 40; *Dialogue,* 35; on primary and secondary qualities, 54–55; persecuted, 235; theories of motion, 41–42, 216; *Two New Sciences,* 29
Galison, Peter, 169
Galton, Francis, 109, 110, 168, 229
Gassendi, Pierre, 46
Geertz, Clifford, 228
Geiger, George, 178
Gender vs. sex, 252
Geocentric cosmos, 21–22, 39–40
German universities: apolitical, 105–106; class composition, 104–105; destroyed in Napoleonic invasions, 72; growth of philosophical faculty, 76–77; research

324 Index

German universities (*continued*)
 ideal, 71–74; science in, 76–78, 96; student enrollments, 76–77; student revolts, 72; women in, 111–115. *See also* Technische Hochschulen
Gestalt psychology, 211
Gierke, Otto von, 114, 119
Gilligan, Carol, 251
Glenn, John, 260
Glockner, Hermann, 172
God: after empiricism, 60–61; Bacon's service to, 25; continually acting in the world, 46; first cause, 41; great watchmaker, 23; Hobbes on, 34; humans to perfect creations of, 21; in all things the same, 40; Kantian conception of, 80; problem of evil, 46. *See also* Natural theology
Goethe, Johann Wolfgang von, 50, 78, 234, 304n6
Goldscheid, Rudolf, 110, 111
Gorbachev, Mikhail, 259
Gordon, Lester, 268
Gossen, Hermann Heinrich, 295n9
Gottl-Ottlilienfeld, F. von, 90
Gould, Stephen Jay, 279n18, 291n5
Gouldner, Alvin, 179–181
Graham, Loren, 218
Green revolution, 237
Green Run, 239
Greenberg, Daniel, 271
Gren, Friedrich, 73
Grotjahn, Alfred, 111
Gulliver's Travels (Swift), 234
Gumplowicz, Ludwig, 106, 119, 143, 284n4
Gunpowder, 1, 236

Habermas, Jürgen, 162, 179
Haeckel, Ernst, 110, 132
Hakewill, George, 28
Haldane, J. B. S., 269
Hallie, Philip, 300n23
Hampshire, Stuart, 61, 206–207, 299n9
Hanford Nuclear Reservation, 239, 259
Hans, Nicholas, 27
Harberger, Arnold, 267–268, 309n2, 310n3
Harding, Sandra, 9, 226, 253
Hare, R. M., 206–207
Harvard University, 267–268, 298n42
Harvey, William, 22, 42, 235
Hauser, Arnold, 82–83

Hayek, Friedrich A. von, 121, 212, 218
Hegel, G. W. F., 73–74, 112, 135–136, 248
Heidegger, Martin, 170
Helmholtz, Hermann von, 78, 81
Herbicides, 245, 247
Herder, Johann Gottfried, 71
Hereditarian research. *See* Biological determinism
Heritability, 249
Herkner, Willie, 87
Herodotus, 19, 215, 275n8
Herskovits, Melville, 228
Hessen, Boris: arrest and imprisonment, 219; British intellectuals enchanted by, 217–218; defends Einstein, 218–219; ignored in Soviet Union, 302n27; on Newton's *Principia*, 215–219
Hicks, J. R., 192–193, 297n26
Hightower, Jim, 244–245
Historical materialism. *See* Materialism
Historiography, 12–13
Hitler, Adolf, 170–171
Hobbes, Thomas: attack on metaphor, 36; materialism of, 22; on God, 34; on invention, 213; on primary and secondary qualities, 56–57; theory of value, 44, 57, 279n8
Hobsbawm, Eric, 226, 237
Hoechst Chemicals, 4, 68, 95
Hofmann, A. W. von, 78, 96
Hollinger, David, 302n37, 310n9
Homosexuality, 252
Hooke, Robert, 33
Hooton, E. A., 176
Horkheimer, Max, 158, 179
Howe, Geoffrey, 258
Hubbard, Ruth, 253
Hubble Space Telescope, 5
Hudson, William D., 201, 206
Hull, Clark, 293n37
Human experimentation, 238–239
Human Genome Project, 5, 243–244, 259–260
Humboldt, Alexander von, 93
Humboldt, Wilhelm von, 71
Hume, David, 7, 58–59, 80, 166, 167, 202, 248, 299n9
Husserl, Edmund, 157, 172
Hutchison, Terence W., 196–198
Huxley, Thomas, 109, 162
Huygens, Christiaan, 33, 34

Index

IBM, 4
Illich, Ivan, 242
Index Librorum Prohibitorum, 235
Institutional economics, 194–196, 297n31
Institutionalization of science, 33, 38
Internalist vs. externalist accounts of science, 211–212, 224–225
IQ, 248–249
Island Sun program, 258
"Ivory tower" (expression coined), 83

Jackson, Wes, 247
Jefferson, Thomas, 261
Jensen, Arthur, 248–250
Jevons, William Stanley, 100, 186
Johnson, Samuel, 29
Jonson, Ben, 29
Just-noticeable differences, 81

Kahler, Erich von, 157–158
Kahn, Herman, 268–269
Kammerer, Paul, 110, 288n30
Kant, Immanuel, 36, 77, 79, 82–83, 90, 132, 165
Kathedersozialisten ("socialists of the chair"), 88–90, 92, 283n10
Kaufmann, Felix, 174, 177
Kautsky, Karl, 121, 128, 131
Keller, Evelyn Fox, 253
Kepler, Johannes, 29, 54
Keynes, John Neville, 182
Kidd, John, 49–50
Kirchhoff, Arthur: *Die Akademische Frau,* 112–115, 119
Klein, Felix, 113–114
Knight, Frank, 195–196, 295n56, 297n34–36
Kolakowski, Leszek, 159
Koppanyi, Theodore, 238
Koyré, Alexandre, 39–41, 46
Kracauer, Siegfried, 157
Krieck, Ernst, 171–172
Kuhn, Thomas, 13, 201, 210–213, 274n12, 297n36, 300n9, 301n13

La Mettrie, Julien Offroy de, 22
Laissez-faire, 102, 187, 189, 194, 197, 270, 296n23. *See also* Liberalism
Language: Adam's, 35; bans urged on eloquence, 36–37; Latin rejected as obsolete, 23, 35–36, 71–72; new science words coined, 66; to soften brutality in war, 2
Language, Truth and Logic (Ayer), 202–205
Laplace, Pierre-Simon de, 46
Latour, Bruno, 13, 201, 225, 303n8
Lawrence Livermore Laboratory, 254
Lehmann, Gerhard, 173
Leibniz, Gottfried Wilhelm, 36
Lenin, V. I., 11, 164
Lenz, Fritz, 111, 171
Lenz, Max, 286n51
Leonardo da Vinci, 236
Lepenies, Wolf, 36, 276n12
Levy, Jerre, 251–252
Lewontin, Richard, 249
Lexis, Wilhelm, 114
Liberalism, 101, 105, 152, 212, 266, 298n41
Liebig, Justus, 69, 77–78, 95
Liebknecht, Karl, 102, 123, 168
Locke, John, 57–58, 280n11
Logical positivism: anti-philosophical, 169–170; as "logic of science," 167, 181; Ayer on, 202; borrows from Weber, 152; contrast with Comte's philosophy, 163–165; critique of metaphysics, 167; death of, 174, 209; ignores Nazism, 169; Kuhn and Popper continuous with, 211–212; Nazi response to, 170–173; term coined, 292n10. *See also* Positivism; Vienna Circle
Lopez, Robert, 233
Los Alamos National Laboratory, 243, 254, 259
LSD, 239
Luddites, 237
Ludendorff, Erich Wilhelm, 141–142
Lukács, Georg, 101, 133
Lundberg, George A., 176, 178, 223, 294n51
Luther, Martin, 54, 138, 235
Luxemburg, Rosa, 168
Lyotard, Jean-François, 13, 304n17
Lysenko, T. D., 218

Macaulay, Thomas, 33, 93, 274n9
Mach, Ernst, 81, 109, 131, 163–165, 167–168
Machiavelli, Niccolò, 65, 280n1
Machine breaking, 237
MacIntyre, Alasdair, 61, 206
McIntyre, J. L., 38
McKeown, Thomas, 241
Mackie, J. L., 208

Magic, 42–44
Malinowski, Bronislaw, 42
Maloney, John, 194, 198–200
Mandeville, Bernard, 65, 99
Mannheim, Karl, 225
Marcuse, Herbert, 240
Marginalism and the marginal revolution: Austrian school, 128; comparability of wants, 186–192, 197; criticisms of, 194–196; equilibrium theory, 195–196, 296n23; ideology in, 199–200; marginal productivity theory, 187–188; politics of, 185–192, 296n18; positivism in, 192–193, 197–198; Robbins vs. Pigou, 190–192; Samuelson on, 297n28; socialist views of, 296n17
Marshall, Alfred, 87, 100, 188, 192, 194
Marxism: academic, 122; anti-ethical bias of, 124; ethics vs. science in, 122–133, 266; identified with sociology, 103, 121; in economics, 184–185, 296n17; neo-Kantian, 131–132; popularity of, 103–104, 123; positivism and, 121, 131; revisionist, 127–131, 266; science theory in, 214–215; value-neutral, 121–133. *See also* Socialism; Social Democratic party; Value theory
Masaryk, Tomáš, 106, 131
Materialism, 22, 56, 79–80, 105, 110, 119, 130–132, 214–215, 302n27
Medicine, 241–244
Meek, Ronald, 295n3
Menger, Karl, 89, 166, 185
Menzies, W. C., 218
Merton, Robert: Barnes on, 228; criticisms of, 223; defense of democracy, 222–223; on military technology, 277n43; on norms of science, 221–223; on Nazi science, 223; on Puritan ethic, 219–220; on social functions of pure science, 221–222
Merz, J. T., 97
Metaethics, 206–207
Michels, Robert, 107
Milgram, Stanley, 240
Militarization of science, 1–2, 4, 175–176, 216, 236, 254–260
Mill, John Stuart, 135, 162, 185–187, 189, 295n1, 295n7
MILSTAR, 5
Mises, Ludwig von, 121, 123, 127, 158, 197, 218, 298n38, 298n41
Mitchell, Wesley, 195, 297n31

MITRE Corporation, 257
Moore, George Edward, 204–205, 298n8, 299n9
Moralism, 206–207, 226–227, 231
Moral philosophy: analytic, 201–208; Bacon's doubts about, 32; Hume's, 61; Randall's conception of, 58; scope of, 58, 207; vs. moralism, 206. *See also* Ethics; Value theory
Moray, Robert, 33
Morrison, Philip, 176
Moynihan, Daniel P., 250
Muller, Herbert J., 218
Mumford, Lewis, 279n6
Murdock, George P., 34
Myrdal, Gunnar, 194, 301n13

Nader, Ralph, 260
Napoleon Bonaparte, 70
National Institutes of Health (NIH), 242, 244
National Liberal party, 101, 122
National Medical Association, 242
National Science Foundation, 238
National Security Agency, 240
National Socialism: as "applied biology," 171; attack on value-neutrality, 171–173; Ayer blamed for rise of, 205; Barnes on, 229; Bok on, 268, 309n3; critique of relativism, 172–173; Dewey on, 292n21; ignored by *Erkenntnis*, 169, 173; Jews blamed for liberalism, 172; Lundberg on, 294n51; medical experiments during, 238–239; on quantum theory, 173; philosophical writings in, 170–173, 293n32; science to save world from, 218; Sombart's collaboration with, 127, 158–159
Naturalistic fallacy, 58–61, 204, 299n9
Natural theology, 35, 45–51, 78
Nature: Aristotle's view, 224; Baconian conception, 30; design in, 46–51; mechanical conception, 22–24; natural theological conception, 46–50; Rickert on, 145; value in, 39–52
Naturphilosophie, 77–79
Nelkin, Dorothy, 306n28
Neoclassical economics. *See* Marginalism and the marginal revolution
Neo-Kantianism: Adler's Marxist, 131–132; as philosophy of compromise, 83; divides

science and religion, 80; Nazi distrust of, 172; positivist, 163–164; rejects positivism and naturalism, 101; Rickert and Weber's, 145–146, 151
Neurath, Otto: christens Vienna Circle, 166; cited by Bukharin, 217; flees Germany, 174; lectures at Bauhaus, 169; manifesto, 167; on ethics, 299n18; on physicalism, 155; on war economy, 168; service in Bavarian Soviet Republic, 168–169; translates Galton, 110, 168, 292n14. *See also* Vienna Circle
New School for Social Research, 174, 195
Newton, Isaac, 35, 43, 46, 215–216, 219, 302n27, 306n26
Nicholas of Cusa, 40–41
Nicholson, Joseph, 188, 194
Nietzsche, Friedrich, 63, 83, 170–171, 214
Nixon, Richard, 241, 250
Nobel, Alfred, 236
Novelty celebrated, 28–29, 100
Nuclear war, 1–2, 241, 255–256, 258–259, 273n1
Nunn, Sam, 260
Nuremberg Code, 239
Nuremberg Trials, 238–239

Objectivity: Catholics deficient in, 286n51; compromised by advocacy, 10, 28, 136; different from neutrality, 10; feminist critiques, 253; Harding's view, 226, 253; Nietzsche's mockery of, 83; Popper's view, 212; requires criticism, 6, 226, 253; Tönnies on, 91, 152; Weber's defense of, 105
Oersted, Hans Christian, 78
Ogden, C. K., 202
Oken, Lorenz, 93
Oppenheimer, Franz, 90, 100
Oppenheimer, J. Robert, 238
Ostwald, Wilhelm, 113, 119, 144
"Ought" and "is," 7, 10, 58, 61, 65, 132, 135, 150, 166, 167, 208. *See also* Naturalistic fallacy
Oxford University, 38

Paley, William, 47–50
Pan-German League, 109, 138, 141
Paracelsus, 42
Pareto, Vilfredo, 99, 192–193, 297n26
Parker, Samuel, 36

Parsons, Talcott, 228
Pascal, Blaise, 60, 213, 216
Passmore, J. A., 176, 209
Pasteur, Louis, 69, 224
Perrault, Charles, 29
Perspective drawing, 53
Peters, John, 225
Petrarch, 233–234
Petty, William, 21, 27, 34, 87
Petzoldt, Joseph, 144, 164–165
Philippovich, Eugen, 86–87
Philosophical faculty, 76–78
Philosophy: ancient Greek, 17–21, 232–233; as "science of all science," 73; Ayer's conception of, 202; German idealist, 77–79; Marxist, 121–133; Nazi, 170–173, 292n22, 293n32; of history, 12–13; of science, 13, 259, 267. *See also* Emotivism; Ethics; Logical positivism; Moral philosophy; Neo-Kantianism; Positivism
Philosophy and Public Affairs, 207
Physicians for Social Responsibility, 260
Physics, 4, 6, 147
Pigou, A. C., 183, 192–194, 296n24
Pimentel, David, 245–246
Pinochet, Augusto, 268
Planck, Max, 113, 169
Plato, 17–20, 232, 236
Playfair, Lionel, 94
Ploetz, Alfred, 110
Plutarch, 20
Poincaré, Henri, 166, 300n9
Polanyi, Michael, 11, 212, 218
Popper, Karl, 11, 121, 160, 164, 166, 209–213, 218, 270, 300n7
Positivism: and early sociology of knowledge, 214; defined, 159; equated with socialism, 107, 121, 131; eurocentric, 214; in economics, 182–200; Kuhn's critique, 210; legacy of detached indifference, 231; links with relativism, 11, 227–230; Nazi critiques, 172–174; origins of, 160; politics of, 165–166, 266; Popper's claim to have killed, 210; Robbins's conservative, 190–192; Scheler's critique, 214; social science vs. physical science, 164–166; Weber's critique, 135, 143. *See also* Comte, Auguste; Logical positivism; Vienna Circle
Postmodernism, 304n17
Potoniés, Henri, 96, 109

Pragmatism, 177–178, 207, 209
Presentism, 12
Primary and secondary qualities, 54–62
Principia (titles), 298n8
Progress, ideal of, 23
Proletarian science, 133, 157, 197, 218, 223, 288n30
Psychology, 238, 240, 293n37
Ptolemy, 211
Public Interest Research Groups, 260
Pure science ideal: functions of, 221–222. *See also* Science
Purity, celebrated, 100

Rabinowitch, E., 294n42
Racial hygiene, 109–111, 175
RAND Corporation, 256–257, 268
Randall, John Hermann, 58
Rape, 248
Rawls, John, 207
Reagan, Ronald, 2, 258
Realist accounts of science, 225–226, 303n5
Reichenbach, Hans, 169–170, 174, 209, 293n32
Relativism, 11, 172, 227–228, 231
Religion, 46–52, 150–151
Revolution of 1848, 105
Ricardo, David, 100, 102, 184
Richards, I. A., 202
Rickert, Heinrich, 145–146, 148
Rifkin, Jeremy, 240
Ringer, Fritz, 106
Robbins, Lionel, 8, 189–192, 196, 198, 296n23, 298n38
Rousseau, Jean-Jacques, 36–37
Royal Society of London: activities of, 26–27; exclusion of morals and politics from, 33, 264; exclusively male, 27, 37; language of, 35, 68; Puritan Calvinists in, 220
Royalist compromise, 7, 33, 65–66
Ruskin, John, 194

Saint-Pierre, Abbé de, 29
Sainte-Beuve, Charles-Augustin, 83, 282n19
Salz, Arthur, 158
Samuelson, Paul, 196, 295n3, 297n26, 297n28, 298n42
Saturated fats, 245
Scheler, Max, 158, 214–215
Schelling, Friedrich, 73–74, 78
Schiebinger, Londa, 277n39–40, 286n35

Schleiermacher, Friedrich, 71
Schlick, Moritz, 165, 166, 169, 174, 202, 205, 292n10, 293n32, 293n37
Schlözer, Dorothea, 112
Schmoller, Gustav, 89–90, 105, 283n9
Schulze-Gävernitz, Gerhart von, 106
Schumpeter, Joseph, 88, 176–177, 225, 296n17, 298n44
Science: ancient Greek, 9, 17–21; and prosperity, 94, 238; and religion, 79–80; Baconian, 6, 24–33, 93; calls for moratorium on, 238, 305n13, 306n26; censorship and self-censorship in, 38, 182; critiques of, 151, 232–261; demilitarization of, 259; Descartes's conception of, 27–28; "for its own sake," 18, 70–71, 76, 94, 96; industry support for, 4, 95–96; methods key to progress of, 26–27; militarization of, 1–2, 4, 175–176, 216, 236, 254–260; Nazi, 170–173, 223; norms of, 221–223; perfectibility of, 28–29, 276n13; political nature of, 267; practical origins of, 23, 73–74, 214–217; pure vs. applied, 3–4, 18–19, 68–69, 121, 129; Puritan values and, 220, 302n30; secrecy in, 2, 43–44; social origins of, 23, 174, 212–223; specialization within, 77–84; social responsibility of, 5, 176, 231, 260–261, 294n42; social vs. physical, 146, 161, 165–166, 176; unifying force, 92–93, 96–98; useable for good or evil, 2–3, 233, 268, 273n6; values of, 29–30; Weber's conception of, 136, 147–148
Science fiction, 36
Science for the People, 258, 260–261
Science policy, 1–13
Scientific growth, theories of, 145, 209–215, 297n36
Scientific Knowledge and Sociological Theory (Barnes), 226–229
Scientific utopias, 218
Scientism, 149, 162, 266
Scientist (word coined), 49, 99, 284n1
Scientists: attitudes towards women, 112–119; class background of, 27, 37; persecuted, 45–46, 218–219, 235, 302n27
Second International Congress on the History of Science (London, 1931), 215–219, 302n27
Seitz, Frederick, 309n80
Seneca, 20, 275n9

Senior, Nassau, 295n1
Shadwell, Thomas, 234
Shakespeare, William, 236
Shapin, Steven, 225–226, 303n2
Shaw, George Bernard, 296n17
Sica, Alan, 289n4
Siemens, Hermann, 109
Simmel, Georg, 99, 116–117, 119, 145, 172
Skinner, B. F., 293n37
Slavery in ancient Greece, 19
Sluga, Hans, 292n19
Smith, Adam, 61, 93, 102
Social constructivism, 223, 226, 252, 303n5, 304n18
Social Darwinism, 70, 109–111, 138. See also Racial hygiene
Social Democratic party (SPD), 79, 103–105, 106, 119–127, 143, 168, 266
Socialism: as applied sociology, 121, 129; opposed by sociologists, 92; popular support for, 103–104; pure and applied, 130; scientific and utopian, 104, 124, 128–130; suppressed, 102–103; Weber's distrust of, 142
Sociology: branded Jewish discipline, 158; confused with socialism, 104, 106–108, 121; early use of term, 284n4; German professorships of, 107, 285n20; origins of, 86, 90, 99; suppressed by Nazism, 158–159
Sociology of knowledge: Barnes's strong program, 226–231; bibliography, 301n15; coercive, 226; Marxist, 214–215; Merton on, 219–223; origins of, 158, 214–215; realist vs. moralist, 226; relativist, 231; Scheler coins term, 214
Socrates, 5, 6, 17–19, 32, 206, 233
Sombart, Werner: academic career, 99, 107, 122–125; break with Social Policy Association, 88; conservatism, 105, 123–126; cooperation with Nazis, 127, 158–159; Engels's praise for, 121–123; excludes value judgments, 88; on dialectics, 288n25; on Judaism and capitalism, 126; on Marxism as "value-free science," 124; on science as unifying force, 92, 97; "red professor," 122; rejects Marxism, 126; value-subjectivity of, 153
Soviet Union, 179, 218–219, 239, 241–242, 250, 302n27. See also Proletarian science
Sprat, Thomas, 26, 27, 31, 33–36, 46, 93, 213

Stalin, Joseph, 219
Stamp, Josiah, 238, 305n13
Star Wars. See Strategic Defense Initiative
Stern, Fritz, 120
Stimson, Dorothy, 220
Stocking, George, 12
Storer, Norman, 223
Strachey, Lytton, 204
Strategic Defense Initiative (SDI), 5, 256, 258–259, 309n80
Strauss, D. F., 78, 81
Strauss, Leo, 178–179
Strong program, 226–231, 303n5. See also Barnes, Barry; Sociology of knowledge
Structure of Scientific Revolutions (Kuhn), 13, 210–212
Stubbe, Henry, 35
Stumm-Halberg, Karl von, 106, 141
Subjectivity. See Value theory
Superconducting Supercollider, 5
Surplus value, 184–185
Swammerdam, Jan, 148, 264
Swift, Jonathan, 234
Sympathy and critique, 12–13

Tachistoscopy, 238, 305n15
Taoism, 236, 273n6
Tartaglia, Niccolò, 23, 29, 216
Technische Hochschulen, 68, 76, 79, 107
Technology: and war, 1–2, 236; critiques of, 8, 236; not distinct from science, 3
Teleology, 41, 90
Theodicy, 61, 135–136
Theory and practice: ancient Greek ideals of, 6, 20–21, 24, 262; Baconian views on, 29, 263
Tolstoy, Leo, 148, 151, 262
Tomlin, E. W. F., 205
Tönnies, Ferdinand: denied promotion, 106; on eugenics, 101, 110; on objectivity, 91, 152; on specialization, 82, 91; on value-neutrality, 82, 90–92, 229–230; positivism of, 100–101; women led by feelings, 115–116
Torricelli, Evangelista, 216
Tower of Babel, 35, 93
Treitschke, Heinrich von, 112, 138
Trials against animals, 44–46, 278n11 and n12
Tucker, Abraham, 50
Turner, Stephen, 290n43

Tuskegee Syphilis Experiment, 239

United States Air Force, 255, 257
United States Army, 239, 254–255, 257
University of Berlin, 71–73
"Usefulness of useless knowledge," 96
Utilitarianism, 185–187, 190–193, 297n28

Vaihinger, Hans, 80
Valla, Lorenzo, 28
Value-neutrality: and ideological neutrality, 198; as reaction to Nazism, 159, 175; defended, 90–98, 159, 166, 198–200, 228; feminist critics of, 253; functions of, 103, 119, 150, 175, 180, 198, 230; Gouldner's critique, 179–181; in Barnes's strong program, 226–231; in economics, 183–200; in Weber's research, 139–140; Nazi views on, 170–173; origins of, 67–70, 120, 149; ridiculed, 9; Robbins asserts, 189–192, 196; sociologists' support for, 86–90; Schmoller's critique, 89; Strauss's critique, 178–179. See also *Werturteilsstreit*
Value theory: ancient vs. modern, 21, 39, 263; Galileo's, 40; Gossen's, 295n9; Marxist, 184–187; objective, 175, 178; Samuelson on, 295n3; scientific revolution in, 41; subjectivist, 8, 56–58, 69–70, 152–153, 183–200, 205, 270, 295n3; utilitarian, 186; Weberian, 135–136. *See also* Devalorization of being
Vavilov, S. I., 302n27
Veblen, Thorstein, 194–196, 227, 230, 297n31–32
Verein für Sozialpolitik (Social Policy Association), 85–98, 105, 137, 158–159
Verstehen, 119, 145–146
Vico, Giambattista, 12, 29
Vienna Circle: and economic theory, 196–197; Ayer on, 202; founding of, 166–167; Jews in, 174–175, 293n32; manifesto, 167; Nazi critique of, 173; politics of, 169, 217; Popper and, 210; view of science, 266. *See also* Logical positivism
Villard de Honnecourt, 23

Wagner, Gerhard, 171
Warnock, G. J., 205
Watson, James, 243
Weber, Alfred, 97
Weber, Ernst, 81

Weber, Marianne, 117–119
Weber, Max: American reception of, 177, 181; attack on racial hygiene, 110–111; critique of Ostwald, 143–144; distrust of Poland and Russia, 138; dueling scar, 118; early career, 99, 137, 143; English translations, 181, 295n56; Frankfurt School critique of, 179; influence on Robbins, 189, 196; invited to join Heidelberg Soviet, 143; letter to Marianne, 118; Mises and, 298n38; National Liberal party, 101, 141; Nazi critique of, 173; on democracy, 141–142; on limits to science, 136; on materialism, 142; on positivism, 143, 146–147; on Puritan ethic, 88, 105, 219; on rationality, 142, 148, 159; on religion, 150; on social Darwinism, 138; on socialism, 142, 149; on Tolstoy, 148; on value-neutrality, 88–89, 139–140; on *Verstehen*, 119, 145–146; on violence, 138; pessimist, 140, 151; "Science as a Vocation," 82, 147–149; Strauss's critique of, 178–179; support for Anti-Socialist Laws, 141; support for First World War, 140–141; support for Sombart and Michels, 107; values irreconcilable, 151–154, 201–202, 266; women's rights, 117–118
Welfare economics, 192–194, 296n25
Wells, H. G., 218
Weltanschauung, 171–172
Werturteilsstreit, 6, 70, 86–98, 115, 135, 198
Wesley, John, 47
Whewell, William, 49, 284n1
Whiggism, 12
White, Andrew Dickson, 47, 279n15
Whitehead Foundation, 4
Wicksteed, Philip, 194, 298n38
Wiener, Norbert, 176
Wilhelm Friedrich III, 70–71
Wilkins, John, 35, 36, 46
Willowbrook experiments, 239
Wilson, Edward O., 248
Wissenschaft, 76–77
Wittgenstein, Ludwig, 166, 169, 203, 205
Wöhler, Friedrich, 94
Wolff, Christian, 71
Women: barred from political activity, 111; debates over access to education, 112–119, 287n52; exclusion from academies, 37; Greek philosophical views on, 18–19;

Hegel compares to plants, 112, 248; ovaries damaged by study, 114; Ph.D.s in physics, 251; separate species, 115; Weber on, 117–118
World War I, 4, 98
World War II, 4, 238
World War III, 255
Worms, René, 108

Wundt, Max, 172
Wundt, Wilhelm, 113

Xenophon, 20

Zeller, Eduard, 80
Zetkin, Clara, 111
Zilsel, Edgar, 23, 174, 213